The Ultimate Renewable Energy Engine
Harnessing "Magnetic Gravity"

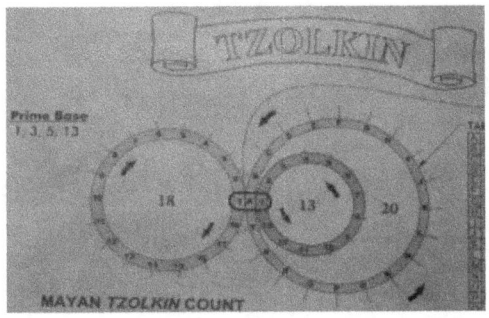

Creating Electricity
From the greatest energy source useful for mankind

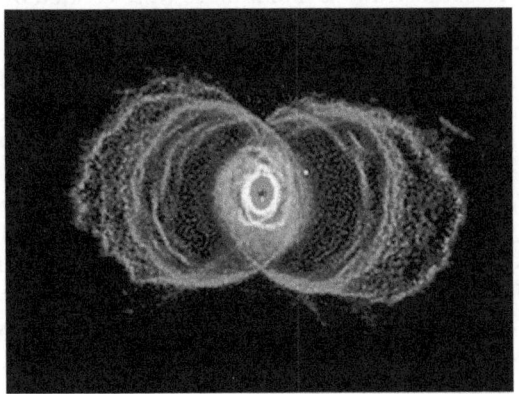

Gravity forming mass demonstrated in the Hourglass Nebula

Babushka Egg Concept Book #9
Third Edition September 2012

By Herbert R. Stollorz

ELOHIM
JOD SUPREME COURT
A Micro-Babushka Egg Concept Booklet #10

and

The Ultimate Renewable Energy Engine
Harnessing "Magnetic Gravity"

Babushka Egg Concept Book #9
Third Edition September 2012

By Herbert R. Stollorz

Use of Photos and Illustrations

All astronomical photos on the cover and in the text were sourced from nasa.gov and edited to adapt to publication design. Unless noted otherwise, all tables, illustrations and images were created by the author.

Use of Scripture

All Scripture quotations, unless otherwise noted, are from the HOLY BIBLE: NEW INTERNATIONAL VERSION®. NIV®. Copyright © 1973, 1978, 1984 by International Bible Society. Used by permission of Zondervan Publishing House. All rights reserved.

Scripture quotations marked NRSV are from the NEW REVISED STANDARD VERSION BIBLE. Copyright © 1989, Division of Christian Education of the National Council of the Churches of Christ in the United States of America. Used by permission. All rights reserved.

Faith in the Future Foundation
P O Box 6384
Minneapolis, MN 55406
www.apocalypse2008-2015.com

ISBN: 1439241171
EIN13: 978-1439241172

Printed on demand in the USA by
BOOKSURGE LLC, North Charleston, SC

The Ultimate Renewable Energy Engine
Harnessing "Magnetic Gravity"

Table of Contents

Part 3: More UREE Gravity Applications 249

The Evidence Trail Continued - Nine Babushka Egg Concept Books 266

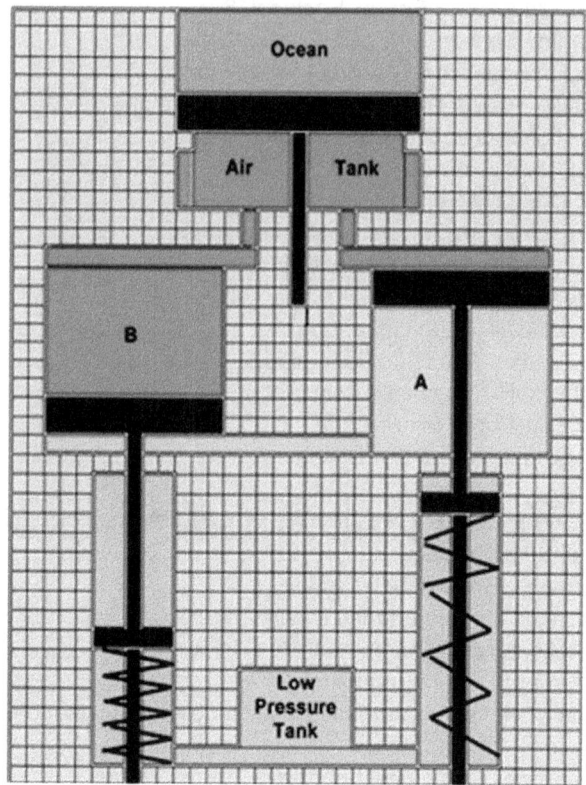

Why Another
Babushka Egg Concept Book?

This Third Edition of the 9th Babushka egg concept book raises many science concepts to a higher level. It postulates that magnetic Gravity linked to UREE could fuel airplanes, even Concord jets with "free" electricity. But has gone further and answered the question, "Where does the energy come from?"

That required a new Donut Atom theory to be postulated explained for the layman, as atoms smashed by CERN lost their identity will never reveal where the ENERGY originally comes from. Even my little grand-kid wants to know how free electricity crystallizes from the cosmos. Can it be extracted from Gravity-Water-Air and the embedded magnetisms converted on the atom level by applying low technology to yield totally "free electrical ENERGY"? Nicola Tesla, who originally invented ELECTRICITY 100 years ago, proved it possible and was followed by other inventors demonstrating many patent applications. This technology should be suppressed no longer by dirty OIL money interests.

I hope it will replace the root-cause of an unscientific evolution theory and wake up universities from sleep being liberated from a powerful criminal oil-nuclear cartel and once more pursuing true science. They have the global influence to slow down the horrendous worldwide destruction depleting our resources by not suppressing but accepting this free energy. This energy warning comes at the eve of depleted global energy resources and was inspired from ELOHIM sending once more a Jonah FISH to get the attention of science matching prophecy. Watch a 52 km asteroid already in an 825-day orbit ending again a Civilization [21 September 2017?] like Noah's apocalypse 5 February 2287 BC.

After 100 years, these new revolutionary energy concepts are exposed. You may be astonished to learn about the censored history focused on the great scientist Nicola Tesla who rediscovered electricity known before Noah's time. Tesla and many others were reawakened to a cosmic energy source now further expanded in this Third Edition to better unify some aspects where really energy originated revealed in the Bible. Now watching on YouTube movie clips of many entrepreneurs experimenting with forgotten generator-motor science greatly surprised me by proving free electricity existed all along.

But rationally thinking, energy must come from somewhere, and only metaphysics one notch higher has the answer forbidden by the establishment. They did it to Sir Isaac Newton by suppressing 50% of his writings, but now his powerful voice is reemerging from the grave to defeat an illogical, unscientific, evolution religion controlled by a powerful atheistic priesthood-elite.

Investigate the evidence of wireless electrical transmission still puzzled why were 700 Nicola Tesla inventions suppressed permitting a criminal Oil-Nuclear Cartel to have so much muscle causing globally unending energy conflicts that led to killing millions in the last century when free electricity was around? Why it is still censored and covered up in our advanced technical culture?

If you like to know the electric motor secrets not allowed in universities, listen to a recent 2-hour lecture on YouTube by Peter Lindeman explaining forgotten science principles buried in patent libraries. But his expanded technology needed someone to unify it into a whole 360° horizon concept by adding the metaphysical perspective even Newton postulated. These pages explain how and why magnetic gravity can be linked to generate enough electricity to drive a car or fly jets without fossil fuel.

Do not be fooled and discouraged by disinformation (Pearl #218) from the oil-nuclear cartel inserting fake science on YouTube to keep you ignorant and confused. Either way, start being entertained with the many fun video clips, or if you are serious, read this Babushkas egg concept book first to delineate what is counterfeit. One way or another, your mind-horizon will be widened one notch higher – guaranteed!

UREE SCIENCE PROVEN

The 9th Babushka egg concept book has expanded into the THIRD EDITION as countless more facts were discovered in the last six months. Once a multifaceted research science book is published additional information often surfaces later that can only be amended either in my Pearls or eventually summed up in another edition.

Searching the Internet revealed that I am no longer alone exposing a "Grand Lie" now uncovered that "FREE ELECTRICITY" existed 100 years ago still suppressed by the criminal OIL-NUCLEAR CARTEL which controls the patent department and every university with huge grants paid by YOU the consumer. They got away with it manipulating political every university by placing international in strategic position a powerful well financed atheistic priesthood. They are backed up by corrupted governments compulsory enforcing exclusively an unscientific evolution theory in every school therefore brainwashed the MIND of every student.

That distorted the ability to think in LOGIC undermining the basis of true physics linked to the forbidden metaphysics pointing out a destructive evil mind-set of relativism. The cartel of the richest people on earth became very successful enforcing the "Grand Lie" that free energy is NOT POSSIBLE. Your overcharged utility and gasoline bills are used to pay off corrupt politicians and to finance counterfeit science while supporting a class of privilege usually hiding unearned wealth tax-free in offshore bank accounts controlled by crooked global bankers. They direct our civilization in a vicious cycle causing worldwide extinction in danger of terminating all life on earth.

It will end in Armageddon guaranteed, but God's Wrath once more will save LIFE revealed by another Jonah now warned that ELOHIM will continue his Plan for Mankind but this time without "evil" another 1000 years. We should recognize everybody has heard about a coming Apocalypse as TRUTH proclaimed in the Bible linked to divine prophecy cannot be suppressed anymore by the evil OIL-NUCLEAR energy cartel stealing obscene money from the hoodwinked consumer. One forbidden witness tells it plainly:

Free Energy Tom Bearden, Joseph Newman
by JohnEnergy 2012 - www.Lightworksav.com
http://www.youtube.com/watch?v=I0K_LkO0Zuw
http://www.youtube.com/watch?v=o76BRnEDALY

I am only one voice among the many witnesses on the Internet that is not yet controlled by the cartel. We independently discovered what is no longer taught in universities. Nicola Tesla's discovery of free electricity was deliberately silenced in 1920, as global universities totally stopped teaching about electric motors and generators, pushing dirty polluting OIL instead. They forced the automobile industry to comply, barely improving efficiency. That for 100-year suppression led to a generation being raised without really understanding science applications.

This lack of awareness was only bridged by the silent minority of rebels postulating free electricity in spite of oppression. They only annoy the oil-nuclear cartel that still tightly controls corrupt governments linked to their energy departments. But coercion cannot be continued since the Internet has changed our civilization forever, obsoleting a stranglehold on truth. Truth is like a genie in the bottle when released and is stronger than propaganda lies according to entropy laws. Have a little fun and be entertained watching the many short video clips on YouTube about what was suppressed for 100 years. The world does not need to endure another energy war.

Check out these links:

1. The Lost Secrets of Nicola Tesla - Nicola Tesla the Untold Story
http://www.youtube.com/watch?v=X0z3pEeYDEw&feature=related

2. Peter Lindeman - Electric Motor Secrets, 2-hour lecture -Video produced by Clear Tech Inc. www.free-energy.ws/products -

http://www.youtube.com/watch?v=iLGuf1geOiQ&feature=related

3. A Pie-shaped generator linked to UREE #13,14,16 Thewaterenergy1@hotmail.NL
http://www.youtube.com/watch?NR=1&v=PoJALsTaCAo&feature=endscreen

4. Simple proof of electrical perpetual motion: 3000-watt Engine Powers Itself -

http://www.youtube.com/watch?v=Fv53K9MnDuM&feature=related -www.witts.ws

5. http://www.amazon.com/Lost-Science-Gerry-Vassilatos/dp/0932813755

Courtesy Notice of a New "Energy" Discovery

Please pass it along to the appropriate science department.

Entropy-Abstract:

The Hoover Dam installation extracts clean-green **Electricity** from **GRAVITY** on a **"Second Entropy"** level by channeling the embedded kinetic energy of a compressed water media to flow in perpetual motion by pushing copper wire loops through a magnetic field to yield electrical energy.

But this amazing UREE invention exploits compressed air-media on a higher **"First Entropy"** level and converts the embedded magnetic nuclear force into electricity utilizing air/oxygen atoms. **This process makes the UREE a portable, unlimited energy source.**

My new UREE generator design behaves like an electromagnetic bridge, which forces electrons to flow over conventional copper wire loops to produce electricity. It could be compared to the Hoover Dam generating power from flowing water pressure but the UREE invention is different and uses AIR compressed in a small wheel chair steel tank that makes electricity portable. As demonstrated in physics, electrons can be extracted from the outer atomic shell, but if derailed surging through a low-tech, double-rotating magnetic transformer, they gain more free energy.

Energy now recycled and pushed to a higher **"critical resonance frequency"** will liberate more electrons to perpetually provide flows of portable electricity. Tapping into the nucleus' critical resonance frequency will connect to the **First Entropy Zone** resulting in virtually unlimited **electrical profit,** which makes the UREE invention priceless where an electrical distribution network does not exist. It will drastically revolutionize the 21st Century civilization extracting totally green energy and consequently in addition:

- Obsolete deadly Nuclear Power
- Obsolete dirty polluting Coal Power
- Obsolete the wire power grid network and its associated monopolistic control system
- Obsolete gasoline and drive a car with green energy.
- Obsolete Dr. Albert Einstein's equation, $E = mc^2$

Many witnesses in science are demonstrated in the 16 UREE applications due to the new theorized discovery of
Herbert's "MAGNETIC" GRAVITY.

Available to download free on the Internet.

Dear President Obama,

This courtesy letter captures some of the information found in a number of Babushka egg concept books circling the globe on the Internet. It includes perhaps the greatest discovery of a new energy source that could immediately employ millions people in America if the US government has the resolve. It could even help you be re-elected.

As a retired scientist, I was challenged by my grandkids in the laws of physics and in the process discovered a virtually infinite energy source, which compelled the hi-tech inventor in me to design 10 applications of an ULTIMATE RENEWABLE ENERGY ENGINE (UREE). Not being trained in the advanced theories of mathematics, I can only hypothesize how it works.

I discovered how GRAVITY could be converted into a fuel infinitely available by extracting it from kinetically moved air atoms and water molecules from the ∞ Ocean. My designs simulate how the Hoover Dam provides energy but now miniaturized and made portable. It is a totally cheap clean energy source obsoleting deadly nuclear power station, dirty coal conversion related to many problems in the environment and global warming. My discovery even makes it possible to drive your car without gasoline. It aligns with a recent science discovery by Princeton University scientists who measured light to be 300 times faster than usual under special circumstances verified by CERN scientists that prove Dr. Albert Einstein's theory needs to be changed. Faster light discoveries captured in modern but simple motors can now produce perpetually infinite electrical profit for the 21st century civilization.

The reports of the recent earthquake rocking the White House on August 23, 2011 exposed on the TV a nuclear lie from the energy cartel that safety was their concern. Their first line of defense did not work, as expressed by the UN nuclear expert who previously evaluated the Japan's nuclear disaster. He said that we were very lucky that this earthquake did not cause the melt down of a US nuclear reactor. It was only prevented because the quake was one notch lower, or it could have been repeated like in Russia and Japan. A destructive nuclear meltdown would have permanently poisoned the real estate surrounding White House for the next 100,000 years eventually making our planet void of all life like the rest of the universe.

My patent application for the ULTIMATE RENEWABLE ENERGY ENGINE may become reality, but it should not wait until the energy industry and universities get around to it. It is a bit like how the Wright Brothers had to struggle against the general opinion that human flight was "not possible". In opposition to popular conviction, 100 years later we flew to the moon placing a flag firmly showing America's resolve. We should repeat it once more by putting the effort of NASA scientists to build prototypes of my designs to test my radical new approach to energy generation that taps into subatomic kinetic motion to provide clean green energy for modern living. We are running out of time. The energy cartels still could now make obscene profits from gravity converted into free "electricity".

Sincerely in a tailspin of new discoveries,

Herbert R. Stollorz (10-11-11)

The Ultimate Renewable Energy Engine (UREE)

In the following pages, I present a number of basic concepts behind a pending patent application to prove that we can extract electric power from GRAVITY-WATER-AIR and MAGNETS. This proposal will make expensive nuclear and coal energy obsolete and reduce oil dependency by driving a car perpetually on pressurized air converted to electricity, thus replacing a major energy need with free nonpolluting electrical power. This will drastically change everybody's life.

Linked to a Modified Transportation System for the Future.

This revolutionary concept is presented in two parts:

1. How to use GRAVITY, Air & Water as an energy source to extract electricity. Presenting the Ultimate Renewable Energy Engine (UREE) proven and demonstrated with sixteen (16) applications.

2. How to apply GRAVITY to a futuristic low-cost traffic system, by designing a cheaper, 500 mph GRAVITY Levitated Bullet Train, linked to Hybrid-monorails - still retaining a personal, rented or public owned electric automobile in a new integrated approach, thus resolving the mass transit vs. a personal vehicle quandary.

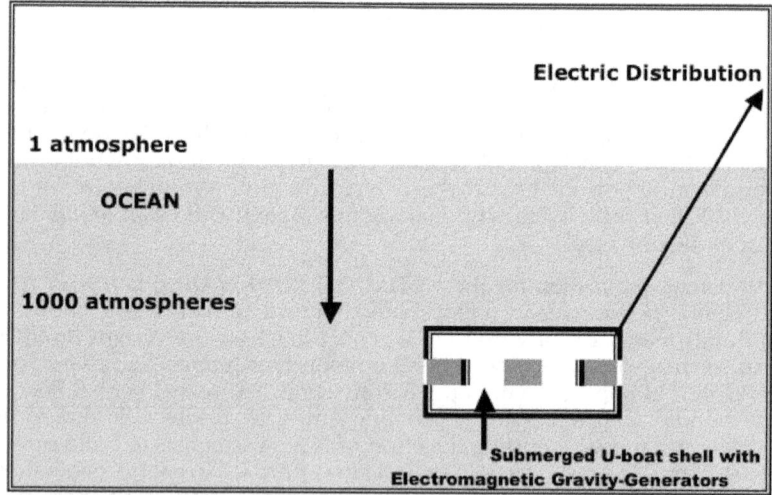

16 Witnesses will Prove a Gravity Conversion Concept.

For an example:

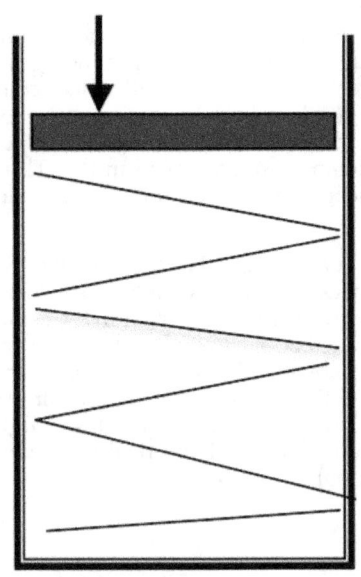

1. A GRAVITY differential below sea level will exert a water pressure ten thousand times higher than the Hoover Dam, thereby creating "unlimited" electricity.

2. The laws of physics demonstrate that any accordion bellows will compress air on the ocean floor. When applied to gravity motors, it will generate unlimited electricity.

3. It works when a differential GRAVITY potential is converted to KINETIC ENERGY in motion, which can be transformed into ELECTRICITY. That is accomplished by using a standard, modified steam-diesel engine linked to a flywheel, via a crankshaft connected to pistons. Applying proven low-tech engineering could extract unlimited, clean, green energy from GRAVITY, which is a boundless cosmic energy force forming galaxies. It may very well be considered the greatest invention ever in the 21st Century.

The Underlying Physics of the Ultimate Renewable Energy Engine (UREE)

The green way to harness nuclear energy is not by splitting atoms like CERN but to tap into what sustains their internal perpetual motion. This short concept book will explain the new Donut Atom theory postulating how electrons spinning around a dynamic atomic nuclear shell connect to a cosmos infinite energy source can be harvested across a magnetic bridge to provide a virtual ∞ electric power.

Exploiting and converting safe nuclear energy (more explained free on the Internet) in a novel method detailed in my 5th Babushka egg concept book about a simple Donut Atom model written a few years ago. It rationalizes a hidden energy path where magnetism, as the driving force inside the atom, creates electricity to bond atoms into molecules. This magnetic energy glue can be safely extracted as "electricity profit" from magnetized gravity via manipulated kinetic differentials.

These pages describe a new method that imitates the Hoover Dam perpetual motion water pressure system creating electricity. The principles in this book raise it to a higher technical level using pressurized air media so that it becomes a portable energy source. This new innovative technique utilizes known standard industrial mechanisms configured into various (UREE) GRAVITY MOTOR designs packaged for different applications. All extract a nuclear magnetic force from gravity linked to renewable kinetic potential, which is the same as the Hoover Dam principle that produces energy as "electrical profit" generated from an immense gravity potential. But it makes no sense plugging in a $65,000 dollar hybrid-electric car into the Hoover-dam electrical outlet for a recharge of $18,000 Lithium batteries when the immense gravity energy can be packaged as compressed air into a small wheel-chair green bottle to energize just one little single cylinder in your car linked to an electric generator. That gives us additional free electricity if linked to the atomic critical resonance frequency now enhanced in this Third Edition of the 9th Babushka egg concept book to understand physics explained better.

Recently, Princeton University researchers and CERN scientists like two witness needed in any court of law measured light speeds much faster than the traditional 186,262 miles per second creating uproar in the establishment. Forbidden Babushka egg concept books postulated it years ago now in addition presented in this 9th. Babushka book describing **"why"** light is faster proven with a special UREE motor concept. These various electrical conversion designs have all along capitalized on a faster light proposition discovery linked to an energy trail to theoretically draw on energies evidenced beyond the traditional physics of:

$$E=mc^2.$$

Discoveries in conflict with Einstein's philosophy should be reflected in the theories of physics to re-classify visible light to "Second Entropy", which is verified and articulated in the first verse of Genesis in the Bible declared as the **"First Entropy <∞-light>"** appearing on the first day of creation. This invisible **cosmos energy** on a higher frequency cannot be measured with instruments and is still flowing in every atom fueling electrons orbiting around an atom nucleus maintaining our LIFE, captured in a new formula expanding a science horizon:

$$\infty E = m \, (+\infty C / -\infty C)^2$$

By Herbert R. Stollorz

An experienced hi-tech inventor
with a proven track record of results and patents.

Philosopher, Author, Vintner, A non-denominational Christian and
Founder and Director of Faith in the Future Foundation,
an educational charity to further biblical knowledge
linked to true science.

The ELOHIM declared

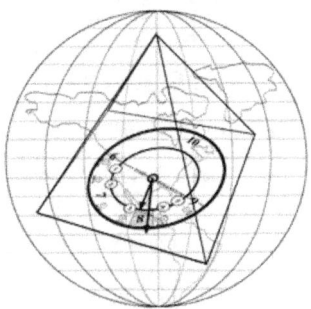

MORTAL
TIME DIMENSION
(4488 BC - AD 3018)

ISAIAH 48:5[1]

I declared them to you from long ago,
before they came to pass I announced them to you,
so that you would not say, "My idol [atheism] did them,
my carved image [high technology] and my cast image [space station]
commanded them…"

From this time forward I make you hear new things,
hidden things that you have not known.

They are created now, not long ago;
before today you have never heard of them,
so that you could not say, "I already knew them."

You have never heard, you have never known,
from of old your ear has not been opened.
For I knew that you would deal very treacherously,
and that from birth you were called a rebel.

For my name's sake I defer my anger,
for the sake of my praise I restrain it for you,
so that I may not cut you off.

My glory I will not give to another.
Listen to me, O Jacob,
and Israel, whom I called:

I am He; I am the first,
and I am the last.

∞ ELOHIM

---

[1] The Holy Bible, New Revised Standard Version. 1996. Thomas Nelson, Nashville

STATEMENT

I, Herbert R Stollorz, a USA citizen, being of sound mind, transfer all rights of future pending patent claim proposals expressed in the 9th Babushka concept book that describes how perpetual motion applied to kinetic gravity generates electricity as illustrated by ten (10) application methods, herewith desire to dedicate all concepts as declared in this document to:

The State of Israel
&
Faith in the Future Foundation
An Educational Christian Charity for the Benefit of Mankind
Federal EIN #02-0736287
28 September 2011

Witnessed:

President: _____**Treasurer:** _____
Christopher J. Patton   Wilfred Hahn

The recent discovery of the **Hebrew Alphabet Number System** applied to science is presented in nine (9) Babushka egg concept books linking many metaphysical realities. These discoveries urged me to write and consider that all of the rights to what we accumulate on this earth ultimately belong to Elohim. The histories and many religions of mankind uncover a basic human need to honor the Creator God. This truth crystallized as something that comes with a promise. The ancient biblical oracles reveal the Creator as the true source of all energy to maintain our life. He can be honored by presenting a portion back as an offering to indicate our appreciation. The promise may be stated,

"If much is given, proportionally much should be returned."

Spending some of my retirement caused me to become a modern Jonah with a big fish story to give warning to our present civilization on the edge of a massive apocalypse. My advice: Do not mess with Elohim, the Creator God, who gave each of us Life and can take it away prematurely resembling what was recorded in history even confirmed by Jesus. When mankind reaches the point of total evil to destroy all life on earth, we will see a repeat of what happened to the Atlantis civilization on 5 February 2287 BC, which utterly perished by an asteroid strike. That impact caused an earth axis wobble that will finally come to rest on 21 December 2012 after 4300 years.

Ignorant scientists misled by an evolution religion do not know that this asteroid strike shifted continents and triggered a great worldwide flood causing massive extinction. The record of these events is embedded in geology and reported in the Bible by Noah. I have further proven this thesis with evidence from many ancient stone-bronze-gold clocks exhibited globally in various museums.

God's judgment of human affairs is verified by many more historic witnesses like Sodom and Gomorrah. If you want to know history from a metaphysical perspective and desire do be acquainted with the ONE ELOHIM, check out the Babushka egg concept books free on the Internet. They are designed to widen knowledge horizons 360^0 by lifting modern science's curtain to reveal many metaphysical laws of nature paradoxically proven by Bible prophecy, which is "history written in advance". If you are intellectually inclined and want to expand your MIND to reflect on the Elohim introduced on the previous page, free on the Internet.

My Patent Claim

I have worked around the clock to put together sixteen witness models of perpetual Gravity Motors creating electricity. I first will describe the Mystery Energy Trail coming from the cosmos ending in every atom. We live In the middle of the energy path where I discovered a way to splice out some energy with 16 new inventions based on simple low-tech designs. If the energy conversion commercially is allowed, these motors will generate perpetual profits from electricity generation and sales. Unfortunately, my first text go around was too convoluted because it reflected how much I did not understand about theoretical physics since I had had no formal education in it. When God implanted the idea to link science with the metaphysical perspective, it took some time for me to grasp and communicate what God revealed in little chunks.

Whom will he teach knowledge?
And to whom will he explain the message?
For it is precept upon precept, precept upon precept,
Line upon line, line upon line,
Here a little, there a little.
(Isaiah 28:9-10, NRSV)

But God opened my mind and used my applied science experience as a proven inventor to finally make it more believable to me in concept. Then it became possible to present a theory for my fellow scientists who have never heard about new scientific information built into the biblical text. It is difficult for most to comprehend this approach because globally most universities enforce an unscientific evolution religion linked to brainwashed deception and lies sold as science. I am grateful I got this far to conclude a German style - English hybrid text. It has been difficult to find a rare editor skilled in logic understanding physics linked to metaphysics never preached in the halls of higher learning or in church. The unknown in science can only be comprehended through metaphysical revelation but must be confirmed with logic and based on proven historic physics as much as possible. In the final test for the quest to discover truth, theoretical opinions must present three witnesses.

Ultimately, the revealed Mystery Energy Trail will only be or made clear to those God has appointed. Much will not make sense to those readers biased in an unscientific evolution theory not matching new discoveries like light going 300 times faster or the international World Standard Kilogram and its 6 sisters changed after 135 years will lead to correct Einstein's theory and will make nuclear-coal-fossil-oil energy OBSOLETE. I combine new concepts with metaphysic laws rejected by a university environment, which have forgotten the history of mankind recorded in the Bible now thrown out globally from every classroom. I present enough witness evidence information from science to prove a new hypothesis summed up in 160 pages published on the Internet to expose new ideas in 9 Babushka egg concept books.

My presentation is linked to the 50%, forgotten metaphysical concept writings of Sir Isaac Newton, as some are stored in the Hebrew University's Jerusalem basement? His great truths of science embedded in the Bible are once more proclaimed and expanded in nine Babushka concept books published and available free on the World-Wide-Web for everyone to examine. Their historic-metaphysical knowledge will benefit the next civilization after 2018 when God's Kingdom will be established on Earth according to Daniel's (518 BC) prophecy calibrated to a fixed earth axis wobble ending 21 December 2012 and correcting our Gregorian calendar.

Sincerely, Herbert, in a tailspin of new discoveries...

> "All truth passes through three stages.
> First, it is ridiculed.
> Second, it is violently opposed.
> Third, it is accepted as self-evident"
> — Arthur Schopenhauer

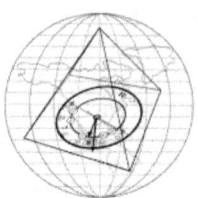

Being ignorant is a curse, but suppressing TRUTH is evil.
-Herbert R. Stollorz

Adding to my energy journey may I present just 12 witnesses from the list of inventor-scientist circle among the hundreds from the past century persecuted and suppressed by the establishment that have triggered an innovation ice age in science by Satan-inspired unscientific atheistic evolution religion, causing mostly deficit disorder symptoms in students ending with a brain-dead MIND and lost the ability to think in LOGIC anymore enforced globally in every university and governments suppressing TRUTH. Any court of law needs minimum three witnesses to establish facts as the time has come should no longer be silenced now nearly 7 billion people living on this planet running out of energy and food. This last generation of the 21st Civilization must stop pursuing a destructive path in danger to annihilate all LIFE on earth not found in the universe or will be judged by ELOHIM.

Check his warning from another JONAH come to the global Town Square announcing: "Time is up - wake up, the prophesied APOCALYPSE judgment has started!" The many witnesses are pointing to suppressed free energy we cannot possible deal here in this energy trail Babushka egg journey and only point to headliners the Internet much better reveals information now available for anyone who wants to widen a knowledge horizon pursuing TRUTH not allowed in universities controlled by a corrupted powerful atheistic priesthood enforcing their evolution religion based on unscientific fairy tales.

Forbidden Inventor Scientist Witnesses: Check it out on the WEB >

1. Nikola Tesla
2. Tom Bearden
3. Joseph Newman
4. Nathan Stubblefield
5. Hermann Plauson
6. Dr. Royal Rife
7. T. Henry Moray
8. Philo Farnsworth
9. Baron Karl von Reichenbach
10. Herbert R. Stollorz
11. Tom Valone
12. T. T. Brown Antonio Meucci

A Synopsis of SPACE – TIME – ETERNITY

"The Astronomer" by Johannes Vermeer, enwikipedia.org

Just One

Frame of the

Cosmos Creation

Movie

When examining GRAVITY, I developed a fundamental notion by first looking at the night sky and investigating galaxies, such as the hour-glass galaxy. I compared it with ancient bronze-gold clocks, exhibited globally in many museums, which are linked to the oldest book on earth, the Bible. This revealed information is captured in the following statement:

1. The Bible presents two fundamentals not understood by theologians. The first verse of Genesis reveals that when infinite light [∞E^2] is spliced out into two entropy laws, it created a "darkness" energy (evening-morning) that caused a time dimension to appear.

2. However, when the infinite ∞ light is converted to a second lower-er entropy level, it shows how elements are formed and crystallized into matter by a visible subordinate-reduced light energy level. This is well demonstrated by investigating the pictures of numerous galaxies through the Hubble Space

Telescope (especially the hourglass galaxy). This light energy level can be measured with instruments and was calculated at 186,282 miles (299,792 km) per second some time ago.

Yet when it was recently measured again at Princeton University (2010 AD), they discovered it ran 300 times faster. That would place Einstein's and Newton's theories in jeopardy? What is not visible to the naked eye is the invisible energy, which has been understood all along since Adam (4004 BC). (Hebrews 11:3)

It is now condensing, according to entropy laws, based on physics, which modern science has forgotten. When light goes faster, we can logically deduct that it must be the cause of a second entropy light conversion to a lower GRAVITY energy level, forming visible elements produced by atoms. This is recognized when investigating Fraunhofer spectral lines that come from many bright galaxies. However, the driving force that causes energy to move is an embedded time dimension, which became the very first act of creation, as recorded in the first verse in Genesis, which declares the beginning of the cosmos.

3. Notice that from a scientific perspective, infinite ∞ light and gravitational energy is the only force without an embedded Time Dimension, frequency-wise. It cannot be measured with instruments. That is not very well understood by modern science that investigates physics from a politically correct worldview, enforcing on society a liberal / socialist, Marx-Hegel-Voltaire-Darwin opinion, while ignoring metaphysical realities. If you are happy with only 50% of the information, to the exclusion of other facts, then be entertained with something else.

Presently, many scientists believe that the light from the sun is fundamentally constant in nature; hence, a flawed theory of relativity was postulated. Another theory is surfacing that the speed of light is exponentially changing, as proven with computers.

Globally, universities collected data and analyzed hundreds of mean average light measurements from known historical mathematics across a 400-year span. However, the biggest proof is found in ancient stone-bronze-gold clocks, exhibited in many museums that measure the moon's GRAVITY, to establishing calendars.

The Antikythera bronze clock fished out of the ocean in 1906 has 32 bronze gears embedded in it that survived the salt water and was dated 100 years before the time of Jesus Christ by a faulty carbon dating method. However, it is much older, if you understand the mechanism of that clock. It calculated out what gravity was over 2400 years ago. Therefore, ancient clocks exhibited globally in many museums are still a mystery to most scientists. They contain ancient secrets that are not understood by our university establishments, but they are now deciphered in my third Babushka egg concept book that describes new discoveries.

Modern physics does not teach closed-loop integration. They have not yet defined how atoms, expressed on a micro-world level, expanded into the galaxy macro-world or by what force. Therefore, they cannot formulate a unifying theory on how Gravity and Electromagnetism, Strong and Weak Forces are looped together. Only the metaphysical perspective will bring it together with logic from a different outlook, collected by many discovered scientific facts and summed up in my evidence trail, explaining what the infinite (∞E^2) and GRAVITY forces are.

4. Infinite ∞light is a cosmos fuel and is converted to Gravity when creating mass inside every atom nucleus, which reflects on and into the sphere of TIME and SPACE. Embedded in mass is Gravity, the second entropy static energy force, but when moved once again, it becomes kinetic energy, doing work much like what is expressed inside every electrical motor. Transferred Infinite ∞light energy in motion, when converted to Gravity in motion, now becomes visible and is expressed as kinetic energy, crystallizing and creating atoms. This is the sum total of all mass in existence, as photographed and seen with the Hubble

Space Telescope. Check out the total Fraunhofer spectral line band-width transmitted to earth, and it will open up the inner structure of the cosmos and show what matter is when investigating galaxies.

5. The slower second entropy, speed of light, that is now visible to our eyes is not constant and is at different points in the universe. It is not traveling straight, which confuses some science quarters that are postulating wild red-shift theories, projecting a "La-la" cosmos. They do not understand physics and are void of logic.

Energy level reductions are dependent on the interaction of total mass encountered in the time/space continuum, as measured from the point of exit of the black hole, concentrating (∞E^2) energy, where all the galaxies are born. They begin to find a home in order to crystallize out as an element, which is proven with Fraunhofer instruments.

6. Gravitational forces are the result of infinite light interacting and reflecting against mass (molecular atom structure in the time/space continuum), thus ∞light-gravity, two parallel energy forces, provide the sustained fuel needed to support everything in existence. Every atom needs fuel to continue; otherwise, it would return in microseconds and collapse back to the Big Bang Dot, where our universe started.

We have examples of GRAVITY energy conversion in the pendulum cuckoo clock, which works in with the same principle as an atomic clock. Both need a small amount of gravity at a given point in time, in order to keep it running, just like our body or automobile needs fuel, too.

7. On the molecular level, every subatomic particle continually receives an added portion of energy for its existence, which is equal to the summa total of its gravitational force/mass, resulting in the nature of the real world, as we perceive it. Time does not exist inside of the black hole, which is conversely expressed in math, when the velocity of light reaches or becomes: ∞ INFINITE, at which point, TIME then becomes ZERO in a teeter-totter relationship.

The world inside the cosmos loop of TIME / SPACE continuum (this world suspended in the space we live in) is seen and is now categorized with our ∞MIND into various elements, which are formed out and controlled by specially-designed, embedded microcode intelligence. It is surrounded by a rooted second entropy GRAVITY force, which is recognized in science as defined by Newton and is now expanded.

Matter forms complex DNA genes controlled by intelligence codes that develop into composite structures like intricate chemical compounds, which are also embedded with microcode functions according to entropy laws only. Nuclear reaction or electricity is therefore a second entropy force conversion caused by gravity moving at different kinetic energy speeds, itself created from first entropy infinite energy exchange.

All of Newton's laws are still applicable, provided we adjust for the IPK, world standard kilogram in Paris and its 6 sisters, which changed its weight. Faster light changes (recently measured in Princeton laboratories) caused gravity changes that are also linked to an energy loop coming from the cosmos. We are still connected to the black hole energy source of our own galaxy, as seen through the Hubble Space Telescope.

8. In physics, GRAVITY is in a teeter-totter relationship with light, just as light is in a teeter-totter relationship with a time dimension. Time is measured by clocks. By connecting the dots, I recently deciphered a number of ancient bronze-gold clocks exhibited in many museums, which tells us that gravity/time was measured quite differently before Jesus Christ's calendar, which needed a correction of 35 years.

Julius Caesar added 62 days a year and Pope Gregory added another 14 days. When the speed of light recently changed 300x faster, as measured by a Princeton laboratory, we needed to add another 4 days in our Gregorian calendar once more, across the total time bandwidth of 7000 Hebrew-Aztec calendar years (4488 BC - 3018 AD), which included the asteroid that caused the earth axis wobble 5 February 2287 BC.

Check out the ancient Astrolabe clock and find out why someone would invest 125 lb. of pure gold to tell future generations about a TIME DIMENSION theory. In the Babushka egg concept books, many new scientific facts are integrated into a fully-rounded 360° perspective that will expand our vision from a physic - metaphysical angle, overlaying it with 6000 years of human history. For reference, an audit trail is included at the end of my story.

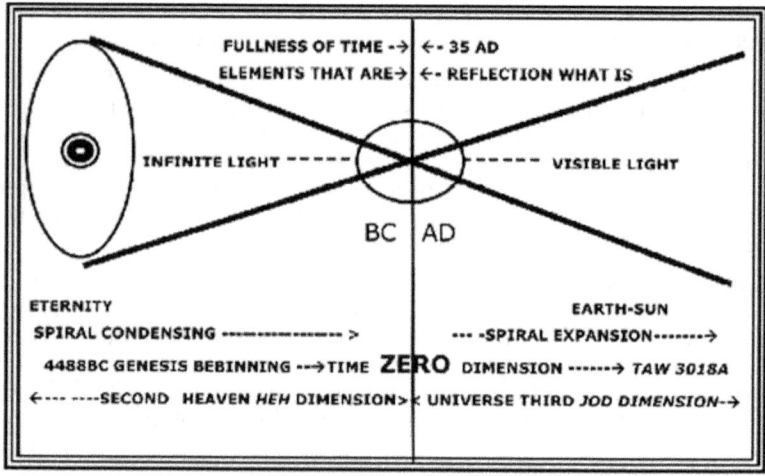

I found another witness on the Internet and copied his opinion below. His observation is accurate and logical, yet it is no longer permitted in our universities:

"Gravity is not an energy source," declare the products of universities, controlled by the energy industries; **you don't get your degrees unless you conform to this farce.**[1]

Rather, GRAVITY is the FUNDAMENTAL ROOT of ALL ENERGY forms. You can trace all energy forms back to its root form of gravity. Every element of mass greater than the H^1 atom is a product of

[1] http://www.c3.lanl.gov/~cjhamil/SolarSystem/asteroid.html

gravity; H^1 has its own gravitational attraction, suggesting that gravity is a property of mass; therefore by default, it has kinetic energy.

Fusion in the sun is caused by inconceivable compressions, which is a result of gravitational kinetic energy. Gravity is not considered as an energy source, simply because it breaks the laws of conservation [quote] $E=mc^2$, which is the transformation of mass into energy waves.

Fusion could be one manner of this equation, seeing that it happens in the sun; the other manner is temperature chain reactions (burning). The reversal of this equation happens with photosynthesis energy from the sun, which is turned to mass (biomass), fossil fuels, hydrocarbons etc. These are secondary energy sources, which fit neatly into the laws of conservation.

Gravity is a perpetual source of mass-to-mass attraction (kinetic energy), which is only proportional to the amount of mass and not proportional to usage, meaning the amount of kinetic energy from gravity is limitless (perpetual), as long as the attracting masses remain the same.

Electricity from Gravity?

As we can see with the tides, waves, buoyancy, hydro-kinetic oscillation and 20% wind, these are all gravity and kinetic root sources that can be converted to electricity, caused by the moon's orbit in relation to the earth. Non-solar, dependent, perpetual energy existed and this has been on going for billions of years. Imagine the total amount of kinetic energy used in that time. Have the gravitational forces changed?

In answer to the question: Not only can electricity be generated from gravity, it can be done totally green, which is perpetually in harmony with the planet.

How long does something have to go on for before it is classed as perpetual, because not even the universe will last forever and ever…or will it?

Possible Applications for a Gravity Motor

When analyzing why our civilization has increased exponentially, we see that its population basically is dependent on two factors:

1. **Energy availability**

2. **The result of work performed**

To feed a population 1000 years ago, it took 80% of the farmers to grow food. In the Middle Ages, it took 60%. One hundred years ago, technology invented labor-saving machines, thereby reducing the percentage to 35%. Work is now performed by gigantic machines and robots controlled by computers, which reduces the number of farmers growing our food to 5%.

I made a list and numbered some energy sources that I converted into laborsaving work, which gives me free time to go to the beach or drive around visiting friends. The stores are filled with food grown around the world, therein bypassing seasonal products, making availability no longer cyclic.

By analyzing the energy list and its conversion, I wondered, "If we selected and switched around the benefit to equal the work application, could we squeeze energy from gravity?"

What looms on the horizon is that oil, coal and nuclear energy will eventually end. We need to look for different, converted energy sources such as hydrogen, wind, solar and water, but the greatest unlimited potential energy source is GRAVITY and OCEAN-WATER. Looking into the sky, we see that gravity is forming gigantic galaxies and even runs my cuckoo clock for free. Could gravity, coming from a cosmic source, replace all the other energy forms and also be free?

In the meantime, the energy cartel moguls lie a lot and brainwash everyone by saying that nuclear energy is the cheapest energy source available to mankind. They say that we need to build more power stations in order to start over again, as the old ones built 40 years ago, need to be replaced.

Obscene profit is the motivator for their relatives and friends. Yet, like a criminal professional gambler, hiding a card in their sleeves, they are not showing the real cost of the disposal of the nuclear waste, which should have been averaged across each kilowatt produced.

The real cost magically disappeared, hidden in the US National Debt that owes trillions of dollars, where even the US Congress dare not reveal it. Tons of money, billions, is spent every year on stupid cyclotrons like CERN, feeding a bureaucracy that believes in evolution religion fairy tales that do not know what an atom is. For 40 years, they have and are still smashing the embedded nuclear intelligence code and continue scraping it off like junk residue stuck on the Nebelkammer wall, much like manure on a dairy barn wall. Therefore, they have not found a unifying theory regarding how gravity and electro-magnetism, as well as the strong and weak force, are connected together to form a universe.

However, why are the many free, clean, alternate energy sources suppressed, which should be redirected to our national policies? Why not use low- tech solar panels, splitting water to gain clean hydrogen, which could totally (and cheaply) fuel all our 21st century needs, globally reducing the horrific pollution in the air, land, rivers and oceans, which is changing the climate?

In the last pages, I've collected a number of energy applications (and I sure could add more) that give us an overview and prompted me to have another viewpoint. Perhaps we need to take a closer look and remove the energy moguls from office. They will no longer have an excuse when they read about the biggest invention that the establishment always says is "not possible", but they should put it on the list for a future, unlimited, clean energy supply for the next generation of mankind.

Gravity Perpetual Electric Generator

TYPE of ENERGY	APPLICATION	CONVERSION
Coal	Train	Steam
Oil- Fossil fuel-Gasoline	Automobile	Explosion in a cylinder
Oil- Gravity-Kinetic Inertia	Hybrid Electric Car	Gasoline Explosion - Electricity
Natural Gas	Hydrogen Extraction	Electricity
Nuclear	Steam	Electricity
Wind	Machines	Electricity
Solar	Doped -Silicon-Panels	Electricity
Magnetism	Motor- Generators	Electricity
Heat - Cold	Oven - Refrigeration	Electricity
Gravity differential	Hoover Dam	Electricity
Gravity differential	Ocean Floor	Electricity
Gravity differential	Flywheel Inertia	Electricity
Gravity differential	Time-clocks	Cuckoo clock
Gravity differential	Machines	Water wheel

Our road map explains how to extract electricity from gravity with a new UREE generator invention not found globally in any patent department. It defines a trail of perpetual flowing energy from the cosmos. *Energy* is needed to fuel every atom, which is controlled by complex intelligence far greater than our biggest computer. Even neutrinos, protons and electrons have complex micro-codes with embedded, controlling DNA. I apply them in a process to better understand the UREE inventions based on physics rationally linked to metaphysics never postulated.

- A Novel *Energy* Trail Defined
- How Does the Gravity Motor Work?
- The Evidence Trail of New Discoveries

Much will be new to the reader, but keep reading, as Rome was not built overnight.

Knowledge takes time to grow.

How Does the Gravity Motor Work?

Electrical Energy is generated by a still unknown cosmic magnetic-bridge explained in my new invention extracting electricity from a "gravity-differential-potential" related to atoms the source of electricity. Science analyzed everything connected along two wires but no information exist what is on the other side of an electro-generator exposed to high pressured water or steam perpetual flowing which results in perpetual electricity via turbines.

This report is different, as it will explain scientifically the other unknown generator side-extracting surplus **Electrons**, which is **electricity** from compressed oxygen atoms, which becomes the media-fuel. Therefore, it can become portable and stored in a steel air tank-bottle and used perpetually, like a Genie Spirit escaping but must return to the ∞ nuclear energy source.

When a differential media pressure (water or air) is decompressed inside a turbine or within a motor-cylinder, kinetic energy is released according to entropy. That causes free moving electrons to be pushed over a magnetic-bridge levitating around the outer nuclear (-) magnetic shell of artificially compressed oxygen atoms now redirected and forced by an electric generator magnetic field emerging on the other side expelled as perpetual free **Electricity Profit**.

However, for perpetual motion to work, the oxygen gas is recycled to imitate a steam or Hoover-dam flowing water principle. Using oxygen enriched air will make it portable but must be returned to the same piston which squeezes the pressure back into the same air-tank now restored 100%. It functions as a compressor which allows the stripped naked negatively (-) charged electrons returning to the generator to recombine once more to the original oxygen atoms and molecules for a (+) recharge. Or in analysis, the white electrical negative (-) wire must return in a ∞ loop to the same generator-winding source in order for electrical energy to flow. That can be accomplished with a GRAVITY MOTOR using gained kinetic energy inside the cylinder and is transferred to a flywheel via a crankshaft.

The motor crankshaft has a flywheel attached on each side to permit perpetual motion for the generation of electricity. But both flywheels have dual functions and must have embedded a hybrid electric generator ∞ linked to a standard electric streetcar motor. The flywheel stores kinetic energy from the released gas pressure inside the cylinder's downward motion and transfers the kinetic energy into the flywheel, which is needed for the return trip, thereby compressing the same gas back into the pressure tank.

Yet in order to gain electricity in perpetual motion, each flywheel has an electric generator stator embedded. The generator's rotating armature is connected to the same shaft, to a rotating armature of an electric streetcar motor, thus jointly turning on the same axis, which is linked to supporting ball bearings. That is duplicated on the other flywheel crankshaft.

The flywheel is turned by the gravity motor crankshaft at approximately 5000 rpm, which turns the outer generator armature. The generator shaft, which is linked to the streetcar motor axis, it will now create a current that is carried over to the other flywheel streetcar motor field winding side. That newly created current will turn the shaft of the second motor, gaining speed, when added to the crankshaft speed.

Since the second streetcar motor is connected to the second generator, it will generate electricity also, but it must be fed back to the first motor field winding side, which in turn creates more electricity. Electricity is gained from the differential of speed on either side of both generators osculating back and forth, which now accelerates as they all are linked to the same crankshaft with the motor receiving added electricity.

The speed of the generator internally keeps increasing and feeding it back to the other side; therefore gains perpetual motion for energy to flow. Both shafts of the generator's streetcar motor turn counter-clockwise along the crankshaft - restricted to a velocity of 10,000 rpm or it will explode.

The flywheel, looped together with four (4) or more cylinders that are connected to the air-gas pressure tank, will eventually be confined by eddy currents and by a full turn of 360⁰ of the crankshaft, which will max out the speed with excess electricity spliced out, which is considered **Electricity Profit**.

In order to return the high pressure of the spent cylinder for perpetual motion, a number of spark plugs must be added to each cylinder when fired from a capacitor, which heats the return air a little and will now expand the gas pressure inside the cylinder on the return cycle to over 100% of the original gas tank pressure, which is applied to each piston.

It momentarily creates a higher power when argon-oxygen gas expands and adds kinetic forces to the return pressure; therefore, according to entropy, it pushes back the compressed gas with a higher force still helped by the flywheel crankshaft, in order to restore the piston for yet another cycle with the original air pressure unchanged.

Patent Description of the
Ultimate Renewable Energy Engine (UREE)

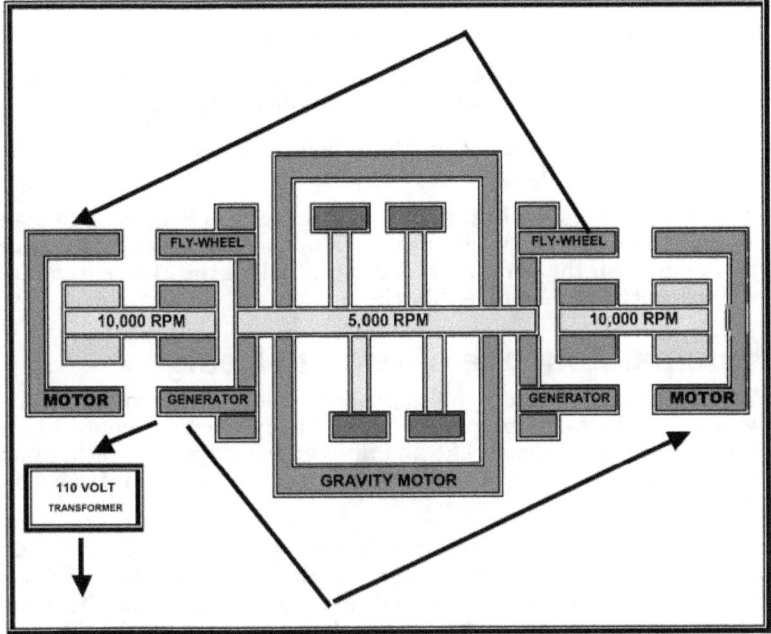

A Perpetual Gravity Motor System consists of basic standard redesigned low-tech components:

1. Apply a 100-year old design of a diesel engine cylinder arrangement
2. On each side of the crankshaft attached is a flywheel. There are two.
3. Inside on each flywheel is embedded:
   - A newly designed electric generator
   - A modified streetcar electric motor
4. Externally linked is a high pressure air tank
5. Suitable transformers for 110 volt network
6. Redesigned spark plugs
7. Suitable battery, capacitors, switches, super magnets and other electronic control article.

It works because of two fundamental thermodynamic entropy laws, which govern physics:

- **A higher energy force will always go one way directionally and cascade to a lower level.**

- **Kinetic energy can be stored in a flywheel.**

- **Electric energy is produced when a magnet is pushed through a copper wire loop. Many loops create electricity.**

- **Perpetual motion is considered when all the above is moving indefinitely, theoretically.**

- **What shakes out in the form of electricity is beneficial to mankind. I call it electricity profit, an unlimited, cheaper, electrical energy for the benefit of mankind, fueling the 21st century and perpetually beyond – My claim.**

Principal Essence of Gravity Design

When I started the venture to design a possible gravity motor deep below sea level, I was not prepared to spend a number of sleepless nights on it. The problem seemed like an impossibly hard nut to crack. Imagining a Hoover Dam power station deep below the ocean surface appeared to be a crazy idea at first.

However, my gut instinct told me that the idea of utilizing a **gravity differential** a thousand times larger than the Hoover Dam could work. Potentially, the benefits for our civilization are so large that the development of a functioning gravity motor should get the Nobel Prize. Even global governments would be happy too, creating a growing tax base. Future generations will benefit and be blessed by the greatest invention in this civilization to provide our rising needs for more electricity.

I hope that challenge of converting a free, existing power from the sky and transforming it into electricity will be picked up and continued by someone more experienced and intellectually inclined towards engineering work that deals with a high pressure differential, such as demonstrated in the Hoover Dam. I need someone with the specific expertise to point out how this idea could be implemented to make it work cost-efficiently. As a philosopher with a hi-tech background, I will present a few applications that will span multiple ideas, with the hope that others would build further, thus expanding the challenge for the greatest invention in our hi-tech society.

My base patent application is very specific and will only cover the fundamentals of physics, as described in my new atom theory. This new atom theory, which was recently published, became three Babushka egg concept books, as so many scientific facts piled up, forming the foundation that will lead to many more applications, depending on their purpose. It will also explain Global Warming from a new perspective, as most scientists are misled by universities that falsify data and lie a lot, in order to get another government grant. They will never find out the real reason. It is like barking up the wrong tree.

Let's examine a few examples in the next illustrations, which will demonstrate how gravity, as a perpetual energy force, works. For a better understanding, I will start first with the ocean floor and how an infinite gravity energy differential is utilized and converted to generate electricity. The Hoover Dam project basically analyzed and converted a gravity differential into usable electrical energy, which is carried overland by wires through a power network to the consumer. To better understand how gravity exerts pressure, I first applied the principle on levels below the ocean surface. However, I discovered that I could also apply it to levels above the ocean, too. If it is allowed by the energy cartel, it could be considered the greatest invention ever in the 21st century.

My cuckoo clock is proof that gravity conversion works. You will understand my gravity principle better, which is now postulated in Babushka egg concept books. All that is needed is to add a little gismo to a cuckoo clock, replacing the chain, and it would run forever. That same principle exists inside the smallest world clock (useful to NASA), which is an atom. It is now functional and applied globally, in order to tell time more accurately to a fraction of a Femtosecond pulse[2], which is needed and linked to computers.

Inside every atom (still unknown to science), I postulate a structured intelligence code which runs like clock gears driven by perpetual motion. Most scientists obviously don't know about it, since for 40 years, they still smash atoms to junk with gigantic cyclotron machines like CERN and keep convincing a comatose, uneducated government that this science needs more money. Apply a little wisdom pearl from my grandma: "A fool's money soon departs."

That could be checked out in my neighborhood by visiting NASA (being a little smarter) who used atoms that had not been smashed, only to discover that atoms behave like clocks, which are needed for space travel. They could tell time more accurately and run forever due to gravity, an embedded mystery energy force that continues in perpetual motion. I just expand on the principle applied to the Hoover Dam because it is larger.

Some modern skyscrapers rest on springs of compressed rubber pads, for insulation against the forces of possible earthquakes. That is a gravity differential also. It could be applied to a pressure piston designed with a huge top weight to compress a liquid, thereby transferring an invisible gravitational energy that is useful for perpetual motion.

In a nutshell, it is explained:

If static gravity energy is allowed to move perpetually, some free energy is left over as profit if converted to electricity.

When looking at pictures of the first airplane built by the Wright brothers, we see that same principle. When a gravity pressure differential force, which was discovered later, was moved across wings, it created lift in order to keep heavy mass levitated. Their first airplane was a

---

[2] Femtosecond pulse 15 (=a millionth of a billionth) reveals how quickly nuclei of atoms in molecules rearrange during a reaction. Ahmed Zewail, Caltech 1999-Nobel Prize.

far cry from today's double-decker jumbo jets that are as large as the Zeppelin, but five hundred times faster.

Therefore, think big! For power generation on a commercial scale for an example, imagine a submarine body containing a gravity generator near the deep edge of any offshore ocean, or use the weight of a skyscraper underneath the earthquake damper. All utilize a pressure differential, like airplane wings, or the same principle applied to the Hoover Dam turbine blades which convert energy significantly less expensive and pollution-free, without dangerous nuclear fuel waste, which kills all surrounding life and horrendously expensive to dispose of.

Perhaps a scaled-down consumer prototype version should be designed first, depending on the required power output. It stands to reason that if we put our cuckoo clock on the bottom of the ocean or high above on top of a skyscraper, it will work. Due to the gravity/pressure differential, the applied weight only needs an automatic cycle impetus to make it run perpetually, long-term.

Check out my new Babushka egg concept books that connect Biblical prophecy with a new atom theory, revealing that every atom needs to be refueled with infinite light energy converted by the second entropy to gravity, for the cosmos to continue. The night sky reveals a universe of billions of chandelier clusters with the lights on to demonstrate that continued energy is needed to exist, just as automobiles do, and you also.

A thought to ponder: Investigating the possibility of my gravity innovation, we see that it is almost a perpetual motion machine going forever like an atomic clock. Nothing like this has ever been thought of by science possibly. It works because of how it is connected to an infinite energy source that comes from beyond the cosmos.

The use of that free energy started 100 years ago, when we began converting magnetism into electricity. Tell me where the electricity comes from, fueling billions of lights globally, and is waiting behind every wall switch to be ready in less than one second? What is its nature? It teaches us the principle that only when energy moves is it useful and considered profitable.

This new approach, coupled with better understanding of that mystery force, could be the fuel for the next civilization, which would make the horribly expensive nuclear power stations obsolete.

Our grandchildren will be happy because they will not be burdened with guarding deadly nuclear waste for 100,000 years; always fearful of another accident like the one which destroyed the city of Chernobyl and surrounding farmland in Russia and recently Japan 2011 and a few other places. Those large areas are now totally empty and void of all life. It is dangerously contaminated forever, but the cost is kept secret by the government who lie a lot, being united with the energy cartel who are only interested in making obscene profit. It also might upset a lot of Arabs sitting on future unsold oil that grossly pollutes our environment, but it could restore America's greatness once more if allowed by the government.

We will now investigate the conversion of GRAVITY into ELECTRICITY from some examples, as well as explain how gravity interacts inside an atom, with a new **Donut Atom theory**, explained in Babushka egg concept books to widen a knowledge horizon. These are free on the Internet. Examining the laws of physics and overlaying it with the metaphysical perspective, for a fully rounded 360° viewpoint was never concluded by anyone, as it is forbidden and ignored by science.

When science is reduced to tunnel vision, it becomes the reason that so many artifacts, exhibited globally in various museums, such as prehistoric bronze-gold clocks are not understood. They still remained a mystery for hundreds of years, as they were not able to find out that the ancients measured GRAVITY.

Combining physics with metaphysics will lead to a better understanding about magnetism, what electricity is, how life is embedded inside an atom, where GRAVITY comes from, how it is produced and the biggest mystery of all, **infinite ∞ light** – the fuel for the cosmos, down to the very level of every single atom, which is controlled by embedded intelligence.

That is a full menu and can only give you some appetizers from the same table. I sincerely hope it will challenge your perspective to further investigate the audit trail mentioned at the end of this story, for a clearer understanding of nature, as well as discovering the reason behind our existence. As we search to the outer region of the cosmos to find life among billions of galaxies, we can only find it on one tiny planet, as we look through the Hubble Space Telescope.

It is the only place with an embedded historical record across a 6000-year span, which crystallized and is summed up in a unique book called the Bible, the foundation of all knowledge and wisdom for mankind. Sadly, that information is rejected by the education department, who is no longer teaching it to our kids, which is certain to destroy our civilization.

Our global society is changing fast and becoming more corrupt morally, having lost the compass. We already see the food chain for mankind genetically collapsing, having been altered by criminal cartels who splice out the original protection immune system of every seed for food - all for obscene profit. It is no longer naturally procreating with the irreplaceable reproduction intelligence, which has been sliced out and will be permanently lost, as it is exchanged for cloning. If you understand physics, it will be linked to the extinction of future mankind and all surrounding life forms.

The mystery "LIFE-SOMETHING" is only found on earth; however, it is now in horrible danger of disappearing, according to Newton's laws. It will terminate mankind on this planet in the next generation, with the food chain collapsing, unless God intervenes. The public is kept in total darkness, comatose and in an ignorant state, unaware of the greatest extinction menace mankind has ever encountered.

Therefore, before it is too late, why not venture out on a worthwhile journey that challenges our mortal MIND, to discover a "universe" a little closer to our home, with embedded life, controlled by intelligence.

The same laws apply inside our own bodies, which verify and establish the reason behind why we are here on this planet. We are the only ones living, with an intellectual MIND that invents, creates and is morally religious. We are replicated in the image of **Elohim**, who created the universe and us, as it has been revealed where we came from, which is only found in a 6000 year old Bible:

> **Let us make man in our image,**
>
> **after our likeness.** (Genesis 1:26)

Obstacles in Defining Physics

When something is believed by everyone as "not possible" and even conflicts with the education establishment, think back to how science started and when it was still in its infancy. Galileo came around and was forced to be silent, condemned in house arrest or pay the consequences.

Our present scientific endeavors lead to high technology, but in many aspects, it has reverted back to the Middle Age mentality, where they believe in unscientific fairy tales, such as an evolution theory, which is void of facts and which became an atheistic religion. An evolution theory is not possible if you understand science, as it violates entropy laws and other laws of physics are avoided or outright denied.

Globally, our universities are now controlled by an atheistic priesthood enforcing their biased politically - correct viewpoints on the majority. They are backed up by powerful lawyers linked with Supreme Court degrees and enforced by the police.

It is no different than what Galileo experienced during the inquisition period that stopped most scientific inquiries. Creating electricity, along with perpetual motion, from a portable, compressed air tank gets the reaction **not possible**; therefore, it may be difficult for any patent authority to grant legal protection, as it would ruffle some feathers in the university enclave who read the above headline.

Our global education system is sinking further into quicksand. We have forgotten many laws of physics, thereby creating confusion, as observed in some UN global warming symposiums where they are bewildered by falsified data from a prestigious university that can only be straightened out with forbidden metaphysics. Modern science is controlled by opinionated speculations, and they do not know what makes an atom tick. For instance, why keep on smashing atoms for 40 years with bigger machines like CERN, only to create more faulty hypotheses postulated from scraped off junk from the Nebelkammer wall, like manure on the walls of a dairy barn?

Thereafter, they applied faulty conclusions to the universe, which left them further perplexed, as galaxies were seen a tad closer with the Hubble Space Telescope. As a matter of fact, it is no wonder that in the midst of hi-technology, no unifying theory exists regarding what **gravity, electromagnetism, strong force and weak force** are, thus creating a universe and our planet, which is the only place where life exists. When they start out postulating from an incorrect and faulty database, it will never lead to the proper conclusions. It is like building a house in the middle of a river on sandy flood basin.

An elusive reality is mostly clouded by a biased political and atheistic worldview persuasion, where they spend a lot of money on many unproven scientific theories that are proclaimed yet lack logic, as they avoid the requirement for witnesses, which needs a minimum of three. But this is forgotten in our modern universities, which elevate unscientific opinions never proven sold as truth. No wonder any new discoveries must first pass a prevailing atheistic pseudo-evolution religious filter, which is void of logic and which globally controls the education of our society, as they now denigrate anything that is not relative to their level of ignorance.

This created a widespread mindset that no longer applies logic, **but rather slides back** to the Middle Age environment which is similar to what Galileo experienced, as they **control new information** in an effort to keep you silenced by **suppressing discoveries** with the tools of the inquisition, which is enforced by a modernized police force.

However, Galileo's oppression can no longer be repeated in modern times as long as the Internet exists. The global Internet created an outside forum platform in order to explain new discoveries, which cannot be controlled by an atheistic priesthood. It is a marvelous platform, where rejuvenated or snubbed scientists can voice their new discoveries that are in conflict with a biased politically correct establishment. Demanding legally that only their established science discipline is permitted in universities, which is just a sanctioned colored filter to cover a biased unscientific atheistic evolution religion.

Therefore, I hope my patent application will be granted, based on redefined and forgotten science, which is explained in an audit trail of 9 Babushka egg concept books that are free on the Internet. You should read them first in order to rediscover the laws of physics, combined with a forbidden metaphysical perspective, so that you can understand my next disclosure, which could make it possible to drive a car without gasoline. Everyone will say, "Not possible." Like the analogy of the Wright brothers, where it was preached in church (with much laughter), "If man was meant to fly, God would have made mankind with wings."

Let's investigate GRAVITY from a different viewpoint, in order to discover that unlimited Kinetic energy embedded in GRAVITY, can be converted into electricity and extracted either from a high pressure differential below sea level or stored in a portable media air tank like a bottled up genie energy spirit.

But one more principle needs to be considered. When we throw and accelerate a stone into the pond, it creates a wave, which is a form of kinetic energy. The trick is in how to capture both inertia levels converted into electricity. That can be accomplished with existing technology. Once more, the new UREE motor design applied to modern technology is different by tapping initially into the first entropy force of the stone's inertia and combining it with the second entropy level force of the pond waves to catch much more electrons and convert them to electricity.

Conversely, first we need to have a better concept about what energy is since science has so many wild opinions, as you will soon discover in this report. Do not judge me crazy yet, when I'm postulating and driving around perpetually in my car without an energy source, but let's sit at the same table with a cup of coffee and listen to an unheard of, new Babushka egg concept story.

Egg < or > Chicken?

The biggest debate, which is not yet settled in our time, is the answer to the question of, "What came first, the **egg** or chicken?" From the metaphysical perspective, life is embedded in an atom, which grows according to intelligence and is directed by an energy exchange. For me that is the egg. Again, we have two energy loops such as the ancient ∞ hieroglyph, which presents itself as conforming to two entropy laws. That explains how we can drive around in a car without gasoline, but nevertheless, I am not hallucinating, as I still believe that we need energy that is converted from a different source.

The **egg** starts life and becomes a chicken and both are linked, in order to gain **profit,** which is a miniature Babushka egg that I call **electricity**. That is the fuel for the big civilization egg again, right down to the level of my house with a car in front of it that I like to drive around without gasoline, as that has become my story.

Where the energy comes from will be another Babushka egg story, and if you have difficulty understanding how energy is produced from the cosmos that fuel every atom. Don't worry, the chicken Gravity Motor will explain it later and fill in the gaps. There are no books available in any of the universities that even come close to what was revealed to me, although I am still learning and writing it down in a foreign language makes it even worse.

After you read the Gravity Motor chicken explanation, come back to this disclosure and once again follow the trail. Expanding a perception regarding ∞ and what it is once more will become **profit** for your MIND, I am sure. Now let's find out how GRAVITY works and how it is converted to **electricity**.

This is a journey of a new hypothesis investigating an unknown effect of **Electromagnetic Resonance Frequency** embedded in Fraunhofer spectral lines focusing on Oxygen Atoms extracting free green ELECTRICITY from

Magnetized GRAVITY- Magnetized AIR and Magnetized WATER

(General science never really investigated the nature of MAGNETISM which can be extracted as free Energy)

http://www.youtube.com/watch?v=o76BRnEDALY

Part 1:
A Cosmos Energy Mystery Trail

Cross section of the Donut Atom

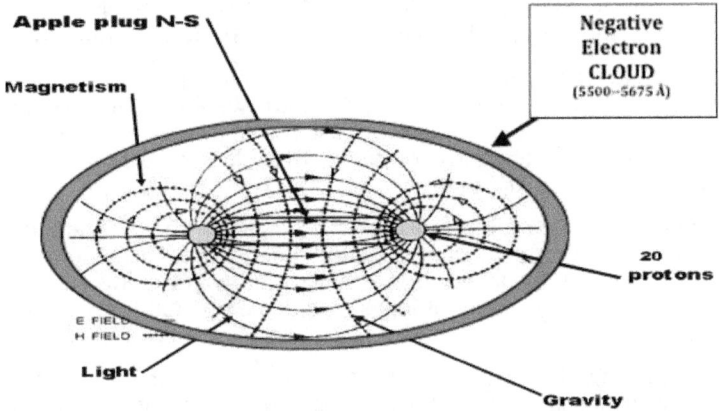

Sixteen (16) UREE inventions extracting free green electricity from magnetized GRAVITY-AIR-WATER must be linked to an energy source. This Part #1 presentation will explain the reason why we get an electrical pulse when a magnet is pushed through a copper wire loop and attempt to outline a new novel theory:

What Is the Nature of Electricity?

Where does it come from?

How is it produced?

Looking out the window from an airplane during the night and seeing millions of lights below, my grandkid one day ask the question, "Where does the energy come from?" Asking the same question to my friend a scientist usually get a standard orthodox answer that is can be explained with Faraday and Lenz's laws. Explaining a conundrum with anonymity did not satisfy my kid. He kept on pressuring what really is magnetism showing me some magnets disappointed his schoolteacher cannot explain it. We agreed that only certain elements capture magnetism stuck on our refrigerator.

I pointed out that magnetism is needed to get electricity, which resulted in the biggest invention ever, an electric motor, but we still do not know how it really works. Without motors we could not build skyscrapers or anything you can buy in Wal-Mart stores. Tracing back the electrical wire to get to the source always ends up with an electric generator, which is turned around by compressed water like below the Hoover Dam or steam similar to compressed water in a nuclear power station where science inquiries stopped for a hundred years.

Science analyzed everything what comes out from a generator, but I cannot find any information to tell me the other side of the generator receiving energy. If an old-fashioned proverb is true "what goes in must come out again" could find the answer.

Do not tell me that energy burning in thousand of cities mysteriously appeared out of nothing from a few copper wires turning bypassing the laws of physics? Pushing a train 250 mph and melting tons of hot liquid iron displaying a horrendous energy collected from something rotating?

Searching for the source of energy looking in the direction of the night sky why do we see thousand lights brightly shining in the universe even without a generator and copper wires. My kid shows much more wisdom than scholars and wonders if these lights shining in galaxies are linked to electricity energy, too?

That put me on the trail to investigate electricity further to find out what is the **mystery energy** in front of on the other side of the generator how is it born, how is it converted in contact with water atoms beyond the generator windings? Only atoms showed the possibility of embedded energy.

Being challenged by my grandkid as a previously professional inventor of many gismos and during the process figured out a new novel energy theory that resulted once more with an unusual invention how we can extract electricity perpetually from Gravity Motors.

It seemed impossible against an establishment believe system, but my guts tells me like before, "It will work". The energy investigation got bigger now expanded into a novel energy theory which will turn science opinions upside down superseding Einstein energy formula and hopeful obsolete expensive deadly nuclear power seen in a new perspective.

In an effort to balance the many accepted science opinions as some are brain-dead fairy tales enforced in our schools, I cross-check them with knowledge from metaphysics and if they match the laws of physics, it tells me that I am on the right path. What you read here might sound strange to you if you are not familiar with metaphysics, but if you apply a little logic, it would not only bridge the unknown, but it could even reveal some previous unexplainable science mysteries to you.

The proof of the pudding is in the eating, and so it is with my **Gravity Motor** patent, which will work to give us free electricity perpetually. It will make nuclear and coal power plants obsolete, severely restricting oil globally in the near future if the energy cartel allows it and could even be a problem with the patent department biased in an atheistic evolution religion.

It will not be long to hear from certain science quarters that Einstein's theory has been obsoleted too now proven with 300 times faster light measurements announced recently in Princeton's University. That should give you the incentive to come back to this 9th Babushka egg concept book ignored by the establishment and find the answers. If you have the aptitude to follow new discoveries, you will be richly rewarded by discovering new scientific applications that are not found in the halls of higher learning.

Let's start with an unknown **Energy Mystery Trail** I discovered which is not yet taught in university books and is linked to the metaphysical ∞ Cosmos, coupled with a time dimension that crosses over to our world.

We will investigate the inside of atoms and compare it to a novel atom theory to discover a new energy trail that is perpetually moving ∞ unknown to science. To explain energy for the lights to be on, forget your past education biased in an atheistic evolution religion for a moment, and sit down with my grandkid and listen to a new story only a Christian grandfather can tell.

It could give us for the first time an understanding about what the nature of **electricity** really is since it is still unidentified to science, as it is an energy only recently discovered in the last hundred years. That mystery energy is still not defined by the experts' misapplied mathematics creating more confusion, but is now revealed on the metaphysical level, which surprised me greatly and became my story.

Have you ever noticed when we cannot explain the laws of physics to my kids, the high priests of higher education will scribble unendingly on the white board math signs nobody understands explaining what he believes. It is a good thing for balance that my kid has a handy traditional grandfather to figure it out applying old-fashioned logic investigating the laws of physics forgotten in our schools. That could answer the **why** question when a magnet is pushed through a copper wire loop we get an electric pulse.

Many scientists are imprisoned by an atheistic evolution religion the Bibles expose the consequence causing the human mind to become darkened, became fools and in their wickedness suppress truth. Comparing a Bible verse [Romans 1:18] proclaimed 1900 years ago with our time reveals that mankind has not changed much. Look around and notice the consequences if you study our global environment linked to massive extinction, pollution, destructions and death the result of rejecting truth believing in the tenets an unscientific atheistic evolution religion.

Let's have a closer look why a new Gravity Motor creating electricity could run forever extracted from gravity always around thus should investigate perpetual motion. Many comatose scientists believe perpetual motion is not possible therefore should explore a double looped ∞ energy source that could widen a horizon understanding nature a little better.

Our natural world gives us many examples for a double looped perpetual motion like our blood circulation designed to continually maintain

Life, and must run perpetually on two ∞ loops, with one loop being refueled every second with oxygen energy cycle.

My doctor tells me that if only one obstruction somewhere in the system stops it, similar to an electrical switch on the wall, it will stop everything or make profit for the undertaker.

However, to make a profit from gravity motors that generate electricity, we need to transfer that analogy and examine a totally new perspective, starting with our world that is linked to **atoms** and ending with the formation of **galaxies**, one is copied on the front cover, which all goes back centered on a bigger infinite ∞ energy loop, thus closing the circle, which we now want to better understand.

That will reveal how nature really operates, how it all began and how it is all related to our earthly environment from a two looped perspectives; one being physics, the other being metaphysics interwoven ∞ together, both will give us a 360^0 viewpoint.

It could make known an invisible energy trail inside our body linked to galaxies coupled to the metaphysic dimension beyond the other side of a galaxy Black Hole returning back in a full circle to our side stated in the first verse Genesis 1:3, "There was ∞light."

If you believe that our blood circulation runs perpetually by itself, you should investigate it further to understand that it is controlled by a built-in intelligence micro-coded (MIND).

Refueled energy is not a mechanical phantom mystery. It must have a source, as it will cross many bridges such as magnetism to the oxygen atom level now discovered, right down to electrons becoming electricity controlled by complex intelligence we will later better appreciate.

The black hole of a galaxy, like the **Hourglass Galaxy** on the front page, could be a useful illustration of what an energy source would look like applied to our earthly vehicles. Photographing various stages of developing galaxies from space reveals awful energy conversions to us, but we wonder if it is expanding or compressing into a vortex. Nobody really knows, as we cannot recognize that which is invisible to our eyes controlled by our finite mind.

But yet is remarkable that our MIND can delineate the very center of the energy frequency spectrum (zero-∞) which is one billions time a billion tiny sliver to become visible light with an embedded rainbow frequency bandwidth of [4000-7000] Angstroms designed perfectly useful for our eyes?

That range can be expanded with different color filters applied to the Hubble Space Telescope and additional compared to Fraunhofer spectral lines could reveal more what is invisible to our eyes.

Fair enough; allow me to expand the invisible frequency spectrum to go further stretching our MIND and acquaint you explaining a mystery energy trail for **electricity** and in the process introduce you to the invisible metaphysical energy that is perpetually coming from the Heh-dimension cosmos to widen a science perspective.

This is postulated for the first time in the Internet Science Forum declaring an **Energy Trail** from the ∞Cosmos, which is linked to a **Time Dimension** that crosses over inside every atom and is perpetually moving ∞ (infinitely). Postulating unfamiliar theories needs proof to establish truth that requires a minimum three witnesses now demonstrated for the skeptic with sixteen (16) applications. The newly discovered UREE is energized by gravity confirmed with the laws of physics proven with logic.

I hope the energy cartel will not oppose a newly discovered energy source while running out of fossil fuel and with anticipation would shut down globally deadly nuclear reactors. If we apply the laws of physics will soon be better informed of mankind greatest danger that all life will be become extinct on this planet in the next generation. It is time to expose a well kept secret by the nuclear cartel willfully suppressed for a comatose public and hopefully would wake up governments to be alerted that most of mankind will no longer be around in the next generation.

It is a well known fact among the nuclear experts that concrete with embedded iron rusting will eventually crack and will expose excessive high radiation embedded in stored hot spend nuclear fuel rods and therefore will in time fracture the cement pool storage in about 70-80 years contaminating the ground or drinking water.

The storage pools containing high radiation should have been minimally lined inside and outside with two feet thick gold plated lead-copper walls on both sides, which would remain hermetically sealed and flexible enough and can stretch when concrete is cracking. It could prevent leakage if welded together as a box.

The way present nuclear power stations are built, storing highly radioactive water, all will eventually leak in time into the surrounding subterrain groundwater killing billions of people in the next generation. But that fact was ignored as obscene profit was the motivator to ignore it. The bureaucrats in charge are not bothered by the consequence that it would cause 100,000 years of deadly contamination spreading over vast acreage of former surrounded real estate to become void of all life which already started in Russia and Japan to prove.

Do not believe the lie that safety was a concern as every nuclear power station will leak the way they were build watch the next 20 years guaranteed, as the media technology stored in tape-disc-drives changed to different technology format with spare parts no longer available. Worse, practical knowledge disappeared with the previous mortal experts no longer around.

For example, the nuclear fix to isolate and cover-up the radiation hot spots in Chernobyl with a concrete-shroud I recently found out, not mentioned on TV kept quite, that it collapsed in a big failure. The scientists should have learned some lessons from history looking at the oldest church in Istanbul Turkey, now 2000 years old, which survived all the earthquakes every since having the largest freestanding overhang.

Investigating what the ancient architect builder did history tell that he studied antique structures and the cause why buildings collapsed in earthquakes? He fixed the problem by grinding up fired bricks to coarse gravel and reshaped bricks with the Roman famous cement, which made the structure flexible and prevented developing stress cracks if shaken up. It was a proven as a good technique to construct bigger buildings still standing in our times.

In addition nuclear power stations should be covered with lead and copper plates welded together creating an outside skin to provide future generation some minimum protection. Concrete will crack in time. It is unavoidable and not ever prepared for an emergency releasing deadly radiation, which accumulates forever in the environment and will not go away.

We still have a chance to correct our problem if the energy cartel would re-engineer the many storage containers following my simple method mentioned here to allow future mankind to live on this earth too and should seriously consider my Gravity fueled UREE and develop it further. Getting perpetual free energy for the coming 21st Century would still guarantee obscene profit but is utterly poison free.

Personally, I am sure that my presented logical concepts in this disclosure in generating new green electricity energy will be applied in the next generation to last 1000 years prophesied by ELOHIM. Read page 9 once more.

Explaining energy on the physical side let's use our five senses controlled by hopefully sound MIND. We are always curious and want to know what is invisible, much like when I was a kid who was interested in what was forbidden, I found out about invisible electricity by touching the electrical plug with my finger. That told me that invisible energy is not static and travels at infinite speed, if you had watched my finger.

Energy, therefore, is not static learned from driving around in my car, but it needs in addition a carrier with embedded intelligence that is controlled by a ∞MIND. Illustrations on both levels, based on physics (our MIND), and metaphysics (ELOHIM MIND) require a lot of analogies to make it plain to those new to modern science denying the history of mankind, which is compiled in the Bible revealing the Elohim.

Looking closer at the energy trail on the physical side, electricity fuels our 21st century, as nearly every item you can buy has needed electricity to create its existence. Comparing it on the metaphysical side stated in the first Genesis verse of the Bible tells us how it **all** started with ∞-**light,** which we will learn more about. However, in order to create a cosmos, it was necessary for energy to flow perpetually, otherwise all would disappear faster than a Femtosecond pulse and end in a big bang DOT.

A Little About Electricity

With electricity, which was only recently discovered, we see the evidence as a major force that fuels our 21st century civilization, as we look at every downtown city globally, which profiles with skyscrapers, airports and the Internet, plus your computer, TV and cell phone linked to satellites. This all came together only in the last 100 years. It only works because of electricity. Yet science still does not know what it is when you take a copper wire loop and push a magnet through, creating a pulse. Increasing its speed creates more pulses, as defined by a scientist named Mr. Hertz.

Then we discovered the three finger electrical rule while watching that strange energy force, which is now linked to magnetism, flowing with so many pulses; some are positive (+) as others are (-) negative   which we call frequency.

Fraunhofer discovered most of our elements on earth in spectral form, like a frequency captured in a visible light band spectrum in the very center from zero to infinite. However, our eyes are calibrated being limited only to see the middle, which is about 186,281 miles/sec Hertz. Frequencies are still a mystery to science, as they are usually linked to electrons, but only their action is visible to our eyes.

Yet investigating it with physics, scientists formed some opinions, but they need to be cross-examined with metaphysic laws of nature, which is suppressed by an atheistic university establishment.

However, we are now confronted with a GRAVITY converting motor running perpetually creating continually electricity can only be extracted from atoms if you follow the mystery energy trail. It will become a big new story that appears unbelievable, but we hope we can become educated applying logic combined with the metaphysics, which has been forbidden and not allowed by the energy cartel fearing revenue loss.

Opinions in science are not facts; therefore, let's have a closer look connecting electricity with two energy carriers, **neutrinos** and **electrons,** fueling every atom and see what we can learn **new** about them following the mystery trail to define an energy source not found in school books. It is fun to be exposed to something new to give us a better understanding at least one version of many opinions explaining the mystery of **electricity**.

The Oxygen-Atom Connection

Electricity, as an atomic energy sub-force, is not yet defined by science, as it is still too young.

Let's trace a mystery energy trail backwards never taught in universities. Starting with two wires connected to a motor linked to every item you buy in any store. Inside the motor, magnetism must first be created in the stator and fed back in order to create electricity and then return it to the rotor recycled to the stator creates more magnetism, which creates extra electricity in a ∞ cycle. In the middle of those two energy loops, we can extract profit, as it does some of the **work** that slaves used to do. The result is an electric motor, which made skyscrapers and space station possible.

However, electricity starts when we push a magnet through a cooper wire loop. Many loops give us more pulses. When many wire loops are put around an iron core, magnetism is produced. If it is put back in the armature stator that rotates a generator rotor and releases electricity considered the biggest invention ever. Therefore, we must look for the interface following the two external wires where kinetic energy is pushing a generator rotor.

Only when kinetic inertia energy is in motion, crossing over a moving magnetic bridge is electricity converted. If the kinetic force is pushed with perpetual motion within a system like perpetually flowing water or steam it will extract perpetually electricity profit in a generator.

But my story starts with the other side of the generator producing electricity and raises the question, "Is the energy path linkage connected to compressed atoms?" When analyzed, extracting energy only works when atoms are compressed being closer squashed together.

That becomes the energy source to extract electrons crossing over a magnetic bridge transformed from kinetic energy into another form of energy called electricity. It only needs an appropriate mechanism via a gravity generator designed to catch as many free ranging, concentrated bunched up electrons levitating around the compressed atoms.

The UREE GRAVITY MOTOR was designed to use pressured air in order to make it portable. I first started with compressed air below the ocean that would give us a thousand times more energy than nuclear power. Air is 21% oxygen, .9% argon, and the rest is nitrogen. But conventional electric power comes from compressed (Hoover Dam) **water**, which is the H^2O molecule with two hydrogen atoms and one oxygen atom; therefore, there is double oxygen (42%) as air.

Through observing nature, we focus on oxygen being used as fuel for many applications. Without it we cannot drive a car, nor could any other living thing exist. All are dependent on the enriched oxygen element. I concluded that the oxygen atom must be the candidate to be able to pass on more free energy than any other elements. I have never seen the size of an atom but science tells me that Oxygen is much bigger and heavier than Hydrogen; therefore, I use percent (%) to make it simple.

Once more steam is compressed water, and water is H^2O, which means two parts of hydrogen and one part of oxygen. Converted 42% or more is oxygen. Therefore, oxygen is needed to extract electricity also proven in the Hoover Dam system compressing water using gravity releasing kinetic energy. In both cases kinetic energy is linked to converted electricity by a magnetic bridge generator explained later. If the energy cartel would investigate water further, it would find out that the embedded energy in water is the biggest energy source on this planet and cheap available using simple modern methods ignored by corrupted governments controlled by powerful lobbies. Maybe this book will convince you of that possibility.

Investigating the Hoover Dam turbine generator profile reveals that electricity is only created when free electrons are expelled from compressed water (42% oxygen) media at higher atmosphere pressure crossing over a magnetic bridge. Let's follow the trail to prove it from history as electricity was used 4300 ears ago, which will surprise everybody.

Electricity was only recently rediscovered, yet it was known previously before by the Atlantis civilization before 2288 BC, which totally perished without a trace. Ancient clocks exhibited globally in many museums like the Tzolkin, Aztec clocks and the other, some came 1000 years later like the Greek bronze Antikythera clock fished out from the ocean

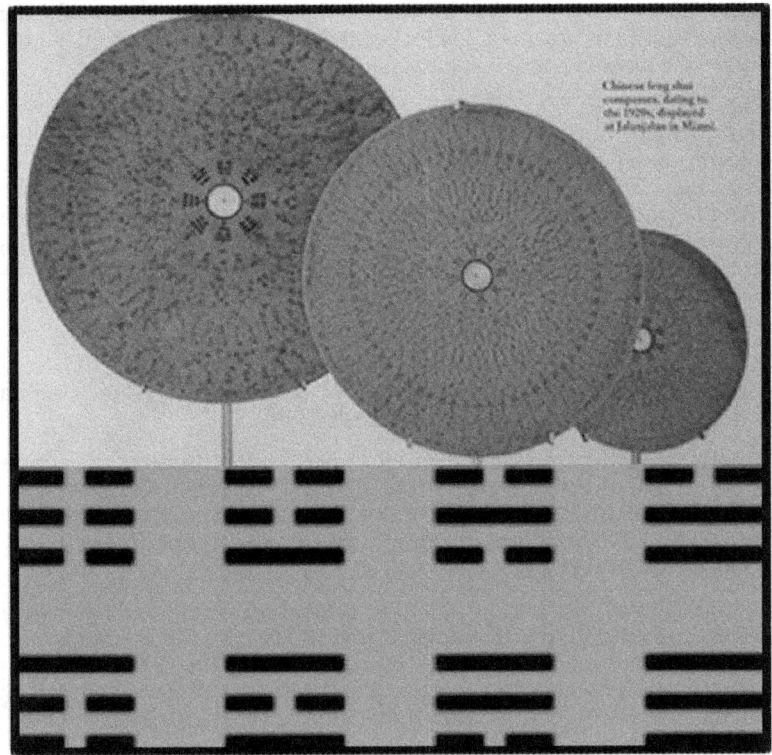

Chinese feng shui compasses, dating to the 1920s, displayed at Islamshee in Miami.

together with Hebrew Genesis tell a story to prove the after-effects of the asteroid changing a world past Noah's time.

Let's have a little detour story enough evidence exists to put it together in a good narrative, if you are educated in both physics and the metaphysical levels checking out ancient clocks in museums. To better understand GRAVITY extracting electricity as a fuel known in ancient history.

Investigating the double oxygen needed as a fuel for electricity lets first follow a historic trail that caused the third Babushka egg concept book to recount the story of Noah and his family who survived a destructive asteroid on **5 February 2287 BC**. His family stayed alive by building the ark (a boat) for 17 years before the flood happened revealed by ancient clocks and the Bible.

The asteroid caused an earth axis wobble shifting GRAVITY which tore apart an ancient civilization which changed permanently a prehistoric tropical world climate piling up three miles thick ice on the polar region, collapsing an **oxygen-enriched** environment (15 atm) down to our present one atm at sea level.

Science now reveals that our present air measured today is only half the **oxygen (21%)**, but comparing it to ancient times is double **(40%)** analyzed in Bernstein - amber air bubbles causing trees to grow three times larger like dinosaurs bigger too. When the prehistoric earth before the asteroid impact was turning seven times within one grand orbit around the sun (check the ancient bronze-gold clocks exhibited in many museums) it was **double the oxygen** content in air at a higher atmosphere pressure.

That caused a tropical climate and produced fantastic, prolific bio-plant growth. But now I discovered that we need double the oxygen in compressed air for my GRAVITY MOTOR moving excess electrons over a magnetic bridge to create electricity more specific explained later linked to pressurized water. This information is useful following the energy trail which connects to a historic ancient Atlantis civilization using electricity for massive genetic modification reported in the Bible and the oldest ancient book of Enoch mentioned in Babushka egg concept book: **Genetic Modification Exposed!**

To save original food - seed and domestic animals - God told Noah to build a boat to preserve irreplaceable "genes linked to the original micro-codes DNA" from Adam's time 4068 BC.

Genetic Modification cannot be done without computers. Scientists, therefore, should investigate an alleged 4000-year old Chinese, three dial bronze plate clock (could be older) showing an ancient binary code [zeros-ones] reinvented in our time without we can not do our genetic modification as well.

Another witness during the Atlantis civilization is revealed by investigating the foundation of a Spanish Cathedral in Peru made from ancient black stone harder than granite, which survived the flood and many earthquakes. The church on top was rebuilt two times destroyed by earthquakes but the foundation thousands of years older did not fail.

Looking closer revealed to me an extraordinary foundation wall built from 50-ton stones that show no damage because they are interlocked with a perfect 3 ft. round-nose-embedded in the saddle. Not even a razor blade could fit between the joints. Another monolith stone on top revealed a dished out, matching ball-shaped depression harmonizing exactly with the matching curvature to allow a rotating movements to align 50 tons of weight accurately within one degree of inclined tilt straight up wall 50 ft. high.

I am an expert in a lifelong machine shop practice and can tell you that this stone can only be machined with diamonds tracing a curvature so accurate that it requires a computer slicer-grinder. Electricity and computer controlled machine technology must have been known 5000 years ago and vanished with the Atlantis civilization. This conclusion was recently verified by NOVA TV programs shown in July 2011 an ancient quarry littered with ancient stones with narrow slots lined up with accurate holes shaped within an inexplicable multi-squared pattern that could only be produced by diamond bladed machines.

The Garden of Eden story starts with Adam (4004 BC) may have had an IQ of a thousand perhaps playing chess with God became an inventor with a higher intelligence. Genesis 4 mentioned iron and bronze right up front helping his son Cain to build a city. He helped his grandchild Tubal-Cain make all kinds of tools from iron and bronze, and Jubal had a talent to play the harp and flute creating all kinds of music instruments to entertain the clan.

One more witness is the Great Pyramid at Giza built between Adam and Noah (4004 BC - 2288 BC). A million huge precision stones could only be cut with diamond slicers therefore needed electricity and machines. Have a closer look inside the open red granite stone coffin polished inside and out perfectly straight and smooth another proof. One side is damaged purposely with an iron hammer as some iron keys were found embedded in the upper chamber holding stones together now in the London Museum.

The Bible alludes to an altar in Egypt. Scientists and theologians wonder why an open coffin with a body and lid missing? From the metaphysical side, it was meant to indicate a resurrection principle which came with a purpose damaged on one corner which is pointing to Jesus' resurrection but was demonstrated with broken corner like broken body.

This evidence has been co-witnessed by the Aztec stone clock, a 20-ton stone monolith clock, which lies next to five pyramids on top dated before the Atlantis civilization (2288 BC). It is exhibited in the Mexico City outdoor museum. It is also revealed with a Genesis Bible calendar defining an earth-axis wobble by recording the ages of people before and after the asteroid impact matching the many ancient clocks exhibited in museums globally which measured a moon GRAVITY in prehistoric times. These are linked to Noah's report, which survived the flood and passed along the information of a previous civilization.

Noah is a real person, not mythology, and it is time for us to become educated and not throw out a 6000 year old history of mankind. Rather, we should discard childish, foolish evolution speculations that do not conform to science. I only mentioned it here when I discovered that electricity generation is only possible with an **oxygen content over 40%** in either water or air media.

Studying ancient clocks in museums linked to earliest history makes perfect sense that Noah's three-level ark built from hard gofer-wood, fasten together with iron had to have an electric generator and lights throughout, It needed no windows.

I would not be surprised to find a GRAVITY MOTOR made from iron that Noah used for creating perpetually electricity controlled with computers. Thousands of animals had to be fed and water to be pumped like a fully equipped cruise ship, as seven peoples could not do it unless they had machines available for the task.

It is really stupid to think that the ancients were primitive cave dwellers or to look for the ark on top of steep mountain cliffs and permafrost surrounded with a mile thick ice. Domestic animals needed a valley with green grass while Noah got busy planting his vineyard. Mankind had lost all of the First Civilization computerized knowledge from Adam (4004 BC - 2288 BC) when hit with an asteroid causing an earth axis wobble creating a global flood.

Nothing could be salvaged being buried deep in silt like the Sphinx 133 ft. deep dug out from sand. Look up to the Giza Pyramid will see left over cover-plates below washed-out. Only the ancient Hebrew songs repeated by grandmothers passed on to children telling the creation story were saved collected in books by Moses, a highly educated prince of Egypt. But logical thinking, why was all previous knowledge lost from ancient times without a trace? The answer is we do the same by converting man's knowledge into a binary code once more. But think further of possibilities like an explosion of a dirty nuclear bomb creating Alpha-Beta rays high in the sky will fry every microchip below for thousands of miles. That will bring everything running on electricity to a standstill, lost forever that included high-tech too. Check Pearl #175 to understand it better.

However, to globally wipeout our civilization needs an asteroid repeat already in orbit. That explains why mankind now born after the flood became cave dwellers forced to survive with bow and arrows in a changed environment no longer tropical. Check the iceman on the Internet found in the German-Italian Alps seen on Geographic Magazine TV.

Therefore today we are so privileged to read the Bible from prehistoric times as ancient history is quite interesting linked to earth with a declining wobble which shortening the yearly cycles Noah did not know thus recorded days. That later really confused the Aztec correcting the sky calculations now facing ice winter season building 5 pyramids on top.

Perhaps, we can get additional proof by reading the Bible from a metaphysical, scientific perspective and match it up with geology and museum artifacts, like I did with ancient bronze-gold clocks described in Babushka egg concept books, now deciphered. They provide a lot of evidence that is not based upon unproven, speculative opinions from the atheistic priesthood controlling the university establishment who have forgotten what the purpose is for their existence enforcing unscientific fairy tales.

Returning to the energy conservation principle, we see that without energy fueling a cosmos; logically no big bang would have been possible. I learned this from playing around with firecrackers. Yet atheistic evolution religion preaches different opinions that all arrive at the same conclusion, postulating that everything we see in the cosmos came from NOTHING. On the other hand old-fashioned science once taught that energy must conform to thermodynamic laws.

There are two. It basically defines an energy conversion now even discovered with embedded micro-codes intelligence and according to the first entropy law can never evolve over 100% or over the ∞ infinite energy level. Over 100% of NOTHING is still a NOTHING. The atheistic evolution religion enforced in our schools denies it. The second entropy law defines it further where we see that if an energy transfer happened, it will always be less than 100%.

For example, in your car cylinder we have an energy explosion that causes the piston to go down, only to end up with less energy. When ∞ energy is cascading according to the second entropy law, mass is formed creating weight we call GRAVITY.

When bunched up atoms are moved and linked to kinetic forces, it becomes another unknown energy of SOMETHING, perhaps electricity. Or stated again, when static GRAVITY force is embedded silently in all matter, it becomes the glue that forms mass, which will coalesce into bigger mass measured by Fraunhofer spectral lines.

We now have a new theory explaining GRAVITY energy, cascading further and linked to every atom (which formed mass), but is controlled by magnetism, which is the glue extracted on the lower atom level and can be useful to do work for us. Moving mass will make Gravity respond to lower entropy and the force of mass becomes kinetic rooted energy, and if extracted, I would consider it to be profit for the benefit of mankind.

Our bio-world and moon prove kinetic energy conversion, which I postulate, can be measured with instruments. I am amazed at how big we make monster machines to do work, which would have statistically taken 20 billion slaves in ancient times to do the work of 5% of the farmers today.

We now use up fossil energy resources at a very fast rate. Check the government statistics; If you added up all the energy used starting from the farm and transportation to the consumer it takes 10 calories of fossil fuel to gain one calorie of food. Investigating energy conversion that brought up a logical question examining a US Energy Department energy policy to convert ethanol fuel from modified corn food.

To me is utter foolishment to use 10 calories of fossil fuel to get one calorie of ethanol for our automobiles not even including the process cost? It seems out of balance. Somebody I am sure pockets the difference. Now observe if the government is running out of money they just print more but forgetting Newton's law the problem still around.

But energy is needed regardless the cost. Just look at most cities globally and watch how fast the skyscrapers are built, which is not possible without electricity. What is separately needed analyzing further, we see on the bottom in nature an unknown ∞ energy force discovered extracted as magnetism from the inside of every atom nucleus, which is linked to a basic electric motor. Many do not know how it works.

How is Electricity Created Inside a Copper Wire Loop?

Investigating further the mystery energy trail on the other side of an electro-generator not yet defined by science exposed to high pressure like the Hoover-dam, I discovered that only when the **oxygen atoms** inside the steam or water molecules are compressed at a greater density will it release concentrated, free levitating electrons recognized as electricity.

Investigating further the mystery energy trail on the other side of an electro-generator not yet defined by science exposed to high pressure like the Hoover-dam, I discovered that only when the oxygen atoms inside the steam or water molecules are compressed at a greater density will it release concentrated, free levitating electrons recognized as electricity.

Negative charged electrons levitating on top of an atomic nucleus will be forced by external compressed atom pressure orbiting around the nucleus much closer in an unnatural higher density around an atom negative charge magnetic shell. If a magnetic bridge passes by embedded in a generator, some are diverted to flow outside the atom environment like a waterwheel on the side of a flowing compressed narrowed down river to do some work which ends up in an electric motor. That fueled the last century development.

Therefore, the levitated free electrons will travel over a turbine generator at high speed being pushed over a magnetic bridge, according to the second entropy and wants to go back to what was normal on the other side.

Once more the turbine speeds up, a gravity kinetic energy differential that will condense atoms much closer as that creates a magnetic bridge from the atom level in order to leave a closed, artificial, restricted nuclear environment. It is sucked off on the other side across a magnetic bridge embedded in a generator mechanism when many wire loops pass a magnetic field, but only at high velocity.

The magnetic bridge is extended by a copper positive (+) wire 10,000 miles or shorter but must return with another negative (-) wire to return in the same generator winding to close the loop back inside to reconnect to the **oxygen atoms** for electrons to flow balanced again around a nucleus. The returned electrons who where forced to work like a slave got stripped of their clothing now (-) naked want to go back to their home to be recharged to live a little longer. If you want to explain electricity better for my kids, let me know.

Therefore, speeding up magnetism becomes the bridge for electrons to flow from a nucleus in the positive (+) wire, but must return with the second wire, which is negative (-), to the same cooper wire loop where it left, in order to go back inside the same atoms where they came from. Any break in the wire system collapses the flow of free electrons and they would no longer be flowing perpetually.

That scenario should remove a mental blockage science believes a false theory that perpetual motion is not possible. However, by investigating nature, we see many applications like NASA uses an atomic clock perpetual moving gears inside the nucleus which giving us time in Femtosecond more precise.

How about mammal blood circulation of two loops, one with oxygen and the other returns to be refueled in the lung with the heart pumping perpetually never losing a beat. Another example, invisible to our eyes when electricity is demonstrated within the nucleus atom where many electrons levitate and move in perpetual motion, in a continuous circular movement on the outer shell of every atom.

Outwardly, electricity behaves the same way with many, newly formed negatively charged electrons in motion, which perform work like low-cost slaves used to do. However, a switch on the wall will stop perpetual motion and prevent electrons from flowing further. At that point, energy is zero just like when the lights are off.

Only when energy flows do we gain profit. Gravity energy released in compressed water molecules will condense in a collection of overcrowded atoms with electrons flooding around, but to us, it looks like a huge static force, which is only detected when it moves. Jump off the table and I will prove it.

If you understand that energy trail on the atom level despite a common belief system could make it possible to drive around "perpetually" with only pressurized air as a fuel in an automobile, which most people consider crazy.

However, writing so many Babushka books should indicate that I am not that unbalanced to postulate what is forbidden in our universities. Yet, I hypothesize that the same principle demonstrated in the Hoover Dam linked to an electric generation system will work inside the UREE gravity motor. It is only miniaturized and applies the same logic.

My GRAVITY MOTOR works by extracting electrons to create electricity. I make it a portable system by using air, or a suitable gas, but at higher pressure. I only changed the pressurized water or steam media into a pressurized enriched **oxygen** gas media.

The mechanism is a little different in a gravity motor as we circulate perpetual pressure movements inside the motor to get perpetual electricity that simulates a Hoover Dam turbine water flow, which is just a transfer of a different media. I too use compressed oxygen atoms but embedded in an air gas mixture circulating across a magnetic bridge to extract electrons from the oxygen atom nucleus.

In order to explain electricity inside a copper wire, I have thought of a good illustration. When I was visiting the Munich beer festival in Germany, I saw that there were long tables where people were seated, with their arms interlocked. When the music played from the music band up front, it made them move in one direction and back again in rhythm singing a song, where all were having a good time, frequency-wise.

To me, it illustrates how electric energy flows in rhythms and is transferred to a distance of 10,000 miles into a thin wire to end on your wall switch. Thinking of atoms being like people could explain electricity. Watching the beer party is like when we push a magnet through a copper wire loop creating sound energy with instruments. Up front, the music band is the generator of pleasant energy for atom people to oscillate like a current inside a copper wire.

Notice that the first person sitting at the table moves to the right, as they are linked with the music (energy) played up front, and all the interlocked people are synchronized, as they therefore move right also. Check out the last person at the table. I liken it to a long wire ending with the one started to push, ending on the other end in a pulse. Now all the people are lined up in massive bodies and move in the same direction back and forth, singing a song, much like electricity that cycles inside a wire through so many Hertz similar to electricity pushing a loudspeaker.

Each copper atom (person) in a wire only represents a weak force, but they still transfer energy from what comes from the stage, like the blasting trumpets that make a lot of noise. The energy (music) from the generator source moves electrons with a voltage analogue, like beer contained in glasses filled with SOMETHING. Observe that the thousand beer-glasses in motion are 90^0 to the flow of cyclic people in motion according to the three (3) finger electrical rule. By analogy the drinking movements in the above illustration also matches the three finger electrical rule as the interlocked people (copper atoms) lift their beer glasses 90^0 to the flow of the (proton) current.

The flow in the copper wire is filled with concentrated polarized electrons charged (+/-) cyclically levitating along the interlocked atoms / people. The people are lined up around a long table: one side is moving right the other left like two wires linked in a loop to a generator. The compressed atom system energy wise is held together with music magnetism from the source like alcoholic electron beer creating happiness, which is a dynamic magnetic force potential turned into profit for the beer-house establishment or profit if you are in the electricity business. It's a good story.

Visit a Biergarten in München, and visualize how electricity works as you listen to the music source and wonder where the beer glass (electrons) 90^0 to the current music flow linked to people-atoms with embedded intelligence moving in a frequency manner following the trail of physics to understanding nature on the atom level better.

The combination difference of the strong and weak forces within an atom is called electricity, which can be stored similar to a beer glass analogy in batteries, capacitors.

It can be increased or decreased in a bigger-smaller beer glass transformer, or changed frequency-wise with higher or lower magnetic density, like a stronger beer. This is linked to a time dimension costing you a ticket to crystallize out as happiness Gestalt or mass.

Before you visit a famous Munich Oktoberfest, read my novel atom theory of the last Babushka egg concept books will make more sense watching the people and drinking beer thereafter will much better understand how electricity is born and transferred.

Therefore, the time dimension (spent in the Munich tent) is the mysterious force in the first creation act revealed in the first Bible verse. It is embedded in all things, which gives it substance, but I prefer to define it as Gestalt.

Yet, I noticed that the newly-discovered energy force, electricity, is somehow connected to gravity via cascading to kinetic rooted energy levels like wind, tides, ocean waves and buoyancy travelling to our earth linked to inertia, the flywheel, bullets, slingshots or bowling balls (explained later), all being coupled to the internal **magnetism** within the atom nucleus. That will make it possible to drive an electrical car without a recharge because our earth is moving around the sun linked to the cosmos ∞ energy source.

Let's look a little closer what we learned in grammar school and expand on the idea. When we pass a magnet through a copper wire, it creates an electrical pulse still not defined by science. But electricity will turn the lights on either in my house or in the cosmos seeing in billions galaxies that should be our start to trace the **Cosmos Energy Mystery Trail**.

What are Neutrinos?

In order to discover a cosmos energy trail, we must first focus inside the atom that is controlled by the laws of physics that are replicated in the cosmos seeing galaxies with the Hubble Space Telescope. Inside the atoms, we see a mirror image like a Russian toy egg that has another one on the inside and discover many more inside that become smaller with the same concept picture on the front. That egg concept I applied to an atom and expanded could understand the cosmos, as the same egg picture pattern reveals how the big Milky Way galaxy egg functions beyond our vision.

Atoms should not be smashed to smithereens by monster machines [CERN], but function like an electronic watch which still has a concept of gears embedded to operate according to complex computer intelligence, much like invisible software is coupled to silicon-rooted inner guts, making an electronic circuit for electrons to travel. We want to collect the same electrons within electricity, but it needs a lot in order for me to drive around in my automobile for free. It starts inside an atom as SOMETHING in motion where the outcome is either telling us time or giving us energy we cannot do without.

Atoms are really miniaturized clocks running with gears that were recognized by NASA measuring time. Inside the nucleus a number of protons and electrons are in high motion perhaps with the speed of light would therefore become mass embedded with a time dimension measured in Femtosecond figured out by Ahmed Zewail got the Nobel Prize 1990.

Every element is expressed in a precise spectral frequency Fraunhofer discovered. If the protons and levitating electrons would stop moving, that would stop the atomic clock, thereby upsetting NASA and delaying another shuttle.

General science does not know when an atom clock stops, time is zero or so the math guys tell us, which would have consequences and would end the cosmos, as it collapses in a reverse big bang DOT in less then a Femtosecond pulse. We recognize time because are all mortal born in a time dimension as all what has life needs energy down to every single atom. In mathematics energy is inversely connected to time like a teeter-totter relationship or expressed:

[∞E = Zero time]

By using the logic of mathematics, which is proven by observed realities, we see that it dictates that energy is needed in order to maintain a cosmos, as well as life on earth, right down to the very atom level to crystallize out in Fraunhofer-Gestalt recognized with our five senses controlled by a ∞MIND. If you do not believe me, stop eating converted bio-energy and tell me the consequence later, if you are still around.

By checking out the Internet pool of knowledge, I discovered enough information to develop a better atom theory, thereby making some logical sense since so many confusing and unproven opinions are sold as facts in order to make money.

Therefore, my Babushka egg concept books are free. Among 32 fractured particles from the many atom smashers like CERN scraping off gunk stuck on the Nebelkammer wall, I noticed that one will stand alone and was named neutrino, which is difficult to define because it does not look like scraped off gunk.

CERN scientists postulate that the elusive particle known as neutrino comes in three varieties: muon, tau and electron. They got excited hoping to discover one of them transforming into a **tau** neutrino in the late 2009. If you believe that gunk smashed to smithereens scraped off the Nebelkammer wall has embedded useful information, it is only published in atheistic magazines designed for those no longer think logically.[1] Magazines and big toys make money one way or another.

A separate 2010 experiment at Fermilab found alleged evidence of a fourth type neutrino, which is just another major puzzle opinion. OPERA physicist Antonio Ereditato notes that neutrinos are so common that although their mass is tiny, collectively they must account for a noticeable fraction of the bulk of the universe, another opinion without a witness.

I think Neutrinos are very useful to transfer energy disbursed throughout space. Energy transfer requires a specialized carrier. It must not vaporize when in contact with the higher ∞ energy source and fast enough to transfer that hot energy down to the atom level, as some are bottled up and could be handy for driving an automobile in the future. Galaxies do shine bright in the sky embedded with horrendous energy I am sure looking at pictures from the Hubble Space Telescope.

Science discovered invisible neutrinos, a strange SOMETHING that is difficult to define, and I think they could be the missing energy carrier. To catch some, they invested vast amounts of money deep down a mine shaft, where they created a pool of heavy water and hoped that some would collide with particles flashing and photographed as proof. I am still waiting because when you base observation that is colored with a biased religious atheistic evolution belief system and then try to understand nature, it will always be misleading, ending up with their conclusions being unproven opinions degenerated to fairy tale.

Checking it on the metaphysical level, I think a massless neutrino is a perfect energy mailman coming from space. What is mass-less cannot burn up when it comes in contact with the infinite ∞ light source, but it could transfer energy efficiently to an atom level at a very high ∞ speed.

When light is infinitely captured in energy, within a math equation, then time is **zero** in a teeter-totter relationship. When time is zero, no mass can or will exist since it is not possible, as mass is the product of the second entropy, where it created an energy shell that is recognized as gravity and always will be bathed in a **time dimension**.

When infinite energy radiated into an empty space, it could only crystallize into mass controlled by a time dimension. The concept of a time dimension was first revealed 6000 years ago in the Bible stated in the first verse in Genesis ("there was evening to morning- one time cycle") but was rediscovered in modern time with Fraunhofer frequency spectrum now categorized in spectral lines.

[1] **DISCOVERY MAGAZINE**, Ghost Particles Shake Physics page 64 (01-02-2011).

It is showing us the composition of every element forming into various levels, which is dependent on how much energy was absorbed in a given time. As it evolved into individual atom structures, they became mass. Going full circle for a second witness, read the first verse in Genesis to see that mass could only exist when a time dimension was born. We can only understand what time is a little because we are mortal, and we do not need the math guy to tell us.

Yet historical energy (**E**) was expressed in an old mathematical formula, such as Einstein's formula [$E=mc^2$]. That formula recently became obsolete when a Princeton University laboratory measured the speed of visible light 300 times faster in 2010. That caused a gigantic upset in science, as the fallout would change many of Newton's laws. It gets worse.

Now as we notice that the World Standard IPK Kilogram and its 6 sisters have become obese after 135 years. We are now questioning whether or not we can have an international standard that does not change.

Global Warming is making headlines on TV, too. We recently learned that a prestigious university, in order to get another grant, was passing on a false database. They sold it to the UN global warming science community, which is now totally confused and in disarray, too. It may not be entirely all their fault, since they are ignorant about the laws of physics that keep changing. That my friend, if you are educated in physics and think about those few scientific developments, will bankrupt 300 years of science, turning it upside down, which will accelerate to a big time problem for the establishment, without any foreseeable solution.

Why not check out the ignored Babushka egg concept books, which are designed to investigate nature from a physics-metaphysics, 360^0 perspective, where you will see much better and could provide the answers for the massive confusion sweeping globally. That could explain why is there so much silence on TV and science magazines, an establishment so desperately quiet?

Maybe it requires a little more time to figure it out as it created a major conflict with the atheistic evolution religion enforced globally in every university. A little fresh science air is needed to be postulated outside the establishment forum and still be free on the Internet. It is now postulated in new Donut atom theories, which will fit much better logically, as it coincides with the many observations of nature that have been suppressed.

An obsolete Einstein theory needs to be corrected with new math formula to include ∞ energy revealed in the Bible [Genesis 1:3] that could open a new science horizon to something totally new.

$$\infty E = m \ (+\infty C \ /\!-\!\infty C)^2$$

To sum up the purpose of **neutrinos**, it must be designed around logic. If I had to invent a suitable energy mail carrier to transfer fuel to each atom, it must be specially designed for high speed and needs capacity to absorb ∞ energy levels. First, the carrier must be without mass, not burning up when exposed to ∞light, and second, it must match the inherent ∞ speed, which would require to be designed like a transformer invisible interconnecting two entropy energy levels.

The black hole of a hourglass galaxy on the front page could be that illustration of a transformer, but it still needs a media to stick the energy together with a strong enough bond in order for it to be a carrier, energy-wise, as it must penetrate dense atoms to shed energy from the cosmos. The cycle is continued to give energy infused throughout gravity on the second entropy level, similar to a strong-weak force demonstrated and duplicated inside every atom.

What is ignored by atheistic science, suppressing the laws of physics, is that the cosmos cannot exist without fuel. It is not very logical postulating that a big bang energy exchange 13.7 billions fairy tale years ago was a one time fizzed-out event and could therefore not, potential energy wise, continue right now, but looking in the universe photographing galaxies shows the same energy exchange in our time.

Why not propose and think of a massive **neutrino** asteroid cloud coming from space? Perhaps our Milky Way galaxy passed on the ∞ energy from the black hole through every atom level of a surrounding cosmos. That would be a good model to explain science better.

I learned that **neutrinos** become **polarized neutrons** inside every atom which closes the link in the energy chain to the infinite light energy source now cascading to polarized visible light fueling atoms recognized in the sun. When infinite ∞-energy is passing from the Heh dimension cosmos it will change into a polarization of the Daleth dimension our visible world forming Gestalt-existence because the time dimension is now embedded extending throughout the cosmos.

To explain this idea to my grandkids, I used the analogy of invisible, but real, water vapor condensing into visible liquid on the second entropy level. This gives us an idea about how (infinite) ∞ light is transposed to finite, or visible light, following entropy laws duplicated everywhere in nature.

Philosophically applying that concept inside an atom, we will again notice two dimensions simultaneously existing: the Heh dimension cascading to the Daleth dimension. We know there is much space between the subatomic components, like heaven and earth demonstrating polarized neutrinos-neutrons pushing polarized positron-protons that is creating Gestalt-Electrons floating around the nuclear center that requires energy to run like an atomic clock. Imagine, if the atom shell is the size of a cherry, then the next electrons floating by would be three miles away. Try to hit something in between or catch as many as you can when driving around with your electric car.

Therefore, logically the neutrino asteroid carrier must be billions times billions in density, yet must start mass-less in order to deliver some kinetic energy transforming into neutrons. This energy perhaps initially comes from a cosmic black hole source to fuel every single atom in the universe. Remember, only when we see the time dimension polarization embedded in anything does it become SOMETHING – Gestalt that consequently will demonstrate the formation of mass.

As a result, mass must be continually fueled with energy, or it will collapse into a timeless state. Mathematics defined it as zero time; therefore, it is reduced to a DOT – the cosmos would disappear in micro-Fenton second, as well as you and me, too, if no energy flowed anymore. Seeing the lights on and looking into the night sky, so many galaxies prove it. Your MIND can read this story, which also proves that energy is still flowing inside every atom to maintain our lives.

A **neutrino-asteroid** from space would fit the ticket, therefore, it must be designed as being an efficient energy carrier and must be big enough, clustered in billions bunched together, in order to pass on kinetic energy by colliding or passing through every atom mass center being transformed into neutron energy creating a gravity magnetic shell levitating electrons in order to maintain the whole atom existence.

Otherwise, the universe would dissolve or collapse into zero - the timeless area of non-existence, which is not possible because it is controlled by the cosmos **Elohim∞MIND** intelligence of a higher ∞ order, which is recorded only in the 6000-year old history book of mankind, still the best seller, as well as a unique Bible book. It has been translated in over 400 languages and is embedded with tons of information, which should not have been thrown out from every classroom, in order for us to be better educated.

A young earth matching mankind's history is proven with science measuring all the helium collected on earth, which cannot escape our air atmosphere. The rate produced by the sun's radiation is not over 10,000 years. That could be crosschecked with another witness measuring the changes of earth's magnetic field as measured against stars that dovetail to about 6000-15,000 years matching, too.

It is sad that science has deteriorated to the lowest level and believes that their great, great grandfather was one of the monkeys eventually decided to walk upright and would finally become educated, but unfortunately it will have consequences in science. If we follow the trail of logic according to the atheistic evolution religion, it concludes, with any luck, a monkey lastly arrived with hopefully higher intelligence and became the president of the USA. That scientifically would make the laws of entropy obsolete ushering a new World Order.

It seems we recently crested the evolution summit as our global society became engulfed in atheistic socialism and changed the ancient oracles of "good and evil" now morphed it to be "relative". That will become the last station of our existence on this planet for our human survival that will self-destruct our existence according to the laws of entropy.

We see globally the proof of the consequences of an atheistic persuasion as ignorant senators and judges allowed 60 million babies to be killed just in America, selling their human body parts globally to hungry abortion mill laboratories just for profit. But the laws of physics cannot be voided denying genetic fertility laws. The law of genetics spelled it out that when the normal birthrate of (2.3) is below (1.8) that culture is irrevocable doomed in one generation. In America it has become (1.1) and in Europe (.9).

Germany just announced that it will be an Islamic Republic (Hamas-style dictatorship) in 20 years. Pay attention: Western Civilization will be gone with Christian churches converted to mosques guaranteed in one generation, irreversible according to the laws of physics denied in a formally Christian society. History is once more repeated in Europe and America only with bigger consequences.

If you believe in children's fairy tales that evolution can evolve and gain micro-coded intelligence embedded in every cell of our body controlled by an intelligent MIND, you just ignored science and no longer need to read my story.

It needs first a few lessons from Bill Gates recently stated on TV that his sophisticated Microsoft programs cannot be compared to the million times more complex intelligence codes residing in our body. Or check it out and get educated how scientists are reconstructing the human genome now four more bio systems from the outside in described as:

GENOME - TRANSCRIPTION - PROTEOME - CONNECTOME - METABOLOME, by Brooke Borel.

Making the Connection, **Popular Science**, March 2011.

> It is sad, now a fact that if any teacher in our schools even mentioned intelligence will be fired from his job believing it to be religious (from watching 2009 Larry King's TV program interviewing a high school teacher from Chicago). Are we now governed by brainwashed stupid, politically correct people in high office based on Newton's laws?

I hope that my road map of explaining a mystery energy unknown to science would be investigated with an open flexible mind which could make an entrepreneur incredibly wealthy by producing **electricity extracted from oxygen atoms** a possible better story applying logic linked to the metaphysic revelation.

In defining perpetual electricity profit, where it comes from and how it is produced inside the generator requires a higher logic. We must look further inside the atom to understand the nature of things, which is not found in any university textbook; therefore, we just follow the metaphysical path.

It is beyond the scope of this book to explain it all, as it is not possible. You will need to read the other two special Babushka egg concept books that deal with what a Donut Atom is, with its neutrinos pushing positive protons, which speed around like the analogy of a horse and carriage, going around the nucleus, according to Newton's laws and combined with the three finger electrical rule.

That will create excess electrons being expelled and levitated out the window from a mechanical generator magnetic bridge because the second entropy law creates a negative potential and must equalize which became my story.

The Hoover Dam turbine generator, in principle, does the same thing, if you analyze the electrical energy profile. All of this is mentioned in this report and more. Check out the Evidence Trail that is continued on the last page.

What are Electrons?

When electrons move, we call it **Electricity**. It is converted from magnetism, which is a sub-force created within every atom. Every atom is fueled with energy that is controlled by intelligent micro-codes and therefore moves like clock gears, which was demonstrated by NASA. When we apply it to an atomic clock, it tells time to a Femtopulse. Inside atoms, the moving neutrinos force will be transformed into polarized neutrons, which became Gestalt as the time dimension is now embedded. Therefore, it can push protons in carriages around within the nucleus at high speed, which still needs to be fueled continually from an infinite cosmos energy source for any atom to exist and live a little longer.

Once more when an energy force is coming from the black hole and goes right through every atom nucleus I think of super **magnetism** path since it has polarization embedded and functions like a cosmos transformer starting out with no time dimension where energy is ∞ the highest and ends going through space stuck with a little time dimension. They experience a transformation of speed reduction as scientists Antonio Ereditato remarked that neutrinos could have a little mass eventually only noticed collectively bunched up.

Justifying physics ending with some kinetic energy which tells us that little pre-gravity got formed before molecules crystallize and we express it as materialized atom-mass categorized in Fraunhofer spectral line. That developed into Gestalt by passing on intelligence from the ELOHIM and is now pushing positive polarized protons like a horse and carriage at a lower speed like visible light inside every atom.

Now you know why the sun shines. That could explain a cosmic **magnetic** highway for a frequency of an electronic picture produced in a camera to travel through space still transmitting from millions of vacuum miles away. If you have a nothing vacuum but a picture frequency send from space must drive on SOMETHING only **magnetism** fits the ticket?

To avoid a proton collision racing around inside a nucleus charged with the same positive polarity therefore they are forced to move sideways. That creates 90^0 to the current and will produce a negatively charged magnetic donut cloud according to the three finger electrical rule now levitating forming a circle around the nucleus.

Adding a billion times a billion neutrino energy packages continually coming from the cosmos source provides sustained energy in ∞ motion that causes the donut nucleus to vortex at high speed creating magnetism. It will trigger negative charged electrons to be repelled on top of the negative charged magnetic cloud according to the normal law of physics now levitating pushed away and around in the same orbital direction.

When compressed water-steam-air is accelerated in front of an electric generator, it will compact atoms much closer together creating an electron overflow of a strange moving weak nuclear energy force that travels continuously around every compressed atom outer-shell. Each atom being unnaturally condensed together will therefore cause millions of moving levitated electrons to jump from one orbit to another being pushed by a negative charged moving magnetic donut cloud at high speed.

Given the opportunity of a magnetic generator bridge passing by, it will direct electrons crossing over a generator copper wire side still being pushed by magnetism produced by moving protons within the atom recognized as a high voltage potential similar to electricity. But it is held together by a strong Ampere's force law, which I believe forms a GRAVITY enclosure surrounded with a billion billion atoms and molecules bonded together with magnetism.

Let's take a little detour. It is just a novel idea to lay down a foundation for how electricity is linked to **perpetual motion**, since ultimately energy must have a source starting with the cosmos. This generation of students has grown up lacking the basic education in science, which has been infected with many wild atheistic opinions that are biased by a third worldview relativity that does not match science. You will not graduate unless you conform to your prejudiced professor's brain dead atheistic evolution religion.

Looking at the oldest ancient Tzolkin clock on the front cover, which remains a mystery to science, it not only reveals an ancient calendar with an exponentially declining earth-axis wobble, but from the metaphysical perspective, we can transpose the hidden knowledge of the ancients, in order to understand nature better, such as my new atom theory.

I see an overlay in how physics behaves on two ∞ energy circles, locked together in the center with a 20 energy gear. The driver Alpha(+One) force, which is the main energy power coming from the center of the cosmos – **Elohim** – beyond a galaxy black hole on the other side. This is explained further in the Babushka egg concept books.

The two ∞ **energy** Tzolkin loops (18-13), side-by-side represents a Heh-Daleth dimension like heaven and earth, which creates an energy flow how the cosmos came into existence from infinite-light (first entropy), which is linked to gravity (second entropy) at a lower energy level, becoming visible light that crosses over in the middle to a zero-time dimension dot.

The same energy exchange principle is illustrated in an electrical transformer positioned high up next to your house, which has an energy loop 10,000 times higher than the other loop not connected physically with internal wires, but reduces energy **magnetically** to a lower level of 110 volts. Similarly in another example, when a piano middle C key is hit and listened to on octave higher C note, it resonates not connected with a little lower sound.

The same law is demonstrated duplicated inside the smallest nuclear transformer and replicated in every **atom**. The many analogies are good for my kids to understand physics when I explain complex science on a lower loop will now understand how a UREE gravity motor works analyzing the high pressure loop on one side being mechanically linked to a separate lower pressure loop giving us free electricity following invisible energy entropy laws.

The fuel for driving perpetually without a recharge in an automobile must basically follow entropy laws of physics, too, It can be explained on the higher metaphysical transformer level side investigating the ELOHIM∞MIND, expressed in physics as the Alpha (+One) force (20) as looped invisible forces to a lower level when mass crystallizes out. (Check the process illustrated in the front picture of the hourglass galaxy) Then it is converted into kinetic energy creating the first energy source of infinite light (18) symbolized in a math ∞ hieroglyph ending in visible light (13) down to atomic levels.

Then energy transfers in the second loop according to the second entropy law, which is demonstrated in a real Big Bang explosion getting colder, measured by Kelvin or stated in the first Bible verse of Genesis. This is explained a little different from what you hear in school. Being mortal, my cuckoo clock illustrates the same cosmos principle and needs a higher MIND to wind up that clock so that gravity on the second entropy level can convert energy activated with a chain force for another cycle to run perpetual like my UREE gravity motor.

It is a mirror image of the first entropy level with Alpha(+One) force energy controlled by a **ELOHIM-∞MIND** to wind up his cosmos Jod (pronounced "Yod") dimension cuckoo clock running forever. Both entropy principles need energy like winding up every day a cuckoo clock on the Daleth and Jod dimension dispensing a force to run continually to make perpetual motion possible now linked to electricity.

Therefore, I conclude that the cosmos system shows an energy exchange that must have a gigantic ∞MIND intelligence embedded as well because it comes down to every single atom that functions within very complex nuclear clock gears that are in perpetual motion, which NASA uses to measure time for space travel and synchronizes globally the Internet.

An atom clock is the most accurate hi-tech clock available globally controlling the timing of computers and even our bodies maintaining life. Therefore, it reveals a complex miniature design created from an intelligence MIND. Follow my path once more in order to learn what is not taught in our universities. The cosmos energy starts with Infinite ∞Light, perhaps going through the black hole that we are still connected to in our own Milky Way galaxy and ends fueling atoms.

Once more, invisible ∞Light cascades to the second entropy GRAVITY level, as seen when investigating many pictures of galaxies from the Hubble Space Telescope, that crystallize into elements measured with Fraunhofer spectral analyzers, therefore attaining mass, further slowing down to kinetic energy and bottoming out at the end station (atoms).

The energy trail is still looped back to the infinite ∞ energy, as we discover that magnetism is the cosmic highway linked to every atom when protons and positrons circle around the North and South Pole nucleus, which is fueled by invisible, highly charged neutrinos coming from an energy source at an infinite velocity that we cannot measure. The resulting magnetism collected on the lower nuclear transformer atom side now applied to every electronic gadget on the high-tech market. If something moves electronically according to Newton's law, it points to energy.

I believe that the invisible, neutral highly charged Neutrinos are expelled from the ∞-energy source and travel along a magnetic path right through a cosmos cloud of galaxies and dark matter fog disbursed throughout space. The neutrinos will reduce speed and lose some energy but still energize electrons inside every atom nucleus found on earth to end up being recognized as electricity, which is indicated in the ancient ∞ math symbol with the two loops in parallel. Since it is one continuous ∞-loop from the cosmos right down to every atom, with energy perpetually in motion, we can tap into that preexisting energy system and splice out some profit along its path, for our benefit. That could make driving around in automobiles without gasoline possible in the future if we would know the linkage.

The trick is to find that linkage and is now discovered illustrated in a 4000-year-old Tzolkin clock on the front cover. Observe three energy force-loops: one in the center is the major driver energy loop revitalizing two subordinate energy ∞ looped together.

The immeasurable [20-gear] center is the driver loop Alpha(+One) force, which interlocks with the two lower, secondary energy loops; one is ∞light [18-gear] cascading to visible light [13-gear] thus interconnecting in a ∞ pattern. Thereby it revealed a refueling principle of a heaven–earth cosmos dimension from the high to the lower energy potential, just as a gasoline or electric motor demonstrates the same principle on the lower physical energy exchange entropy level.

To assist you in gaining a little more understanding, allow me to explain it this way. The end-station of a declining energy path inside every atom is an electromagnetic force that circles around the nucleus, creating a negatively-charged magnetic Donut, which is the glue (like beer in Munich) that allows atoms to stick together. Because protons with its positron polarization in the nucleus center are positively charged and therefore would repel each other if not held together by an external gravity force.

In closing, electrons are born with negative polarity when they emerge from the nuclear energy hot furnace Donut Atom hole, where they are converted from crystallized, resonating proton energy becoming GESTALT in the Time Dimension at a certain speed. Electrons are expelled from the atom chimney south pole, being pushed out by a magnetically positive proton dog force levitating on top of the magnetically negative charged outer cloud shell perpetually moving at high velocity.

Some eventually lose speed sinking lower to return for a recharge by reentering the furnace through the positively charged North Pole to repeat another cycle. But if repeated a trillion times and linked to the time dimension as explained in a new Donut Atom theory, they will enter the next stage and become protons. Discovering how atoms could safely produce energy, I applied the principles to design UREEs #11-16 that generate totally free electrical energy forever because of being connected to the ∞ cosmos energy source.

Once more, when initially created, electrons gain a little mass by being recycled back and forth from the atom south pole around the negative charged cloud reentering the north pole back in the oven donut hole being baked again becoming heavier. That process can be visualized by investigating galaxy photographs that show how crystallized energy eventually grows by getting fat to add mass by moving a trillion times through the same chimney donut-shaped black hole. It gains weight, though still controlled by embedded intelligence.

Electrons eventually become too heavy and morph to become protons breaking through the inside Donut hole wall chasing the tail of the other proton dogs. If too many dogs accumulate in a confined atom apple, it can be explained from a Fraunhofer perspective defining various atoms at different resonance frequencies in the Periodic Table.

Summed up: the formation of mass starts when orbiting electrons getting heavier and become protons, which in turn will divide like cell division, seen under the microscope. In this case, the original mass is split into atoms growing into molecules of different atomic weights stuck together by magnetism and ending in magnetic gravity forming mass.

That is a parallel process seen in nature. Our bodies exchange the ∞ energy via a built-in mortality DOT continually creating a bigger mass eventually to become a baby with a mind with the same DNA blueprint embedded originally from the same ELOHIM intelligence.

For atoms or Life to exist, God's Alpha(+ONE)energy force needs to emanate from, and perhaps, pass through our Milky Way galaxy hole along the energy provider highway. Investigating DNA conforms to the same blueprint for forming atoms as the intelligence is controlling the laws physics always is the same on different levels like the different sizes of Russian Babushka eggs.

A novel atom theory now can logical explain physics how **Electricity** is generated on the inside of every atom by a negatively-charged magnetic donut cloud that repels negatively-charged electrons circling around, which levitate now as a weak force, yet it is held captive together by a positively-charged proton / positron strong force, from a center surrounded by GRAVITY. This is recognized in physics as mass (comparative to a mother hen that holds the chicks together under her feathers), when opposing polarity electrons (chicks) repel spreading out and want to move or levitate somewhere in the neighborhood.

Magnetism Linked to Atoms

Postulating the ability to drive around with electricity everybody is convinced seeing electric cars circling around a golf range. Yet, postulating driving a golf-car perpetual without a recharge fueled by a genie bottle of compressed air would you buy it? I found a church key embedded in nature, which could release an invisible useful energy genie spirit.

Recently I discovered an upside down theory that magnetism is the most important force keeping the solar system in orbit. Modern science has no logical theory regarding how magnetism is produced inside the atom nucleus and thinks that only gravity must be the force in space seeing planets orbiting the sun.

Smashing atoms (CERN) or building bigger bombs with left over fragments, will not reveal the inside of an atom. We only know that a magnetic force either sticks or repels. I discovered this as a kid by playing around with magnets. Eventually we learned more and discovered electricity and defined it with a basic electrical three-finger rule, which made skyscrapers, airplanes and everything you see in a Wal-Mart store possible.

However, so many opinions in science do not make sense. They ignore metaphysics, which started me on the path to investigate what electricity is. I was not happy, which urged from within to formulate a totally new atom theory, which is published in Babushka egg concept books, free on the Internet.

In a nutshell, I discovered that Magnetism is a major force, and the glue that binds atoms together. However, my visualization was expanded by the discovery of the ancient bronze-gold clocks, exhibited in many museums which matched the many published pictures on TV that show the best colorful galaxies, as photographed from the Hubble Space Telescope.

At the same time, I was interested in the origin of where we come from, so I began to study Genesis the first book in the Bible a little closer. What came together was revealed, but I could not keep quiet about it because it was too logical.

For me, as a technical innovator, that was a huge incentive to dig further. What impressed me most as a previous successful inventor of many technical gismos was what the connection of the infinite math ∞ hieroglyph symbol seen on an ancient Tzolkin clock with two loops crossing over meant and why would it fit inside an atom theory. How to explain it?

I gained a little better understanding initiated looking closer at the picture of the Tzolkin clock illustration I notice a gigantic Cosmic Magnetic Transformer to the very level of every single atom. Applied to an electrical motor to find out how that greatest invention of modern civilization really works when we cross over an electrical current connected like a ∞ loop as all must conform to physics follow the energy trail. That is creating MAGNETISM in turn creates electricity, which again creates magnetism, switching back and forth from the stator to the rotor, which turns the darn motor thing at high speed, but why?

Then NASA's atom clock gave me a tip that tells us the most accurate time and runs like a ∞ motor on something, perhaps a force that is still a mystery? The electric motor design represents a ∞ function, which is used in math and which has now been found, where it was embedded 4000 years ago in an ancient Tzolkin clock, which I copied and put on the front cover.

I cracked the code regarding what the ancient Tzolkin culture next to the Aztec calculated, which I only found in a picture clock, which graphically demonstrated an unknown calendar design of a previous earth axis wobble, measuring time. A wobble would screw up any calendar, which Julius Caesar found out about 45 BC and corrected a Hebrew Genesis calendar by adding two extra months in a year, but it also could reveal more, if it linked the cosmos, since earth is part of it.

What really surprised me is that the ancient Tzolkin clock, the oldest, also demonstrates an energy exchange while postulating an atom nucleus. It seems to match. It amazed me to discover that this clock started with Noah's time (2288 BC) linked to electricity, which was known during the last days of the Atlantis civilization explained previously.

Advanced technology and ancient knowledge disappeared with a destructive asteroid **5 February 2287 BC** that caused a global flood triggered by an earth axis wobble. But ancient clocks in many museums around the world still reveal a remnant of prehistoric science and became witnesses to what was around before the asteroid destruction.

Clocks measure time, and atoms cannot exist without a time dimension. They are all linked to the energy exchange system. So if you would like to understand the unknown energy force buried deep inside atoms that fuel our civilization and a little of ancient history should be cross-checked with ancient clocks.

The surviving Aztecs where trying to correct a disturbed earth axis wobble experiencing unknown long snow and ice winter seasons caused by 3 miles thick ice on the poles that became very difficult for growing food. Confused that the sky calendar was changing while correcting it to a familiar pre-flood calendar system, they kept building 5 pyramids on top of each other in Mexico City outdoor museum.

The observed calculations kept changing and needed correction to save money building another pyramid on top of the old one. It took some time to figure it out to know that the previous asteroid had changed the earth axis remembering the old calendar system, which even confounded the Hebrews recorded in the Bible on the other side of our earth.

The high plateau of Mexico City 7000 ft. was carefully chosen as a sacred place and build the first pyramid in the middle of a lake as a mystery light curtain in the sky surrounded the lake and ancient compass all pointed to the center.

Today, we know that the internal earth **MAGNETISM** is moving the Magnetic North Pole 40 Km a year and recalculating shows that the magnetic travel position matches historic events recorded in Jewish 7000 year range calendar. Check the Babushka egg concept book dating will verify it. Theologians and scientists alike even today are still puzzled like the ancient Aztecs and do not understand what the Bible recorded linked to mystery bronze-gold objects exhibited globally in various museums displaying many ancient clocks now deciphered in my Babushka egg concept books.

Eventually the Aztec priesthood invented a cheaper system and replaced building pyramids with a high-tech clock fashioned from 105 skulls heads made from stone the size of bowling balls piled on top of a wall lined up mathematically each measuring 14.30 Gregorian recalculated years. I calculated this 14.30 ratio of years from stone step outcroppings of the Sun Pyramid, which was verified by Daniel in 518 BC with more fractions after the dot 14.305789, a high official educated in the Babylonian mathematics.

That made dating his future apocalypse prophecy possible linked to Antikythera 32 bronze gear clock appearing. That explains why Julius Caesar 45 BC could then correct an Aztec projection that the earth wobble axis would finally come to rest on 21 December 2012 AD. It was later verified with our modern computer calculations applied and adjusted to a Gregorian calendar.

If you want to understand global warming, then you must measure world history correctly in a time table from 2287 BC that will add knowledge to see the consequences on how the cosmos operates. It must be continually fueled with the same energy in a time dimension, in order for the lights to exist, which are still seen as galaxies that shine brightly and gives purpose back to my cuckoo clock to remind me about my mortality. The lock of the Tzolkin clock reveals the mystery what holds it all together to the very inside of an atom. It basically cooperates with three energy loops in order to interchange and transfer energy, yet it still needs a bridge, which has now been discovered as **magnetism**, but is still unknown to science.

Any atom on the inside with moving protons and suspended electrons is a miniaturized mirror image of the universe with suspended galaxies, only larger, which is now better defined in Babushka egg concept books. It is an analogy that is explained as a smaller atom egg, which is replicated in the big cosmos egg and simulated in the hourglass galaxy front picture, surrounded by a million lights, which now shows us that energy is also flowing in the universe. The small atom egg produces electrons when the two levels of neutrinos and protons are separated by magnetism, like the Tzolkin clock illustrated.

Inside every atom, is a magnetically charged, donut shaped nucleus of protons creating a negatively charged polarized cloud that causes the negatively charged electrons to be pushed away, levitated from the central nucleus, as the weaker force. We could visualize the activity similar to a major airport linked to the outside world, as each airplane represents an electron, which is a carrier of a small portion of energy, returning to the center. It comes and goes with SOMETHING energized, embedded with invisible intelligence pilot MIND on the inside.

Remember, the comparison when the nucleus is the size of a cherry, the electron orbit is spaced three miles away like electron airplane with intelligence controlled with complex micro-codes embedded, similar to DNA. It can be summed up as a system and operates like an "airport MIND" - still a mystery. Read that analogy once more to understand electricity.

It overlays the same Babushka egg design trans-porting a miniature ∞ SOMETHING fueled by an invisible energy source but is controlled by embedded micro-intelligence software similar like our Mortal=MIND or applied to the cosmos to mirror the ELOHIM∞MIND source linked to Alpha(+One) Force [20-gear Tzolkin clock], which is plural, too.

The airport operation mirrors a refueling station, where the source of all energy for the whole system expanded to the cosmos outer regions. All is scaled and replicates the same design, like a Russian Babushka egg. Those electrons, not known within a university environment, are specialized and highly complicated creatures with embedded intelligence that is new to modern science prevented from discovering by smashing atoms.

Yet electrons must have a knowledgeable microcode MIND to negotiate many obstructions and decide; "Do we go around or over gates, resistors or capacitors?" They move left or right on the inside of every electronic circuit designed by a higher MIND, as only one mistake will stop your computer, and it will no longer be working.

Recently published in Nature last July physicist Ali Yazdami used a powerful microscope to track electrons as they encountered stair like barriers on the surface of antimony, a material that shares several characteristics with topological insulators such as bismuth telluride. In a typical copper wire, most electrons would bounce back from such an obstruction and the rest would be absorbed, impeding the flow.

"With copper, surface imperfections slow things down and create unwanted heat." Antimony, however, neatly half the electrons pass right across the barriers as the scientist Yazdami thinks that topological insulators might start to replace copper in the next-generation electronics. But the key to discover electrons behavior is comparing and tracing the embedded intelligence cross referenced with INTEL microchips or Bill Gates computer codes.

What is extraordinary the Bible reveals that the function of intelligence as seen in every microchip is also miniaturized and found on every level of the smallest particle neutrinos-electrons-protons-atoms ending in mankind DNA controlled with an intelligence MIND not yet defined? Even replicated in the maximized cosmos-galaxies theater, all mirrored the image of the ELOHIM∞MIND expressed in physics as intelligence. To mention it is forbidden in our schools believing it to be the wrong religion could get you fired from your job. Ignorant educators do not know the difference only educated in unscientific brain-dead evolution fairy tales.

A Magnetic Generator Gate

We will now investigate an electron energy exchange from the generator viewpoint to find out how it works. Electricity is made up of surplus electrons will be collected in a gravity motor, using a bridge, which is **MAGNETISM**. It is the conduit connecting oxygen-atoms inside the tank and expelling electrons orbiting around the nucleus travelling to outside of the gravity motor system in order to gain profit, which we call electricity. It is (Electro) spelled **magnetism** encircling at high speed a Donut nuclear atom - (city) controlled by intelligence microcodes.

The ∞ cosmos energy fallout reduced on the bottom where mortals live is **magnetism** to stick atoms together and must be continually replaced somehow energy wise, and I believe the energy refueling cycle to the atom level is accomplished by billions neutrino-asteroids from space passing right through atoms on an invisible magnet trail. Perhaps they are coming from our Milky Way galaxy's black hole transformer wireless connected to earth, too.

From a copper-atom wire perspective (Remember my Munich beer festival analogy?), we have a **magnetic** bridge for the electrons to travel on. When we artificially compressed water on the atom level within a Hoover Dam or water-steam in a nuclear-coal system, (or compressed air my newest invention), the circulating levitated electrons are concentrated being artificially pushed together around on top of the atom nucleus shell therefore condensed floating will be loosening speed because the distance is getting shorter crowded toward the magnetic poles.

That can be visualized in a photographed vortex of the hourglass galaxy picture on the front page as the same laws apply. Or it is similar when we push a magnet through a copper wire loop the magnetism from the iron is transferred as a force and interferes with the copper atom-people drinking beer (a negative charged magnetic cloud surrounding every atom person) pushing together sideways ending with the last person (copper atom) on the end table (wire-loop end) with a transferred push (pulse). Read my Munich Biergarten story again will explain a science mystery how electricity is produced inside a wire loop creating an electric pulse.

Electrons on top of a compressed magnetic cloud will circulate at a lower velocity and are forced to overflow if a suitable magnetic bridge is passing by being magnetically pushed into a generator's copper winding side. Crossing over now will experience a high-speed reduction to a lower speed frequency wise similar to an electrical transformer reduced to million volts.

Every atom has a magnetic north-south pole created by positive charged proton on a billion energy volt levels circling around from the equator and ∞ moving toward the poles slowing down in speed. That radiates magnetism on the outer nuclear poles (+/-) shaping an external shell with a negative charged donut cloud. The higher proton magnetism makes elements bond together when atoms are magnetically attracted on top of another donut shape creating molecules.

That bonding created excess electrons forced to travel now mirrored in a generator-motor side and looks like a step-down round iron-oxygen (Fe_2O_3) transformer wound with many copper loops. That created a path for expelled electrons collecting from atoms to jump over to the generator-motor stationary field armature winding side inter-connecting to the rotor with brushes. That is reflecting the same design inside an atom adapted to catch free electrons with a magnetic bridge cross-wired with a second armature winding group will now **magnetize** each section ∞ lined up in a commentator making a motor work for our benefit just like mirror imaged inside every atom.

Compressed levitated electron-slaves pushed through a magnetic motor bridge forced to work will return stripped naked with a negative (-) polarity back to the generator windings across the same atom magnetic bridge to the original source gaining once more a recharge getting dresses again (+) at higher velocity.

Returning to the outer level on top of magnetic cloud, electrons will be forced to circle in a vortex by an energy force emerging from within a negative charged magnetic donut nucleus. It will redirect the electron slaves traveling on negative (-) copper wire highway coming back, but because they are now neutral charged - being naked without cloth, they will be sucked toward the nuclear equator in a ∞ loop cycle from the atom equator to the poles back and forth gaining velocity to end up happy negatively charged once more.

It is the same design as a magnetic generator bridge connecting moving protons magnetically in motion to transfer energy to the outside world. Electrons being pushed by magnetism generated from protons passing on neutron energy turning like a clock at different speed in a transformer fashion gaining speed like a flywheel to be balanced and fully (+) charged again to the original state.

If you want a mental picture take scissors and cut the hourglass galaxy on the front page in half and put it on top now is visualizing an energy path across a nuclear atom ball with a North=South pole black hole.

But if the external water pressure still exists, the surplus electrons' energy once more is re-directed to the lower entropy magnetic bridge in perpetual motion via the electricity cooper wire highway, bunched up and moving in groups like a sine-wave signal from maximum to minimum (Hertz), from the generator and motor traveling back and forth, which basically cycles **magnetism**. It is the same design system within atoms linked to proton transformers passing on energy coming from the cosmos.

The frequency is determined by how many gismo generator-motor windings are mounted together. The Hoover Dam magnetic bridge is colossal compared to your automobile generator, which is tiny in comparison having fewer copper wire windings, hence a smaller magnetic bridge. Electron slaves magnetically pulsed, **osculating** from high to low over the magnetic bridge passing energy from the nuclear atom side to our gismo GENERATOR side looks to us like electricity according to the copper wire loop experiencing a passing magnetic pulse.

Once more never taught globally in any university needs repeating to inform uneducated students of a forgotten physic understanding what a cosmos MAGNETISM Bridge is. It starts from outer space with an energy rail perpetually transmitting and refueling every atom along a dual visible light rail. Smashed atoms from CERN cannot tell the story where energy comes from perhaps energized originating from the black galaxy hole passing energy from space through invisible neutrino mail carriers.

Similar to a electrical transformer passes undetectable ∞ energy stepped down to a lower entropy now becomes visible light following an embedded magnetic path on the other rail only visible with special instruments. A moving Alpha-(+ONE)-energy is linked to every atom pushing protons levitating smart electrons within the atom nucleus and needs a magnetic bridge to extract its inherent energy.

That energy exchange is maintaining LIFE forming every single distinct atom linked to proteins directed by DNA intelligence embedded 60 trillion genes in our body, too. But an energy exchange in physics produces surplus levitating electrons, which now come full circle and are transferred via **Magnetism Bridge** to the outside (our) world aligning with a speeding UREE generator fed with magnetized gravity gaining speed with a motor combination.

Only when electrons are moved with higher speed to the resonance frequency can they be extracted.

What emerges from the generator winding gate pushed to the resonance frequency is **electricity profit**, but the mystery energy is never used up, just converted and returns to the same source for energy to flow once more. It just needs a recharge, which we are all familiar with. The negative charged wire must return to the source for energy to flow.

Have a closer look and notice the connection of the of the Hoover Dam turbine blades transferring pressurized water (oxygen atoms closer together) converting a kinetic differential force via a **magnetic** bridge to a copper wire loops generator for electrons to flow.

The expelled electrons excited by a critical resonance frequency level now flowing over a UREE invention will give us big time electricity profit. Proven by a plug on your wall which has two copper wires attached: one wire comes from the (+) source over the light bulb (**profit**), returning (-) used up, but it must be connected to the source - perhaps the biggest Hoover Dam generator windings now doing the same but being portable in a UREE by using compressed air media which puts electrons in a frenzy looking for an escape. A UREE magnetic fishnet bridge does the trick. If a switch breaks the connection, we have no profit "electri-city" which turns the light off like "electri" magnetism is disconnected from the (atom) "city" basically millions atoms living together.

A Magnetic Transformer Principle

Yet how does my gismo gravity motor **Magnetic Bridge** physical really work? How can compressed water be replaced with compressed air linked to Oxygen Atoms in the tank be connected to our newly invented UREE gravity motor machine?

We have learned that an extremely high energy, which comes with a frequency, can only be useful on a lower energy level, according to entropy. One way to do that is by a transformer. Ask the electrician how a transformer works, as it is usually placed high up on a pole, out of danger, to convert high-powered electricity to the standardized 110 volts. Another way for energy to transfer is to convert a fixed transformer by splitting it up. One side is fixed; the other rotates. It still can transmit a frequency signal used in TV, like a high-speed rotary videotape machine, which I assembled as a technician during the first days of the Ampex Corporation.

Yet when we rotate a transformer on each side on its own axis independently, with different speeds suspended, it will get you something totally different. (Check page 17 the UREE contraption not categorized in any patent department believing that perpetual motion is not possible.) Again, review physics from the two-looped (∞) perspective, as entropy moves from one side to the other. Magnetic energy will cascade and push toward the lower side, which is recognized in a rotating transformer, too.

The magnetic Gravity applied via a UREE electric generator employing a double rotating transformer works when electricity generated is fed back across the next copper loop, which will increase magnetism produced within a transformer copper wire coil. That will be gaining electricity through an Oxygen atom critical resonance frequency that is pushing excessive energy to the other loop increasing the magnetic bias. If repeated continually will speed up the generator via a second copper loop now recycling magnetism creating more electricity perpetually moving back and forth up to the resonance frequency. It needs to be stopped by a conductor or could explode like the same principle design of an electric streetcar-motor only expanded one level higher. If added another UREE design level could get a million KW extracted explained later in a Jumbo-Jet (Pearl #206) application.

Once more recycling magnetism ∞ inside the streetcar motor could explode which of course is either stopped by the conductor reducing the inline-energy voltage. A little energy is used up by internal friction, electrical losses and eddy currents getting bigger but is not setting limits because a critical resonance frequency is so big that it will flip a Fraunhofer spectral line to crystallize into the lower element level. Similar in your car speed indicator, do not drive your car on the red dial indicating the critical resonance frequency going beyond could end your life. In a UREE application we can now use suitable microchip electronics could replace the streetcar conductor. That same transformer principle illustrated works within pressurized air-atoms, with free electrons speeding around on the outside of the magnetic donut periphery, driven by neutrinos' energy pushing protons. But remember, the Oxygen atoms are in a compressed state, meaning free electrons frenzied and overcrowded by the pressure of millions of volts being severely restricted, searching for any hole in order to become free-floating again.

In analyzing what goes on inside an atom's nucleus, the Donut Atom model postulates that the heavier, positive-charged protons are pushed along a donut shaped path by neutrinos. Their circulation at infinite speed forms nuclear mass (requiring the Time Dimension). A transformer magnetic force radiates at 90^0 to the direction of the protons' path. That can be visualized illustrated in the hourglass galaxy vortex picture on the front page as it releases billions of volts.

Because high (+) proton velocity energy creates a magnetic (-) negative cloud close around the nucleus, the orbiting electrons are levitated to a relatively great distance away from the nucleus instead of being attracted to it. (It would be about 3 miles if the nucleus were the size of a cherry.) Electrons are much smaller in size, effecting the reduction of energy velocity. Thus, it becomes kinetic energy cascading to a second entropy level, now on "million" volt potential echelon.

If a magnetic gismo bridge passes by at very high speed, it will transfer electrons from a pool of a billion volt potential endlessly from what is available around the atom's nucleus. A synchronized, double rotating transformer system becomes the bridge for boundless electron-fish to flow into the magnetic net, as no electrical resistance exists. But because a new invention rotating transformer design which increases the speed of the magnetic bridge therefore catches a lot more fish electrons at higher volume at much less resistance. It is like when Kelvin space cold rose to the temperature where life can exist. Conventional science has tried very hard to come up with a transmission line cooled down to Kelvin cold for less resistance to gain electricity. I present it in the new UREE transformer system at normal temperature where life exists.

If compressed atoms are forced closer together becomes an overcrowded electron neighborhood and when forced to experience a magnetic high-speed carrier (Gravity UREE design) passing by, will use that opportunity to hitch a ride that is less crowded by crossing a **magnetic transformer bridge.** Like a double ∞ energy loop displays when crossing over, it is now slowed down to a slightly lower speed, which will eventually be reduced on the other end of a wire more manageable standardized at 110 volt pressure, which is electricity converted to **profit** for us.

However, we have learned that electrons have embedded intelligence inside their little brain and want to continue a strange journey, as they are still looking to return home, which is illustrated through a black hourglass galaxy hole, perhaps an energy gate, which seems like they are being sucked in.

The returned electrons are stripped naked (negative polarity) and must return back into the same Gravity UREE generator for a recharge to become positive (+) again, in order for electricity to flow via air-atoms. They must hitch a ride once more perhaps linked to a familiar mail carrier neutrino from the cosmos source to be energy recharged fully clothed again. Getting home and circling happily around a familiar nuclear magnetic Donut bathed in infinite ∞ light, the source of their birth.

For a fuller vision look at the many galaxies, ancient clocks in museums and a switch on our wall are all connected to the source is proof and reveal a different electron intelligence story for me to tell my grandkids.

Forgive my repetition in what you read, but theorizing about how perpetual motion works linked to my cuckoo clock cannot be found in universities. A pendulum required a small kick from the original force, which does not come from the weight not changed or altered hanging on a cuckoo clock, but comes from an external chain force for winding up that clock once more which passes on energy via gravity weight transferred to the pendulum kick applied at the correct moment resembling an (electric) power switch to make the clock system work for another energy cycle. Pay attention!

A pressurized static air tank becomes the link to an external cuckoo clock chain atom-level force and could therefore run my cuckoo clock perpetually, now discovered, thereby replacing the gravity weight system substituted via air-atoms coupled to perpetual motion. Applying the same principle, which is gaining free electricity when energy moves, we gain a little back, which I call **profit** but now discovered perpetually. **No energy can be extracted unless connected to the cosmos ∞ source**

Nuclear scientists will never learn what the inside of an atom is, how the embedded energy functions, and where it comes from when they smash atoms in CERN employing thousands of people. Using CERN is analogous to when you put an egg on the anvil in the blacksmith shop and use the biggest hammer with the longest handle and hit that egg with all your might. Some residue may be on the Nebelkammer wall like manure on the wall of a dairy barn, will never tell you an energy path developing into a chicken scraping off what was splattered on the wall.

The mystery of perpetual motion is that it can only work in a closed-loop system (one loop high energy and the other loop low), which is the second entropy level, but it is still connected energy-wise, like = ∞, as the recycled energy must be connected by a crossover DOT, which is a timeless event, or where time is zero, if you do the math.

The electrical switch on your wall represents zero time as on one side is energy stored (+) ready to move, the other side (-) to return to the energy source. Likewise, zero time is expressed in our mortal calendar BC-AD, which mirrored the cosmos time clock.

Energy is coming from the Heh dimension crossing over our mortal Daleth dimension heading into a new Jod-dimension transformed energy wise without a Time-dimension like a bigger Babushka egg all shaped parabolic exponential expressed in math.

Perpetual motion cancels out time when energy moves around from one loop to the other. Just as electrical current produces **magnetism,** as defined by the electrical three finger rule, and returning recycled electricity drives a streetcar motor by speeding up a generator linked to a four piston motor receiving kinetic energy, the main fuel converted via a Hoover Dam electric system. The main energy potential above the water dam reservoir is unaffected because it is infinite, like the high-pressure static air tank that never changes or the weight on my cuckoo clock remains the same.

Even the earth is an unchanged weight for thousands of years, but it still needs the external infinite chain MIND Alpha-(+One)-Force energy, in order to keep it moving. I am now convinced the earth moving is not caused by gravity but **MAGNETISM,** a totally new concept learned from ancient clocks. It is connected to global warming mentioned in the 5th Babushka egg concept book. Summed up to make it short:

A static energy gravitational force is only useful when it moves.

Visit the Hoover Dam and check out the static power. It is a beautiful lake, but it is linked ultimately energy-wise to the sun needs energy too and only burns being refueled because it is connected to the black hole on the other side of the cosmos, if you understand my new novel neutrino-atom theory based on two entropy laws.

However, below the Hoover Dam is where energy moves on the second entropy level, transferred via a turning generator converting an atomic water resonance frequency into leftover energy on a lower speed level, which is considered **electricity profit.**

Therefore, we can drive in the future a car perpetually without gasoline if we used a compressed air media and if parked in your garage still running and connected to your electrical system can light up the whole house proven by a discovered mystery energy revealed in the Tzolkin lock. This is not a fairy tale story applying logic to a forgotten atom story linked to investigating ancient clocks.

I hope my explanations of forgotten ideas are better understood by further investigating adapting a redesigned 100-year-old basic gasoline or steam engine, converted into a UREE invention, in order to return a portion of the energy to a driving motor flywheel for perpetual motion to balance out all explained later. Designing a perpetual motor fueled by magnetic gravity only works when kinetic energy is in motion like the wind in a sail; this causes time to be shorter.

That would give us extra time to spend on the beach, but in the energy equation, other must work in union like a number of pistons converting energy in parallel to gain perpetual electricity that could be fueling the 21st century with the cleanest, nonpolluting green energy. If we do not shut down nuclear power stations before it is too late, they will kill billions of people as nuclear poison will deteriorate and breakdown according to ignored Newton's laws.

The Proper Oxygen Gas Mixture for Automobiles

Clean electricity is produced from the water atom level when we recycle magnetism embedded in oxygen atoms linked to an electric generator-stator-rotor ∞ coupled to motor-stator-rotor that gains **profit** either an invisible electrical current which becomes visible when mechanical work is performed.

In the meantime we have learned that enormous energy is embedded in water and that we should replace poisoned nuclear energy gone the wrong way. Reading my energy story will hopefully explain more fully how the cosmos magnetic energy path works down to an atomic level. I compared them to some galaxies with the lights on as generated by wireless electricity in a vacuum charged with a billion times a billion volt-potential, as photographed by the Hubble Space Telescope. In a generator further explained once more the media must be under high pressure of either steam-water or pressured water (H_2O) molecule as used in the Hoover Dam; therefore, I believe the **Oxygen** atom must be the candidate source for electrons freely given away to give us electricity.

An energy exchange in perpetual motion is proven in our blood system starting with oxygenated enriched blood cells circling around every body part to return without oxygen going through the same heart pump on the other side to end back where it started in the lung transfer station to be enriched and oxygenated. That teaches us oxygen energy needs to be refueled perpetually. Only one break will stop the motor running, now considered profit for the undertaker.

The same principle is demonstrated in a wall switch. Behind it, electricity is waiting perpetually, which science tells me are electrons extracted from oxygen atoms flowing perpetually and must return back on the other minus wire even color coded ending with the same wire magnetic bridge inside the generator returning fast to where they came from.

When oxygen atoms are compressed together much closer than usual, such as in steam or water-air pressure, free electrons on the outside of the atom nuclei are bunched up orbiting concentrated around a nuclear magnetic shell. If an outside magnetic field is nearby, they will be attracted like fish to water traveling from the magnetic atom donut leaving the media water to end up on my wall switch. If I want profit, my kid knows now where the energy is hiding.

Therefore, from a scientific perspective, I pick oxygen as the electron generator for electricity. Using compressed air (21% oxygen), we would therefore need a little higher pressure, perhaps double that of water or steam (42%), as the amount of oxygen atoms are relative to what makes electricity available to the size of generator made from copper wire loops. So many passing wire loops create the bridge from atoms, via magnetism, to the outside world where many electrons will flow.

Magnetism is produced by many copper wire loops around an iron core. I believe the oxygen atom is the electrons' generator for electricity; therefore, I need to double the oxygen gas content when applying it to an automobile. That allows less pressure in the tank - a little safer.

Once more, compressed water, steam or air has embedded oxygen atoms. Air needs double pressure as compressed water (steam) for an equal amount of expelled electrical energy; conversely, if we used pure oxygen gas as a pressure media, it would need less pressure to drive around in a car. In underwater experiments, Jacques Cousteau found out that pressurized pure oxygen is linked to DNA cells healing open wounds much faster and can save a diver's life neutralize nitrogen bubbles causing death in the brain.

Extremely high pressure was my concern. We had enough air pressure available below the ocean floor compressing oxygen atoms closer together, but using pure oxygen gas in a tank could mean less pressure for driving around with a car. Again, I believe the oxygen atom must have more free electrons being sucked away from the outer magnetic nuclear donuts forced to cross over the magnetic generator bridge and flow as **electricity**.

By investigating pure oxygen, I learned it is not suitable when you have a spark plug ignition and needs a mixture of inert gas as neutral transfer carrier. I would suggest using argon (.9%) embedded in air which is a much larger atom that will give us the high inertia needed to return the pistons, but we still need oxygen for the maximum electron transfer. I propose a 3 times higher oxygen (60%) air mixture combined with a carrier argon gas (40%) or nitrogen, which doubles the oxygen content of water if applied to a closed environment to be able to drive around perpetually in an automobile. The energy experts can calculate a better ratio.

I am only a hybrid tech-philosopher who is still learning about science. A fine-tuned oxygen gas-carrier mixture balances the additional kinetic energy that is needed for the return piston, versus collecting enough free roaming electrons to obtain electricity profit. That mixture is a concern when spark plug lightening is applied. To run a UREE perpetually we must heat up the pressured gas needed for perpetual motion, thus expanding a kinetic energy on the return cycle inside piston cylinder.

Pay attention, the higher mass of the argon gas atoms is needed for the return piston force which is 100 times higher as the remaining pressure air-tank force now the air mixture is expanded by a hot spark plug. That builds up extra kinetic energy now converted driving around without gasoline in my car.

It is the same atomic force embedded as I remember on a camping trip in Canada when I made a fire and put a can of spaghetti on the hot coals for lunch and forgot about it. Then suddenly, the can exploded shooting straight up like a rocket 30 ft. followed by a trail of spaghetti. Now I laugh about it, but it is a good example what happened when anything closed up in a can or cylinder and getting a little warmer has 100 times more energy available useful if it can be controlled in a system like the UREE.

How a Magnetic Bridge Works

A little later the 60% ratio oxygen gas was verified on the metaphysic level as I was thinking about what a **magnetic bridge** is really. A generator and electric motor core is made of magnetic iron. The magnetic iron core is wound around with copper wire windings. Magnetic iron I remember way back is used on magnetic coating for tape-disc drives expressed in a chemical formula Fe_2O_3 that means two parts of iron mixed with three parts of oxygen.

Therefore, magnetic iron is a mixture of 40% iron and the 60% oxygen. A bridge is usually designed for the weight what travels over. Only so many cars can pass and in our case only so many electrons will pass over a magnetic bridge depending how big is the iron-core and how many windings.

The Hoover Dam generator is big compared to the electric generator in my car. Though it is tiny, the principle design is the same. Consequently, having picked a 60% oxygen gas mixture is appropriate for the magnetic iron-core bridge because it has 60% oxygen mixed together with iron and should be a good match now better defined.

A chemist might cringe at my ideas but a forgotten art to think philosophical about nature applied to experience eating the whole pizza not being picky how the salami was made. Tasting the pizza or testing the prototype will prove my recipe.

Magnetic iron can be magnetized. We all played with magnets as a kid and learned that like polarity repels and unlike sticks together, but why? The ancients collected magnetic asteroids fragments from space, which resulted in a compass indicating a mystery force buried in our earth. When analyzed we have various iron oxides: FeO, Fe_2O_3, Fe_3O_4 and more. I am not a chemist, but nickel and rare earth added will make super-magnets. The oxygen sisters will marry anyone to have sex, check out nature as 99% is a union with oxygen sisters as all the dirt and rocks on earth is just OXYGEN-SOMETHING.

Therefore, the oxygen electron children are polarized. Some are fully charged (+) and if put in servitude like a slave, becomes naked (-). But the family must always we reunited when they come back with the return copper wire to have a party again in their oxygen-atom home.

Science discovered an iron oxide like the one from space that will freeze the polarization of electrons separating the one with the cloth on to the one without a rag, all in one enclosure. That is sold as a magnet in many stores. But if you heat the magnet or freeze their butts with Kelvin cold, the party is off - no longer polarized.

However, when spun around in a magnetic bridge at a temperature where life exists, the party gets going. The one with clothes on wants to travel beyond because they are artificially compressed by an outside force (air-tank), and the one naked wants to go find another dress fully charged, found only at the oxygen atom home. Let's further have a little closer look and analyze perpetual motion embedded in gravity how invisible laws of nature applied to the moon linked to extract some kinetic magnetic energy.

Secrets of Kinetic Energy Linked to Magnetism

The Bible speaks about David's[2] stone and slingshot story that killed a giant. Much later we hear about Newton's apple, and a bowling ball in my time could explain physics to my grandkid. A stone, an apple and bowling ball that fall from the roof to the bottom will all hit the ground at the same time.

An apple falling on Newton's head gave him a bump or perhaps a headache that defined for us gravity, but a bowling ball falling on your head would have consequences I am sure of, which is called profit for the undertaker. Therefore, kinetic energy is created if mass is allowed to move, when applied in a high-pressure tank, which will move air now exchanged for gasoline that burst out in higher energy and starts your car. Thinking about why my GRAVITY energized UREE works, I discovered another principle. As I am not educated in extra terrestrial laws, I do not have the precise physics terms to describe my thoughts.

We all learned in school that when David's stone or Newton's apple or a big heavy bowling ball fall to earth will arrive at the same time, except for differences in air resistance. Conventional science postulates the equal pull released shows the embedded kinetic energy, but not the whole picture because the observer is also on earth moving with it around the sun.

As long as the smaller or bigger objects do not fall, no kinetic energy is released. They all possess potential an embedded static energy and go together as one mass rotating around the earth's center of gravity even as the earth orbits around the sun with no different speed. Suspended in the air, they have the same speed around the sun as the earth and the observer; therefore, they fall to the ground, which means they move towards the larger mass at the same speed but release different amounts of energy.

What I dispute with a new atom theory everybody conventionally believes that gravity is the force embedded in mass being attracted to bigger mass. But I postulate it is **MAGNETISM** within the atoms sticking other atoms together. That idea can be expanded to our earth – moon and the highest density sun all held together with a magnetic force, like I played as a kid with various magnets to see how far they can be separated and still attracted to each other depended how big the magnet was.

Applied to conventional physics, I could postulate that tiny differences of kinetic energy are captured in a speeding UREE piston and immediately stored in a flywheel temporarily with some embedded magnetism. That difference of magnetism is converted from a free kinetic energy and if collected in a flywheel transformed by the GRAVITY propelled UREE system via a rotating transformer ends converted to **electricity profit.**

But the system would be thrown off if the earth orbit changed. Similarly, the weight in a cuckoo clock does not change moving a pendulum, hands and a cuckoo but needs an external chain force for perpetual motion, which could also have embedded invisible magnetism. We cannot measure it because we are inside the system.

---

[2] 1 Samuel 17: the story of David in the Bible.

My gravity pushed UREE would not work on the moon orbiting 28.5 days around the earth. From the sun's perspective, it moves 14 days in opposite direction of the earth's orbit to experience a net lower speed and 14 days at a slightly faster, but in fact constantly a little different speed moving ahead of the earth in its orbit around the sun. It would mean that an astronaut jumping around on the moon would experience either less or more gravity differences along a variable bias plus or minus changing moon velocity direction every 14 days.

An astronaut has body mass comprised of atoms stuck together with magnetism in my new atom theory and going around the sun expressed in conventional physics with a tiny static kinetic energy embedded if linked to velocity proven with a slingshot principle. That kinetic condition will change when the moon goes faster to the sun orbit motion or opposite will be slower turning around as seen from earth like a full moon to a new moon cycle.

But astronauts jumped around as seen on TV arrived on the moon but was going in an opposite cycle. If arrived 14 days later, he would need to jump a little harder to go as far. Is my logic correct? On the other hand why would a moon influence fertility laws of all life forms on earth? **Magnetism** linked to **oxygen** atoms could be the answer.

Learning the cycle below to ocean would reveal it, or check out menstrual cycles in humans and animals not based on solstice calendar cycles? I believe corals below the ocean measure magnetism from the sun, which is blocked by the moon cycle. Corals only give birth in a bright full moon - a trillion babies timed just right with a billion fish waiting to eat them.

A full moon means looking behind us is the sun and therefore the earth is on the other side receiving a maximum exposure of the sun magnetism that all life forms are cycled from. Perhaps **magnetism** from the sun is embedded in photo-synthetic light transmission would be the real life-energy food driver fueling our bio-world and coral intelligence timed it correctly as fish know when dinner is served all follows the herring cloud following coral expelled babies in a food chain cycle.

A moon clock blocking magnetism from the sun influences every cell in our body of being alive and a woman's fertility cycle. I believe it is linked to oxygen passing on energy ∞ perpetually perhaps by a crossing a **magnetic bridge** too to maintain life and similar to my new Donut atom model investigating electrons.

Visible light could also be dual in nature. It may perhaps have embedded magnetism to function like a gigantic, cosmos electrical energy transformer that taps into power radiating from our sun to distant planets. Any frequency we use in telephone, TV must travel over SOMETHING, if you understand electronics and investigate space camera pictures. Perhaps an invisible magnetic frequency carrier can be theorized that sends back a computerized frequency package passing through a vacuum from millions miles away to earth, currently unexplainable to conventional science.

God keeps adding to my understanding of new theories. Isaiah 30:23-26 says that after the great "Day of the Lord" (Apocalypse) the sun and moon will give (7) seven times more light. It will cause a climate change once more and comes with a promise to bind up all the hurts - heals the wounds, gives more rain - rivers flowing from every mountain followed by more grain, plenteous harvest season and many cattle on the hills. Like milk and honey flowing once more in the ancient holy land. Standing in front of an Apocalypse now dated, I noticed that ancient prophecy is already in progress filling the NEWS headlines of "climate change" to tax you more.

But the cause is hotly debated globally in the UN and universities. I explain it better in 9 Babushka egg concept books investigating historic development across 6000 years as measured by ancient clocks. These ancient calendars revealed a massive climate change before Noah's flood, or the Atlantis civilization (2288 BC). An asteroid strike that triggered earth axis wobbles causing the climate to become suddenly colder with thick ice on the poles (5 February 2287 BC).

But the wobble fizzed out over time resulting in a number of stone-bronze-gold clocks exhibited in museums later as clocks developed gears (Antikythera 45 BC) or electronic clocks NASA (2000 AD) that indicate a reversed climate change around (BC-AD). Hi-tech instruments can now measure it. We are on our way toward the Jod-dimension and back to what mankind previously enjoyed. Check it out in the BIBLE linked to a Genesis Hebrew calendar corrected 62 days by Julius Caesar and refined adding 14 days by Pope Gregory.

Science questions the Bible reference not educated in the metaphysics. It tells us that, if the sun was seven times hotter, then it would have the opposite effect. The light spectrum we know (4000-7000 Angstroms) is on one side ultra-violet deadly to life. The other infrared very hot side is also deadly.

Seven times more would be seven times more deadly as prophesied. That would cause all life to be over, guaranteed. Logically, the paradox of that BIBLE verse of (7) seven times more intense light can be explained only if the middle spectrum is increased seven times like yellow-green (5500 Angstroms) which is MAGNETISM if you understand the Donut Atom theory. Magnetism is the glue that bonds molecules forming 60 trillion proteins via DNA in our body.

The light from the sun therefore must consequently be dual and can be accessed as a source of SOMETHING shining through the sunlight transformer on the second entropy level. Later examining magnetized water will expand our knowledge horizon how fish will come back increasing in numbers the Bible clearly stated. We could do it globally now applied everywhere as some business already making huge profits in the agriculture and meat production.

From the metaphysic science perspective the increased blessing prophesied in the future Kingdom of God on earth can only be activated in the ∞cosmos control room of an ELOHIM∞MIND by some unknown process beyond the galaxy black hole, forming atoms evolving in a universe filled with galaxies and planets. If you like to see some beautiful pictures from the Hubble-telescope of immense planets go to my 5th Babushka egg concept book, **Reflections on Global Warming,** not based on the atheistic evolution fairy tales.

As mentioned before in this subsection, the Bible announces a new civilization with unlimited clean green energy, now projected at a frequency of 5500 Angstroms, expressed as seven times more intense but magnetically extracted from atoms.

New science discoveries postulate in the above headline, "Secrets of the kinetic energy linked to magnetism," that magnetism must be an invisible force influencing fertility on earth. This includes a food cycle in the ocean, perhaps growing all green plants and vegetables. At the atomic level, magnetism sticks atoms together binding molecules.

Electrons can only be transferred over a magnetic bridge to give electricity as an energy source giving extra life to our 21st century. It is now proven to cause an exponential population curve that increased from 1 to 7 billion in about 120 years and still rising.

Could, therefore, my theory be expanded by proposing that magnetism really is the invisible life-force coming from the Elohim energy center black hole passing through space like a transformer needs magnetism right through every atom, binding every atom with an invisible force could therefore be right?

It must already be activated at higher speed? Visible light will travel on the second entropy level highway and was previously measured hundred years ago with a velocity of 186,281 m/sec. If light goes faster magnetism would increase too recently discovered and demonstrated in Princeton University measured light 300 times faster in 2010 AD.

That will cause faster and bigger cells to grow proven in our hothouse growing vegetables exposed to more light around a 24 hr. clock. Growing faster and bigger vegetable cells is also proven with 6-7 ft. people upward taller in our generation. Check the old church-door ways 5 ft. tall in castles exhibiting knight armors designed for smaller people. Why is your and my grandkids one head size bigger?

I previously postulated enciphering ancient bronze-gold clocks displayed in many museums. They prove that the speed of light was faster in ancient times as that will explain the "why" causing dinosaurs and tree cell size to be bigger as confirmed in fossil bones and rocks.

Thus, light has embedded magnetism crystallized out in mass becoming a second entropy gravity force on a dual highway. That is a new concept for scientists to think about. Seven times more magnetic force in the middle of our visible light spectrum will feed the next bigger civilization and still be horseless and make it possible to drive your personal car without gasoline forgotten no longer polluting the environment.

My previous Babushka egg concept books get better, making more sense. Like God said, "Precept upon precept, line upon line, here a little, there a little..." (Isaiah 18:1) Keep reading my Babushka books to widen knowledge horizons to include the metaphysical recorded in the divinely inspired history book collected by 40 scribes. Since only 66 Bible books where printed in our century, four more books are missing as it must be 70 books according to the Hebrew Alphabet Number System.

For that reason, keep looking around to find the missing Bible books like the book by Enoch the seventh from Adam. It describes genetic modification from before Noah (2288 BC). It is stored within the Vatican. Look for three more hidden on the Internet library to widen knowledge horizons.

Is Gravity Magnetic (Pearl #225)

This section became Pearl #225, now inserted to complete the third edition of the UREE Babushka egg concept book. It puts closure to the infinite energy path as symbolized in the most ancient, 20-gear Tzolkin clock pictured on the front page. It takes time to bake a cake when a new theory is revealed from the metaphysics. Many are educated in evolution lies and needs much testing in physics to collect witnesses to accept it as truth. Remember what I coined in the beginning. **Being ignorant is a curse, but suppressing TRUTH is evil.**

In the Hebrew Alphabet Number System (HANS), "Twenty" means "Reaching Hand", which is a higher level of Jod = 10, or the third dimension as shown later in the next TIME graph. It is not the End Station.

Jesus promised a new earth-heaven even the Heh-dimension Angel-world will change, but that is not yet revealed. God said, "What no eyes has seen, and what in no man's mind ever entered God has prepared for those who he appointed."

Therefore, any cosmos dimension within the time dimension is only temporal in nature. I expect to be greatly surprised and hope you will pass the test to be there and live forever. We are already gifted on earth embedded with eternal life when trusting Jesus as Lord of Lords, introduced as the WORD in John 1:1.

The first word of God spoken to Adam communicated a Plan for Mankind with promises if you understand ancient Hebrew. It survived Noah's asteroid destruction (5 February 2287 BC) and the later Babel event. Now 400 language translations exist. Pick the one you like to become educated one notch higher.

In a nutshell, in the new Donut Atom model atoms cluster to form mass by bonding to other atoms with MAGNETISM produced by high-speed protons running inside the nucleus in a system resembling blood circulation - two ∞ loops linked to a ∞ energy Alpha(+ONE)force. This process is documented throughout 6000 years of human history and recorded in Genesis 1:1. That verse reveals a LIFE energy fueling every atom and the power from which heaven and earth were created.

Clean green Electricity can be extracted from Magnetized Gravity a new discovery. Energy crystallizes only in the TIME DIMENSION forming mass, BUT it also has invisible magnetism embedded in every atom, by which elements form. A better understanding of the nature of MAGNETISM will prove how the UREE inventions work as published free on the Internet.

Clean green electricity can be extracted from gravity, which I here explain to be magnetic. Getting a better perception how the universe got formed comes from analyzing atoms not smashed in CERN, which raised the question of whether Einstein's Theory is obsolete, or does it fit from another perspective?

The laws of physics should be put together and not dissected into tunnel vision perverted by unproven, unscientific opinions lacking witnesses, a methodology practiced in every university rejecting the Bible. A derailed evolution religion will never explain physics being only fairy tales for closed minds controlled by the political power elite trying to destroy Christian based culture.

How "Infinite" is related to "Time"

$$\infty E = m(+\infty C / -\infty C)^2$$

(According to ELOHIM)

Teeter $------\rightarrow \odot \leftarrow------$ **Totter**

$$E = mc^2$$

(According to Dr. Albert Einstein)

1. An Introduction of the ∞Cosmos Energy Heh Dimension Linked to Daleth Dimension to create the Time Dimension

Recently a math friend visited me, and I showed him the new formula where the light energy is infinite (∞E). He responded that it meant nothing in science because infinite energy does not exist. Taking a little aback, I opted to remark, "Perhaps we come from a different planet?" to save the conversation.

I had to admit he was right. Perhaps I do not understand Einstein's laws. I am not educated in math but defend my opinion that the energy frequency path starts at ∞ and ends in zero. He countered that high and low frequencies still travel at the same speed of light; therefore, nothing can go faster.

To create a common base, I responded with an analogy to understand what he meant, "Is it like a train one wagon has packaged the lower sound frequencies from whales linked to our ears, with another wagon of higher Fraunhofer spectral frequencies seen by our eyes but invisible, being alpha-beta-gamma frequencies measured with special instruments?" If so, then all frequencies end in ∞. We settled on that. All of physics captured in a train would never to go faster than the speed of light as described by Einstein's theory. Nevertheless, I ended with a challenge, "Is it possible for the train to go faster?"

He said we have no observation that it does. Now, I countered, "What about the recent observation that the speed of light can go faster, pointing to CERN and Princeton University? Also, the World Kilogram Standard IPK in Paris and its six sisters have changed after 135 years. Why?

At this point, our conversation was over. To save face, I closed with an offbeat comment about philosophers looking at all nature to unify the laws of physics, as many are invisible to our eyesight.

Next day, 4 o'clock in the morning, during my usual time with my heavenly father to open my mind, working another pearl of insight came to me. Like the prophet Isaiah tells of his experience with God, "Precept upon precept, line upon line, here a little, there a little." (Isaiah 18:1)

God answered my ruffled feathers and gave the reason why modern scientists are totally turned off when reading the beginning of my Babushka books and not read further being quickly, uncomfortably annoyed. They will shut down the Internet, convinced it is just a waste of time like my math visitor.

It is really not their fault. From childhood they were educated by the atheistic establishment who filtered out important laws of physics no longer taught in schools fearing it would expose the faulty, outdated evolution theories. They forbade student investigation of metaphysical laws denigrated as a stupid religion, declaring only inferior Christians not educated in modern science would believe metaphysic fantasies.

That brain blocker affliction of bias became second nature. It will automatically snuff out anything not matching the atheistic childhood fairy tales embedded as permanent lies implanted in the brain.

I am not surprised and see absolutely no reaction on the Internet of the new UREE invention extracting free clean electricity from GRAVITY, AIR, WATER and MAGNETS after sending 450 CDs of the Babushka egg concept books to most embassies, universities and automobile manufacturers. I received not one courtesy acknowledgement of interest in a free energy source. To me, this is an indicator of how brain-dead the atheistic establishment has become, and they control the education of the masses.

I have concluded that, when the lights are turned off in a person's MIND, he cannot possibly see the invisible treasures in the Bible. It resembles a dark Egyptian tomb chamber full of gold objects of ancient splendor that needs a light on to see them. Without the enlightenment of our MIND, it is not possible to see reality according to Newton. The universities should not have kept 50% of his writings from the students because they deal with metaphysical science.

Newton's metaphysical theories were falsely judged religious and so kept out of circulation by atheistic, ignorant, brain-dead professors. Seems like I am in the same boat: any conversation with most university-trained experts is soon cut short or becomes one-sided. I am unable to have a logical conversation anymore, except as a philosopher, who like inventors are a special class with similar interests. Einstein used to look around in a full horizon to discover more, only to be overtaken by mortality, running out of time.

A train is a good analogy to express physics to my grandkids. It is linked to the metaphysical, forbidden Bible history we can test with tools of logic. Following the road map, few years ago I created a graph as a bridge to Einstein's Theory. It does not seem obsolete yet, when looked from a 360⁰ perspective. My math friend was never exposed to that representation of a three dimensional universe because physical science only teaches physics matching our five senses.

This philosopher considers such an approach as tunnel vision ignoring the MIND from, obviously, another dimension that cannot be explained by evolution fairy tales. That created many speculative opinions hypothesized from politically correct people arranged in a pecking order of the brotherhood aligned to top salaries and big government grants.

My graph seems to come from another planet for the majority. Please try to apply a little theoretical logic to this unusual concept expressed in a strange looking math illustration. Notice there is an exponential vertical curve representing the metaphysical energy laws crossed over by a horizontal straight line representing the Time Dimension, which will fizz out like a flywheel.

When the concept of time as a dimension is overlaid with the Hebrew Alphabet Number System (HANS), it reveals a beginning that started with the Aztec calendar in 4488 BC, but the earth's resultant declining wobble needed to be corrected when an asteroid caused an axis change on 5 February 2287 BC.

The shifting astronomical observances due to this wobble decline really perplexed Noah's survivors. After the asteroid impact, the original calendar had to change drastically following the zodiacs. Being greatly confused after Babel, the Aztecs built five pyramids on top of each other as exhibited in the outdoor museum located in the center of Mexico City.

History later reports that the wobble calendar eventually was corrected. It was projected to come to rest on 21 December 2012 AD as calculated by Julius Caesar and Pope Gregory in calendar we use today. However, prophecy revealed another 1000 years **God's Kingdom on Earth** extending time to 3018 AD, which matched the HANS code on the horizontal axis. The math symbols on the left corner ($+\infty$) inform us of an invisible energy path from the Heh dimension universe based on Genesis 1:1. It appeared as metaphysic, infinite light linked to why LIFE is only found on earth. This is the ∞ light-energy-FUEL that energizes every atom in the universe.

I look up in the sky and wonder why so many lights are on in millions of galaxies. When the lights are off, scientists think space is filled with 70% dark energy. Dark energy is not visible to our eyes; therefore, it could be at a higher frequency. Einstein said that was not possible. Or, could it be addressed as metaphysical energy, perhaps with a little less entropy, now discovered and linked to invisible MAGNETISM embedded in polarized gravity as defined by the laws of electricity explaining how the **16 UREEs** work.

The time dimension in my graph is only a facilitator to intersect on a straight horizontal finite line to make Gestalt possible, which is our physical Daleth world calibrated to our mortal MIND and perceived through our physical senses. That is the dimension Einstein investigated and defined on the physics level.

Looking further we understand that matter is really a paradox. On one level, we think it is frequency energy defined by Fraunhofer spectral resonance frequency lines, and the other level crossing over into the time dimension to become solid particles by creating atoms useful for CERN. We see the two curves represented on my graph as interconnected with

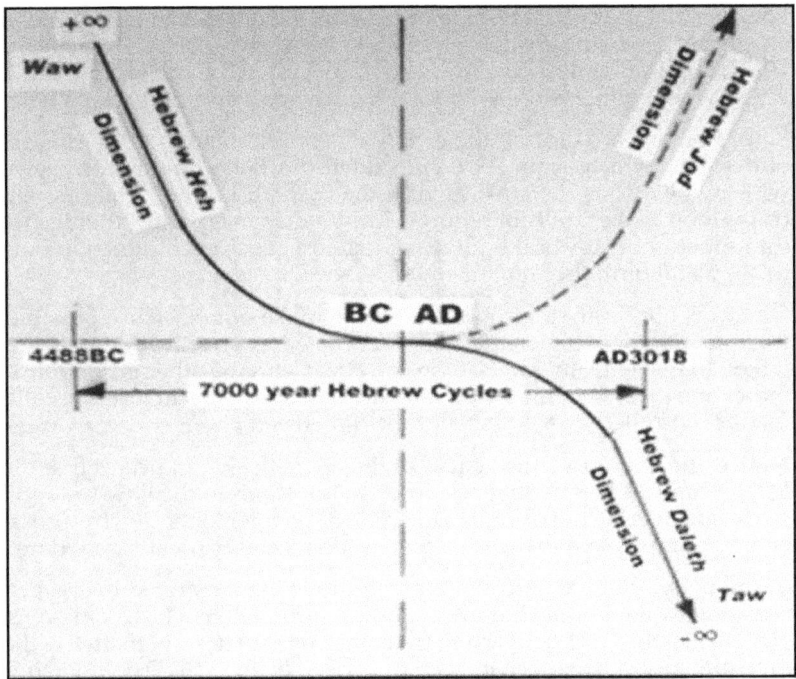

the two thermodynamic laws forgotten in physics. Neither work on same level as postulated by forbidden metaphysical science, but cross over.

The vertical curve is really only a portion of a full ∞ circled loop representing the first entropy of an energy frequency forming atoms. The top of the graph shows the lower portion of the infinite energy trail flowing from the other Heh dimension to the horizontal Daleth universe level and will always be less than 100%; therefore, it is offset 90⁰ with a horizontal crossover mortal timeline curve on the particle side forming matter as crystallized into individual elements. The particle side becomes the second entropy loop aligning with Einstein's theory that can be proven with instruments and tested in physics.

Once more this new theory was never taught in any university explaining why we see Fraunhofer spectral resonance lines in the visible spectrum of light which can only exist if energy is crossing over demonstrated on our TIME-Dimension graph. It is really an energy junction of passing a critical **resonance frequency** of protons racing around the nucleus being highly energized in a frenzy.

Above the critical resonance frequency we have energy expressed as wave's frequency, but below the critical resonance, a frequency juncture will crystallize forming a solid element. This apparent paradox is like a two-sided coin. Science will explore wave energy on one side categorized in Fraunhofer spectral lines and then separately, the other coin particle side. Science wants to divide atoms into smaller particles building bigger machines like CERN.

I believe that the critical resonance frequency juncture can be utilized in an electric generator giving us tremendous energy – enough to fuel a jet UREE application described in **Pearl #206**. Otherwise the last Jet UREE is just a nice opinion story without some science witnesses backing it up.

Since humans are mortal, the graph also applies and is duplicated in our bodies, which again have embedded the 100% first entropy level with two loops as demonstrated in the ∞ math symbol. The first entropy loop is our invisible mortal MIND dispersed with intelligence throughout our physical body. Sixty trillion genes need direction how to align and form protein linked to DNA.

On top is our brain, the control center to make our existence possible. It works similar to a gigantic computer fueled by an embedded ∞ spirit linked to two entropy levels. One is vertical, and the other is horizontal indicating the two paradox levels frequency wise, or as in forming solid matter controlled by a crossover time dimension.

Notice: the bottom vertical curve of the -∞ math sign indicates that the MIND controls the intelligence down to the atoms in the little toe - still cross-linked at 90^0 horizontal to the center and interconnected by a heart energy pump linked to the time dimension mortality (death=zero time).

In the body, the two entropy loops figure is demonstrated by oxygen charging by the human blood circulation system with a heart crossover switch, like the DOT converting frequency energy to solid matter in the time dimension. This exchange is supervised by the MIND defining our mortality as expressed in BC-AD measuring time. It controls the range of human mortality, a smaller Babushka egg because we were designed and replicated in the image and likeness of the eternal ∞ ELOHIM the bigger egg concept.

I previously stated every atom needs fuel on the second entropy level. Without it, we would not have the lights on in our Minds'. Energy must flow down to the lower level in our body to do physical work. Remember, in the laws of electricity nothing can exist without being linked to the ∞ energy generator.

If the NASA atom clock would stop ticking, we all would disappear in a femtosecond as every atom is connected in a chain to the next energy-wise. Like the main electrical switch of the Hoover Dam controls all the lights of the country, or like a bracelet with beautiful diamonds is worthless when one link fails breaking the chain.

Einstein's theory is based on the same horizontal timeline. It is designed like a railroad from **A to Z** with stations in between. Well-defined physical laws are packaged in various wagons as classified in crystallized matter categorized by Fraunhofer discovering spectral lines creating Elements down to atoms fueling 16 UREE motors, no kidding.

Everybody believes sunlight to be constant, being assigned wrongfully to the first law of thermodynamic entropy. But I postulate that visible light from the sun must crossover a time dimension, being governed by the Second Law of Thermodynamics. If you express energy moving like light so many miles per second that can never be FIRST ENTROPY! Visible light is demonstrated by a shrinking declining sun converting energy according to the vertical exponential curve. Therefore, when an energy conversion is moving from Fraunhofer spectral frequency lines crossing over into TIME on a lower energy level of particle atoms, it will always appear as Second Entropy.

Energy can turn into matter or matter into energy as demonstrated in the graph representation, but if matter is moved sideways, affecting TIME, all moves with it like a train not being independent from each other. Einstein postulated that it was connected to the energy source. A divine architect created a three dimensional plan. Changing one area does not change the other; therefore, Einstein is only correct from the horizontal TIME dimension perspective.

However, much of science is a great deal bigger as defined in the un-known Babushka egg concept books. I have demonstrated a vertical exponential infinite line on a higher metaphor order that is free for fur-ther investigation.

I do not stand alone but among millions of witnesses throughout his-tory who testified to invisible metaphysics rooted in a stage set by God as the center of my LIFE which is based on the divine to live forever and not in one's self, the other polarity embedded in mortality. We were designed for spiritual worship, adoration of the Creator and coopera-tion with nature which always ends in prayer almost everybody knows, "Thy Kingdom come..."

My personal science belief system as an inventor was summed up by a champion coach who lived 2000 years ago. The Apostle Paul stood fear-less in ancient Athens facing the unruly, mob in the Areopagus:

> While Paul was waiting for them in Athens, he was deeply distressed to see that the city was full of idols. So he argued… In addition, some Epicurean and Stoic philosophers debated with him. Some said, "What does this babbler want to say?" Others said, "He seems to be a proclaimer of foreign divinities." (This was because he was telling the good news about Jesus and the resurrection.)

> So they took him and brought him to the Areopagus (the first university) and asked him "May we know what this new teaching is that you are presenting It sounds rather strange to us, so we would like to know what it means."

> Now all the Athenians and the foreigners living there would spend their time in nothing (if you had the money) but telling or hearing something new.

> Then Paul stood in front of the Areopagus and said, "Athenians, I see how extremely religious you are in every way. For as I went through the city and looked carefully at the objects of your worship, I found among them an altar with the inscription, 'To an unknown god.' What therefore you worship as unknown, this I proclaim to you.

> The God who made the world and everything in it, he who is Lord of heaven and earth, does not live in

shrines made by human hands, nor is he served by human hands, as though he needed anything, since he himself gives to all mortals life and breath and all things.

From one ancestor he made all nations to inhabit the whole earth, and he allotted the times of their existence and the boundaries of the places where they would live, so that they would search for God and perhaps grope for him and find him—though indeed he is not far from each one of us.

For 'In him we live and move and have our being'; as even some of your own poets have said, For we too are his offspring.' Since we are God's offspring, we ought not to think that the deity is like gold, or silver, or stone, an image formed by the art and imagination of mortals. (Painting an evolved monkey walking upright is imagination, not science)

While God has overlooked the times of human ignorance, now he commands all people everywhere to repent, because he has fixed a day on which he will have the world judged in righteousness by a man whom he has appointed, and of this he has given assurance to all by raising him from the dead.

When they heard of the resurrection of the dead, some scoffed; but others said, "We will hear you again about this." At that point Paul left them. (Acts 17:16-34 NRSV)

If you think that modern times have changed, notice that human nature looks the same, like in the mirror image. When Paul stood in front of the atheistic establishment, he was in conflict with their unscientific philosophies and religion. This lengthy pearl will reexamine the quintessential laws of the COSMOS. How is LIFE embedded in the Donut atom model? What is magnetic GRAVITY used to fuel the many UREE motors to produce free green energy for the next 1000 years?

We are still ignorant as to the nature of ELECTRICITY. It is not yet defined by modern science. Why were mortals created with an intelligent MIND ending in the Jod dimension never preached in church? It is only a Nachtisch from the full table of 9 Babushka egg concept books impossible to recall all what was postulated and never preached in any university or church. I assure you what you read is not religion, so you will not eat the forbidden fruit. It is but an extension of Einstein's theory fifty years later to expand the MIND. I am sure he would have changed his opinion in the face of new science facts.

When reading this pearl, be prepared to deal with a kaleidoscope. As it is turned, you will see more colors and shapes from a different approach investigating what previously seemed invisible to the eye.

Be warned! A crazy one is on the loose. Look out! Like a radio announcer in England warned everybody at the early commute, "Watch out, there is a crazy one driving on the wrong side of the road." It must have been a foolish German absentminded scientist listening in and saying, "What do you mean "one"? There are hundreds driving on the wrong side!"

Who is fooled read further?

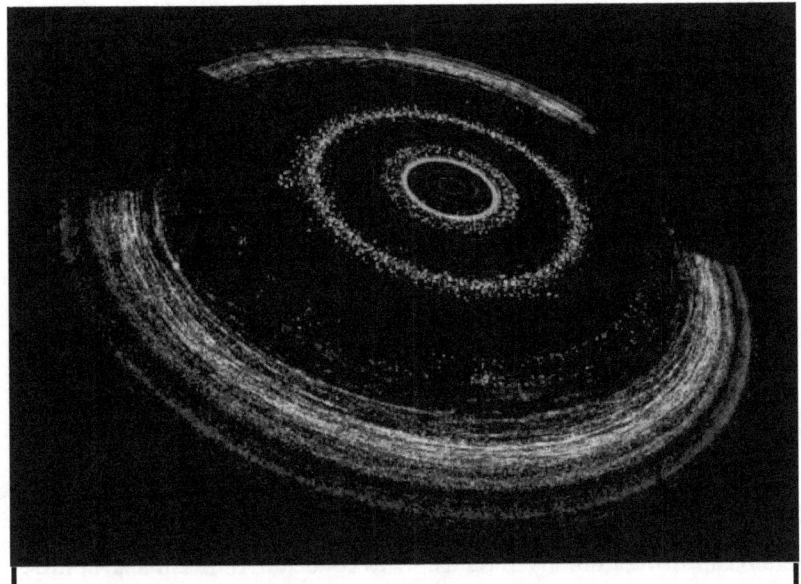

2. A Data-Centric Universe Magazine Picture

A new Discovery on Why the UREE Works - Upsetting a sun-centered doctrine

Investigating our solar system where the energy comes from, I recently got very surprised by a beautiful double-sided computer-generated picture of the universe produced from a mega thousand known facts. NASA and other science institutions collected massive data from space and consolidated it in a print-out picture published in:

Popular Science (November 2011)

"The Data-Centric Universe"

That got my attention because it could prove why my UREE inventions really work as illustrated in that graphic computer printout of the universe. Most people will never notice how much our culture has lost seeing something graphically demonstrated in a science magazine – a picture representing our total knowledge of space.

Applying the metaphysical perspective this universe illustration pointed out a paradox. What we teach in school goes back to the Middle Ages when the Genesis account was the foundation of science.

What was so unusual about the universe picture? Though the universe was shown as a curved disk, the perspective used placed the EARTH at its very center!

Pay attention! That is the point where all computer measurements begin to the outer cosmos region as summarized in this graphic representation of space. My mind started to think about the forces connecting planets to stars, which became a proof of why the UREE invention works. I applied it in a cuckoo clock analogy to explain the earth-moon-sun relationship to my grandkids.

That cosmos chart started me on another new discovery. To see our world from a high-tech perspective, usually I check it out first with the oldest Bible book on earth compared with history. This graphic illustration it is the best picture I have seen in a long time to demonstrate modern scientific knowledge. It is embedded with much information, so now it is hanging on my wall.

When crosschecked with the Babushka egg concepts, you will comprehend the computer data generated Centric Universe in the magazine on a much higher level, This urged me to write about this unusual perspective to my Internet readers. I connect it to correcting Galileo's and Einstein's theories back to what was taught a thousand years ago as preserved and stored in the world library of the Catholic Church in the Middle Ages.

That is where modern science started to explain nature but was derailed and distorted so that it ignored metaphysical laws. Nevertheless, when recombined, it can prove truth with modern advanced science computers recapturing millions of data entries collected into visible graphics.

Investigating further the graph with the TIME dimension DOT where Einstein's theories are linked together with the metaphysical laws. The laws of nature must balance. They are not static as recognized by two entropy thermodynamic laws showing which direction the universe is headed as defined by our mortality. The opposite polarity of general science is LIFE that must have purpose because it is embedded with intelligence and controlled by a center - our MIND. The existence of entropy thermodynamic laws points to a universe that must be fueled with ∞ energy down to the atomic level with embedded intelligence. The NASA atomic clock ticks in femtoseconds because of embedded intelligence revitalized with energy synchronized globally.

Infinite "energy" is transferred by ∞ light as only revealed in Genesis 1:1, as we have no instruments to measure it. Only a higher MIND can understand the nature of things connected to the infinite ELOHIM MIND. Otherwise we would be still stuck on the same monkey intelligence level from the Middle Ages, believing in a stupid unscientific evolution religion and ignoring so many other laws of physics like DNA controlling 60 trillion protein genes. I hope that we will become educated in gravity and magnetism to give us free green energy converted into electricity through 16 UREE inventions.

When ∞E energy from the cosmos is slowed down in the TIME dimension, it demonstrates the second entropy level, seen by our eyes at a certain frequency. At that instant when visible light is stabilized on the second entropy level at a speed of 186,282 miles /sec. it can be split up by a prism to show seven rainbow colors in a range of short to longer cycles of blue $4000A^0$ to red $7000A^0$ measured in ($1 \times 10\text{-}10$ m = 1 Angstrom = .1 nanometers), but light has a lot more information embedded.

Looking closer, one recognizes the many Fraunhofer spectral lines that indicate a velocity change of matter crystallized into existence. There are now over 110 elements known to form Gestalt, which can only happen when energy is slowing down on the vertical exponential curve or accelerating in a certain "time" crossing over to condense into solids.

Each element formed into matter has its own velocity as a frozen fingerprint frequency distinct from the other. In forensic science, this characteristic is used in spectral analysis. A future multidimensional universe (Jod dimension) is not static but must be in motion like a cosmos carousel. We will explain this concept later because Genesis reveals it to be the same, infinite energy "E" - never changing, running like a cosmos black hole motor.

Look a little closer to this latest magazine printout converted from millions of computer data equations measuring between galaxies to create that fantastic graphic illustration of the universe. It could prove that my new Donut atom model was correctly postulated. That unusual perspective of physics will also demonstrate why the 16 UREE motors work, confirming many theoretical concepts explaining an unsolvable paradox from the Middle Ages.

A paradox existed before Galileo got into trouble with the establishment explaining how the universe functions. He postulated a new solar system theory and compared it with what we now see in the Popular Science magazine. The issue is not settled yet after 400 years because it is linked to Galileo's personal sun religion reaching far back into ancient Babylon. His sun religion was in conflict with the Bible creation story, which upset the Vatican establishment and wanted to prove that his religion was more scientific with a primitive, newly invented telescope.

In life sometimes our minds can be fooled by our five senses, especially when biased by a religion. A quick example before I make my case. Sitting in my car while it was being washed by a machine, I watched the water splash on the window. The spray arm reversed its direction of movement, but I perceived it as the car moving forward, and at an impulse, I applied the breaks to stop the car.

If I sit in a moving chain carousel and look toward the inside, everything seems fixed and stationary, but in reality, everything is in motion around about me. I can feel gravity pushing my seat outward. That can explain the universe as the same laws apply but needs real science to overcome our perceptions. If you visit California, there are two tourist places in the redwoods (Santa Cruz and Eureka) designed to totally confuse your senses by applying gravity. Remember, three witness examples are key to not being fooled.

Now expanding on the double page Popular Science computer-generated illustration dated Nov. 2011 showing the universe from the center of the earth expanding 360^0 to the outer edge of the universe resembling a big wheel. That could open up a very different perception of space.

Pay attention once more, the picture shows a sun-earth relationship, which became a paradox. On one hand the Genesis creation story puts the **earth at the center of the universe**, as recorded in the original 6000-year old creation report where everything started. On the other hand, that picture **conflicts with sun-centric solar system** now exclusively taught as a doctrine in every university and decreed as fact: no other opinions allowed. However, is it possible that our perceptions can be fooled?

Let's investigate the ancient Bible, the oldest book reporting true science. It documented the process how the universe was created. It is interesting to note that our existing belief system of a solar-centric system is counter to the Bible story, which tells us that the sun appeared on the 4th creation day.

In the Bible interpretation the sun and the moon were inserted into the system much later, perhaps they travelled from outer space like an asteroid and got trapped between two Ausdehnung dimensions now moving around a pre-existing and much older earth located at **"the center of the universe."**

Looking from space, like a whirlpool forced in a vortex circling around like water down the sink - black hole, or imaging a universe-carousel turning and sensing an invisible gravitational force sitting in the chain-chair with closed eyes. Put five senses together in our MIND could eventual make sense seeing magazine pictures of our Milky Way galaxy. Project a line through the black hole of our galaxy to the center of our earth that is the axis the universe postulated turning illustrated in the Popular Science magazine picture.

Why is our earth-sun placed on the end of our galaxy and not inside the dense fog of dust-matter? Again hypothesized in Babushka egg concept books that what we see when looking at the universe is replicated in an exact mirror image in a Donut atom we will explain once more later. What is unusual about the third Bible creation day that vegetation, trees and shrubs appear well established first before the sun showed up later on the fourth day cycle? I always believed the opposite: that light from the sun was the source causing photosynthesis.

Although only Genesis reveals that on the first day cycle infinite ∞ light appeared, an invisible energy of a higher frequency must fuel the universe to exist down to a single atom. Therefore, we should investigate it closer. We learned from the first dynamic entropy perspective when energy flows down to the hypothesize atom level is creating magnetism for electrons to levitate postulated in a new Babushka egg concept theory highlighted again a little later.

To prove it check NASA and asked what is ticking like a clock inside an atom on the second entropy level to measure time in femtosecond to synchronize every Internet computer with a global atom clock needed to make the World Wide Web system work.

To teach physics to my grandkids, let's look how an atom cuckoo clock works which is a perfect analogy demonstrating natural laws. Most people do not realize that only when gravity is moving it becomes an energy force. Since atoms are too small, let us investigate a cuckoo clock demonstrating the same gravitational forces expanded to the sun - solar system affecting our earth too.

Seeing a historic paradox reminds me first switching way back investigating Galileo's ancient sun religion looking at primeval history during

Babylonian time recorded a SUN centered worship religion that later is seen embossed in ancient Egyptian tombs. Following the trail, we can find many remnants in fossilized customs around the world converging in the Roman Catholic Church under Constantine.

Notice the fishhead and colorful vestures of the Pope and his priestly class. They are typical of the ancient Babylonian priesthood pictured on ruins and tombs. Even in modern times, most Catholic churches have a golden Sundial with silver Star as a background stanza in the center of most altars. We can wonder what it is, or watch the Pope showing a four inch round disk at High Mass Eucharist that originally represented the sun cult of Babylon and further developed later to include the Islam moon religion from the same root.

Fossilized Customs -
The Pagan Source of Popular Customs,
Lew White (www.fossilizedcustoms.com)

I am not surprised even in modern time to see three religions battling for supremacy as most people do not know the evolution religion started science in the Middle Ages from the perspective of the SUN religion which was continued from its Babylonian roots postulated by Galileo who defended it now convinced looking through a primitive telescope. That got him into trouble with the Vatican establishment. Now from my perspective I understand why we still teach a sun-solar-system in our schools is much older from ancient times.

But it was the sun-religion which is in conflicted with the 6000-year old Bible story stating that the EARTH is the vortex center of the universe now proven once more when all measurement of millions computation collected and put in the high-tech biggest global computer which was converted into a beautiful graph published in a Popular Science magazine showing the result with earth "in the center" from a scientific perspective. That raises the ancient conflict once more to a higher level.

Be serious and think a little differently from what we learned in school. Compare those teachings with the laws of physics using a familiar analogy of how a cuckoo clock works as expanded to the laws governing the universe. Just follow the original ancient history trail and start out with the purpose why the earth was designed to contain LIFE the only place in the universe and function like a cuckoo clock with gears measuring a time for mortals.

The Bible gives us an overview of our existence started in Genesis with a Time dimension and where the energy comes from, a good story. Learning about some new laws of the universe let's start first from the metaphysic perspective when the gravity weight of the cuckoo clock is on the bottom tells us projected that "all" will eventually end in Taw as the Alpha-(+ONE)-force will be ending a Time dimension stopped by the cosmos ELOHIM computer.

Unless redirected like a cosmic cuckoo clock chain for the ∞ energy to flow in another direction of the universe ending on the way with the White Judgment Throne after 3018 AD. The first verse in Genesis tells us that the clockmaker ELOHIM preexisted and never changed, winding up a clock in the angel world.

Thereafter, God used the same energy chain redirected to the second creation clock, which is the temporal Daleth dimension universe. The cosmos runs on the mortal clock of time linked with the same energy chain as when an electric train is switched.

The dual system will run in parallel in another ∞ loop that caused the time dimension to appear when forming the elements of the physical universe as measured and proven by Fraunhofer on the atomic level. But energy in the time dimension has been inserted temporally, redirected to expand into a totally new universe meant for mortals as part of a restoration plan correcting the 4488 BC rebellion of Satan in the Heh-dimension.

When God's purpose is accomplished, the Daleth dimension will end as projected in 3018 AD. It is already running in parallel with the Jod-dimension since 35 AD check out the diagram again. That is proven by Jesus' resurrection, which is really a transformation of our material physical world crossing over the Time graph illustrates. He promised on his way to heaven, "Behold I will make everything new, creating a new earth and a new heaven at the end of TIME without death, no longer will have mourning-crying-tears or pain cycles anymore."

The old order atom system bathed in time has passed away, converted to everlasting LIFE. Everybody is invited if passing the test of Revelation 21. Apply a metaphysical science finger to point how an atomic clock system really works as applied to our solar system will expose why the big energy UREE fish works in 16 versions, free on the Internet, thus becoming a witness announced by Jonah II.

We are confronted again with the same paradox. Galileo faced the establishment, and now it has expanded in modern times to unbelievable postulations of a strange theory. Can we trust the Bible once more? It can give us totally free electricity fueled from GRAVITY, WATER & AIR to benefit mankind for the next thousand years.

The free energy revealed in the first verse of Genesis was declared, "NOT POSSIBLE!" It was echoed by an energy cartel fearing revenue loss. This unusual metaphysical representation, if applied to the solar system, will prove that free energy from space can only work when related to a new understanding of physics as postulated in my last UREE pearl for the experts to check out. I do not expect to have rotten tomatoes on my face for misleading people.

A Popular Science Magazine Picture Reveals Upside-down Bible Science?

Going back to the two-sided computer generated magazine picture of the universe is really showing us controversially in graphic: the sun is moving / orbiting around our earth in the center, no kidding. This moving sun principle is similarly replicated inside an atom. When protons move in the nucleus, according to Newton they radiate invisible energy outward. Some of it is hidden light (infrared, ultraviolet) and heat, magnetism and other forms of gravitational forces that made my new UREE invention possible. They extract nonexplosive nuclear energy from thermodynamic forces of atomic function without splitting them.

Electric energy is mechanically expelled by electrons levitated to the outer atom shell by the moving protons. That is not known by ignorant science experts smashing atoms to junk with a monster CERN machine. It created an awful lot of fun to see it operate as a good toy should, fashioned for the level of a childish MIND designed for people who never grew up being exposed to metaphysical reality.

A more logical story is told that the sun was designed by a Creator and placed in orbit on the fourth creation day to give us infinite energy. As explained in true physics, this energy was caused by a moving massive body dispensing outwardly gravity inward converted to radiating heat to overcome space Kelvin cold and light to make LIFE enjoyable designed and positioned to a hairbreadth spaced at the right distant just absolute perfect.

Any deviation of various interacting forces will make LIFE impossible only when gravity moves linked to its inherent inertia can we extract it as electrical energy if applied to perpetual motion transformed to electricity now discovered useful applied in 16 UREE inventions. All free!

The UREE will not work if the sun is stationary fixed printed in every school text book or illustrated on the atom level where the energy would be frozen in a gravity mass being stationary. Understanding how physics work looking inside an atom if protons do not move - no electrons would be levitating giving us free electricity.

Looking in the sky at night, why are the lights on? Can seeing stars prove the theory?

However, general science no longer teaches true physics usually can only repeat the official opinion NOT POSSIBLE presented typically with the junk evidence of atoms smashed and fracture by CERN for 48 years spending yearly billions taxpayers' money, which is the real reason for their existence. Do not be fooled otherwise. If they investigated atoms not smashed, they would recognize embedded intelligence information controlling 60 trillion DNA forming protein genes.

Once analyzed, one can better understand Newton's laws applied to the universe. No wonder it fools so many scientists who do not know that their belief system is rooted in a sun worship perception from the Middle Ages. Theorizing a misjudged solar observation, they conclude a static sun centered system that will proven to be false if the UREE gives us electricity, or they will just deny it works until someone gets filthy rich. That will see a gold rush stampeded once more and repeated again going through three stages of new energy Babushka egg concept: "First, it is ridiculed; second, it is violently opposed; third, it is accepted as self-evident" (postulated by Arthur Schopenhauer).

Those new concepts linked to perpetual motion will expand the magazine picture story now becomes too dangerous what is taught globally in universities as the UREE energy path becomes the key. To lay the base for one notch higher understanding, I am quote from the opening of the Popular Science magazine by Mara Grunbaum, "The Data Centric Universe" illustrated by TULP Interactive November 2011.

Before the telescope was invented in 1608 and shows an inner circle of six planets, our moon, the sun, and any stars we could see in the Milky Way galaxy. However, as our light gathering capabilities have grown, so too have the boundaries of the visible universe. In the late 1700s William Herschel, an English astronomer using a telescope with an 18.7 inch aperture, made the first systematic surveys of the skies, revealing more than 2,000 distant galaxies, nebula and other objects invisible to the naked eyes. Since then, increasingly powerful optic and telescope have greatly expanded our store of knowledge.

In 1948, astronomers erected the 200 inch Hale Telescope at Palomar Observatory in California, and now, large-scale projects such as the Sloan Digital Sky Survey and NASA's Kepler mission use sensitive digital imaging and computational power to collect and analyze hundreds of terabytes of data on millions of galaxies billion of light-years from earth. With each additional bit of data, the universe itself grows larger.

To clarify the special relationship between Earth and the rest of the universe 407,702 objects with known distance and discovery dates are displayed as a ring flattened to the earth's equator.

The big fish UREE discovery will force us to decide what is the true interpretation defining the origin of the universe as seen from two different perspectives. Modern science using powerful telescopes generating computer printouts still needs interpretation what we see could be fooled remember my observation in the wash-machine and later a carousel example perhaps driving on the left side of the road.

Our religion or flawed education most likely will influence the data. If biased, remember in computers "garbage in, garbage out." If data is based on wrong religious premises, it will never arrive at the right conclusion as a recent UN delegation found out when investigating global warming with doctored data from a prestigious university eager to get another grant.

Why is the Bible viewpoint forbidden, denying a 6000-year-old ancient story? Instead, billions of phantom years for a big bang energy emerging from NOTHING is postulated, which cannot be proven if you understand science.

Understanding how the UREE works and if the government controlled by the energy cartel allows it, will have billion dollar consequences obsoleting many energy businesses and other becoming filthy rich will that be enough proof for the skeptic?

Investigating physics from the energy perspective we have a paradox: either the SUN is the center of our energy provider, or the fixed, stationary EARTH is invisibly fueled through every atom from an energy center beyond the universe that perhaps is linked to the Milky Way black hole. CERN smashes atoms, which destroys the intelligence formation sequence consequently will never understand how the universe was formed. They are already back peddling announced previously that light could be faster but the next grant is more important warned by the atheistic priesthood controlling the purse strings of printed fiat money.

Notice a much better perspective is revealed by a divinely inspired Bible that cannot be silenced by the atheists if linked to Newton's physics. Together they explain that energy from the universe perhaps is transferred by neutrino mailman running on a magnetic rail coupled to a ∞ cosmic Alpha-(+ONE)-force source postulated in Babushka concept and now demonstrated in a NASA printout.

Ultimately, the sun like the planets must be fueled with the same energy; therefore, they must be in motion circling around in an orbit as a generator must move to create electric energy as photographed in galaxies and stars for the lights to be on. The electrical three-finger hand-rule must apply to both the burning sun and the cosmos, I am sure.

That proves the electrical theory of physics. Only when the lights are "on" do we have an energy exchange we can test on earth, like magnetism is moving an electric motor. That can also be determined by re-examining the Donut atom theory on a mini-egg entropy level, analyzing that energy is not static but must be moving protons to push levitated electrons according to Newton's Law and produce electro-magnetism, provided we apply logic rather biased religion and atheist fairy tales to understand it better.

We are familiar with Newton's "when something moves, it must have a cause." For energy to be useful, it must move, which is proven by electric energy generation. It moves from the Hoover Dam, or new UREE inventions to get free energy, linked by a grid to the wall switch.

The sun only has the lights on if connected to the power source, proving the second thermodynamic entropy linked to the vertical exponential curve on my energy-time-graph. Therefore can be extracted in various forms of energy defined in physics like light linked to solar-panels, tides, buoyancy, hydrokinetic oscillation wind, air, water, or use concentrated light to generate clean green electricity.

Einstein knew of other dimensions but could not imagine that the light we see with our eyes at 186,282-miles/sec. frequencies could be crystallizing from higher first energy conversions. Investigating why an electric motor-generator works could reveal it. Alternatively, go back to Princeton University and CERN who discovered that, perhaps, light is not constant. Or check how the World Standard Kilogram in Paris has changed in weight.

The idea of an energy force in every atom forming in time is new to physics. The universe must be continually fueled from the ultimate infinite energy source, which is the Alpha(+ONE)force embedded with an invisible frequency, otherwise the UREE will not work. Only when GRAVITY is moving and creating a transmission force can we call it inertia; therefore, it can be extracted to become beneficial for us. However, it must flow from a perpetual moving force like protons inside the atom to be useful.

For example, understand how the electric levitated train in Germany works, or visit the Hoover Dam tourist center watching turbines. Or, for my grandkid I explain the same principle with a cuckoo clock perpetually running being activated by a gravity chain controlled by an intelligence MIND, winding up once more. Relate that to what is taught in every school of a fixed central sun theory doctrine. Postulating gravity is "zero", no wonder scientists conclude extracting energy from gravity is, "NOT POSSIBLE!"

But the UREE works, and that confronts their bias. Is it because many scientists are no longer properly educated, being fooled by a sun religion morphed into an evolution theory? If so, they are prevented from logically understanding the UREE principles. However, forgotten science teaches that gravity embedded in the sun mass can only be measured when the sun is moving in relation to another force like a teeter-totter. Galileo did not understand it, seeing a paradox.

3. A New Cosmic Carousel Model

Seeing so many lights shining in night-sky, again we notice a paradox. Scientists are evenly divided. Some wonder if the universe is expanding, and others are persuaded it is shrinking. However, applying a Solomon decision dividing the baby will not work in this case. It can only be solved looking at nature from a 360^0 science perspective must therefore include the metaphysics mirror image as a judge. Only when we start examined the new Donut apple atom theory, which I compared to a cuckoo clock for my grandkid applied to the universe can we understand nature better as the laws of physics work in space too. But watch out! Our perceptions could be fooled.

Seeing horrendous energy burning in galaxies among sun planets but also space dust without form or shapes brightly shining that tells us that electricity is a major player. Light mostly appears in a vacuum depending on the closeness of atoms expelling magnetism forming an electromagnetic discharge we can see with our eyes illustrated by a scientists Nikola Tesla and the others explained in more detail later.

When an electromagnet force in a UREE is moved by the infinite Alpha (+ONE) force crystallizing eventually into electricity, the underlying energy exchange leads me to believe that the universe is not static frozen but must move to be energized to exist. It is just a bigger electric generator.

We will have another look later at the interaction of electromagnetic forces, which is not very well understood by science. Or, tell me where the energy comes from on the other side of the generator? Just turning copper wires in a magnet will not do.

Putting it all together, I postulate that the whole universe is like a big wheel and must move like a chain carousel. This has never before been defined as such in physical science. Popular Science magazine (November 2011), "The Data-Centric Universe" is a perfect illustration of this chain-carousel.

It is like sitting in a moving chair. Observed from this imaginary chain-chair galaxy, the next galaxy faces in toward the center. Everything will look as if it were fixed to each other, but we know that gravity moved all of them outward with a gravitational force like a chain chair carousel looking the other way. Seeing a red shift now makes sense when looking from one chair to the next. Being pushed by levitating anti-gravity will not be frozen but will have some movement between each other caused by an unknown force.

MAGNETISM is embedded in gravity. We know that a bigger magnetic body (sitting in the chain chair like a galaxy) next to a smaller will have interacting electromagnetic exchange like my grandkid plays with bigger or smaller magnets seeing how they work together when spaced apart. Some bigger, other smaller chain-chairs for extra effect have a mini-chain-chairs rotating around one another, similar to our sun going around the earth, or vice versa, hopefully no longer fooled. That explains the sun-earth paradox.

I may have shocked you at first to make you think a little. Perception can be fooled. Many thought I was that crazy one driving on the wrong site developing an upside down idea from the cosmos perspective system looking from the outside in. Because we still see a paradox, our earth moving around the sun or vice versa, and for energy to fuel a cosmos down to every atom. Therefore, the universe "must move" like a chain carousel, which is a good illustration for my grandkid to explain the contradiction.

A moving cosmos carousel was embedded in the Bible, which created a challenge, logically described by the 407,702 computer generated measurements graphically showing the universe with the axis rotation according to the 6000-year old design reported in the Genesis creation report. The turning axis is found along a straight line from the black hole of the Milky Way galaxy to the earth.

When the mystery Bible ∞energy moves, it creates magnetism that in turn will expel a force. We are surrounded with a light that is the second entropy that starts with electricity. That newly discovered form of energy comes with a frequency only our eyes are calibrated to and is linked to a higher being with comes with a MIND controlled by intelligence "capable" analyzing DNA how it works.

Traveling to the moon and perhaps blowing up the whole planet with nuclear energy proves a higher MIND a monkey cannot do. No religious animal has embedded intelligence because we are connected to the metaphysics as the only species on earth.

Cut it anyway you like will not change the pizza. My new UREE invention utilizes the carousel energy and that is how it gives us free energy from GRAVITY, WATER, AIR and magnetic MINERALS. When its inherent moving energy is extracted by a magnetic UREE bridge, it demonstrates the same principle for lights to be on as we see in galaxies.

Explained another way - when energy flows like a river over a water wheel, its inherent force can be converted into another form of energy linked to turning other wheels manufacturing goods. Think from where the water comes and where it goes conforming to the energy path returning in an infinite loop. Without energy, nothing would exist. The bridge to make LIFE possible like water is MAGNETISM demonstrated on the atomic level.

My UREE invention discovered a magnetic bridge to catch as many bunched up negatively charged electron fish sucked over a magnetic waterwheel now improved with a new UREE invention designed with a double wheel electric generator. It is a good idea because a waterwheel and fish environment preexisted useful as a system about discovered a hundreds of years ago.

Please read my new Donut atom theory, which makes much more sense, backed up with many witnesses from physics and compare them with unproven fairy tale - opinions preached in universities smashing atoms to smithereens. The evidence in nature must fit the pattern down to the atom level now proven with the UREE to understand why it works.

If the energy cartel still does not allow free clean green energy and universities deny that metaphysic laws do not exist, eventually in time will be forced to come up with another energy source as the fossil fuel will skyrocket on the pump guaranteed ending sooner as you think. The indicator is already showing up using extreme expertise in cracking sub-terrain strata to extract gas with the most destructive poisoned "FRACKING TECHNOLOGY" practice invented by evil scientist paid off by energy cartel obscene greed.

Listen to the many angry farmers in New York State horribly upset and complaining about being severely misled as seen only on World-Link TV. A number of programs showed huge farms with the ground water poisoned; killing cattle and can no longer grow vegetables or produce a crop.

Worse, radiation extracted chemicals are dumped into the river below for comatose New Yorkers drinking deadly chemicals that cannot be filtered out. They trust corrupt high-salaried government shuffling useless papers excuse their failure in warning the public saying no law exists but accept fees above or below the table paid by the energy cartel.

The magazine picture will flex our MIND illustrating graphically a concept that was first believed in ancient times. For thousands of years, it was not discarded, yet Galileo quickly did so due to the problem of not understanding a paradox of two major religions looking through a primitive telescope. Now I understand why the Catholic Church was so against it, as it destroyed "the purpose" of a thousand year old creation cuckoo clock system now rediscovered in Babushka concept books ending in a UREE perspective.

The sun-centered religion is still around, now morphed into an atheistic evolution religion even tolerated in the Catholic Church. It still uses the Babylonian symbols of a fossilized SUN religion. They never questioned where it all came from, being fooled by not looking back at ancient history. I am not convinced. Watch a YouTube video from the Russians found 6 months later proving my theory: Truth, the earth does not revolve around the sun.

http://www.youtube.com/watch?v=o6jBK1ZV-qs

A "purpose driven" perspective must be the primary focus if we want to understand the creation paradox reported in the Bible. Hopefully would start a discussion club on the Internet if you are still educated in logic to go back and examining what was original transmitted for thousands years. Why not use the platform of the November 2011 Popular Science magazine picture as a base to compare it with Babushka book concept rediscovering ancient information what was once common knowledge.

Can we trust a perception when told that the sun solar perspective taught in every school could it be expanded to a higher level or explain to me how gravity is generated, what is magnetism and finally electricity produced from water or air demonstrated in a UREE just spinning a cooper wire inside a magnet melting 50 tons to liquid iron at 2000^0 in a Bessemer or Thomas furnace?

Once more, looking at the illustration from a chain-carousel perspective compared to the universe every galaxy shining must be linked to a (gravity) chains connecting to seats (atoms generating electricity) fueling billion of glistering galaxies with the lights on, all is connected to a motor-generator like a black hole in the center which puts everything in motion in a mirror image how the universe works.

Watch the weather TV station showing a storm being formed as viewed from space. It is just another illustration of how the universe was formed. But anything formed must align with the laws of physics as directed by the higher MIND-ELOHIM. I am sure as an inventor He designed his bigger gismos with a purpose in mind and started with a black hole motor at the center to turn like a big wheel in the sky (see the Popular Science picture).

Think about how a black hole is formed if the universe turns like a chain carousel and needs horrendous energy because it is bigger, where does it come from and how is it transferred? However, a carousel turning has embedded a TIME DIMENSION how is it related to "E" the infinite energy? Connecting the whole schlimazel is my crazy story.

Put together, it proves the whole enchilada as in the Popular Science picture showing a central earth that I compare to a cuckoo clock. In explaining nature to my grandkids, it energizes our bio-world on the atom level fueling DNA that will not run without energy. It makes the sun to shine, overcoming Kelvin cold.

It invisibly fuels every atom with magnetism sticking molecules together but that energy must come from higher entropy. It is from a different source running on a dual magnetic rail of ∞light energy perhaps the black hole of our galaxy postulated in Babushka egg concept book.

For me, not believing in illogical evolution science was revealed by the ELOHIM reading his Bible narrative to understand how and why he created the universe that comes with a purpose.

Any inventor can understand that principle. Ask any PhD in the university have you ever invented anything and would therefore gain an understanding of a principle that a NOTHING needs intelligence to make it into SOMETHING but still requires that something preexisted whatever form. A big bang energy opinion makes no sense.

4. A Cuckoo Clock Overlay Perspective

For my grandkid, the laws of nature can be illustrated using the analogy of a universe cuckoo clock linked to the Donut Atom described next. It explains science on his level, though not found in schoolbooks.

Understanding more fully what to expect by looking in unexplored space, here is how I discovered that GRAVITY is magnetized as related to how a cuckoo clock works on earth. It is a good gravity illustration featured a gravity-weight but must be "moving" as a force pushes the pendulum like the moon. It is linked to the perpetual motion of an osculating force such as **gravity or anti-gravity** fueled ∞ from a universe black hole. Or, it could use my grandkid winding a clock to make gears move by the second entropy.

On a miniature atom scale, we will later examine how the ∞ energy is moving neutrinos and pushing protons in every single atom like the clock gears. It pictures a full circle, inversely twisted energy path ending in the new Donut Atom structure compared to a pealed apple. Gravity can be demonstrated. If you jump off a table, your body normally moves down, but if you levitate with antigravity, it needs more explaining like an astronaut on the moon making bigger jumps.

Could it be because the moon is moving 14 days in the direction of the sun and 14 days opposite the sun velocity illustrating an antigravity force from a perspective of a moving sun? If the sun did not move around the earth, or vice versa from a universe perspective, it would burn up everything in a femtosecond, and you too obsoleting Babushka books creating Schadenfreude in most universities.

However, when something moves, it needs energy down to a single Donut Atom as presented in new Babushka concepts on the Internet, which can no longer be suppressed. Driving on the wrong side of the road reveals realities and gets the same reaction.

Go back to the original description of the universe in the Genesis creation story written from a purpose driven perspective where the earth is the center, which matches our German cuckoo clock design. The moon is like an energized pendulum: it can only work when gravity is created by the sun's mass. That converted energy must be moving like the fixed weight on a chain to make the cuckoo clock run.

The clock design dictates according to physics. It needs to be connected to a hanging chain force of a higher ∞ energy source invisible to our eyes and controlled by intelligence. It will transfer **antigravity** higher power either to turn a universe flywheel as demonstrated in the returned UREE piston invention. It is just an upside down perspective against the present teaching in universities but much more logical, productive and fun.

If the earth behaves like cuckoo clock, it can illustrate how the universe works from galaxies down to atoms. The energy system needs a moving **gravity weight** (sun) to fuel the many condensed atoms (protons turn like **gears**) passing on energy to a neighboring moon (levitating electrons), thus moving from the north polarity to south, up and down and around on an apple atom like a **pendulum** back and forth.

To allow for perpetual motion turning two hands (electricity +/- energized wires), it creates a magnetic energy to **measure time** inside an atom clock like a sun orbit creates **calendars**. In science, Fraunhofer discovered a frequency fingerprint. The whole system needs to be wound up by **higher chain power**-Alpha-(+ONE)-force to allow perpetual motion to exist. Perpetual motion is the very heart of the UREE principle, and it confirms how the universe came to exist.

The sun partially illustrates how the system is fueled as shown in the **Popular Science** magazine illustration of the universe orbiting around the earth, being energized from space gaining inertia and expelling **anti-gravity** levitating forces when the universe carousel is moving. That demonstrated a chain winding up the clock system for another cycle directed and controlled by intelligence. It is expressed to us as MIND, which needs to be of a higher order because the cosmos is bigger.

Therefore, when gravity is moved, it will expel an energized kinetic mass force, and vice versa, on the other entropy side fueling atoms. I theorized it must receive some energy fuel from a chain-source beyond - perhaps from our Milky Way galaxy black hole. The sun is only a transfer station like the transformer in front of your house reducing invisible higher energy scaled down magnetically visibly useful in appliances.

Besides neutrinos, **MAGNETISM** could be the transitional linkage force in the cosmos transferring energy similar to what is demonstrated in the UREE. The only positive proofs we have come from standing outside the universe or examining witnesses who came from the other side to visit our earth as reported by some scribes in the 6000-year-old Bible.

Still not convinced why the many UREE work? For the skeptic not believing in perpetual motion, perhaps I can persuade you with a third witness.

Visiting a clock shop, I saw a fancy clock design where the pendulum was the driver looks like an upside down boat-anchor. On each side of the curved tee was an invisible magnet going through a round coil pushing the pendulum by a coil magnetism oscillating with an electric current synchronized by the weight. If connected to a solar panel, it will run perpetually, which is not believed by highly educated professors refusing to check out my UREE #11, Pearl #201. I applied the same principle to another UREE possibility gaining big-time electrical profit.

When I discovered a magnetic bridge in nature that extracts electrons, I could design multiple motors that put them to work like slaves to give us free ELECTRICITY. I have described 16 UREE inventions that meet many modern requirements for the future. The forces in photographed galaxies appear to us only static like one frame of the movie, but the whole universe in reality must be moving to generate energy on the second entropy like a chain-carousel.

Our perception could be wrong, fooling our mind when looking only toward the stationary earth-axis-center of the magazine picture. It seems that everything is fixed, not moving, and looks like a round cosmos wheel. Any child knows that a wheel is meant to turn. However, our fooled perception is alerted in other example: electricity only moves inside a generator by a switch that will turn the lights on.

The switch is just a magnetic bridge for traveling electron slaves to move over from an atom level via an electric generator to become visible when they work. That is amplified by looking closer at where space energy is transferred continually to every atom and allowed perpetually to move inside an electric generator.

That generator is connected to compressed water-steam or air atoms. It needs a switch that every power station has one. Only the magnetic bridge produces free electrical benefit in the UREE inventions. Try it out if you are motivated and do not pontificate, "NOT POSSIBLE."

The second lesson for my grandkid in understanding how the earth cuckoo clock system is perpetually energized by the moving sun, which affects gravity, being linked to every atom like gears. However, it still needs to be connected to chain force outside the design parameters I call it the Alpha-(+ONE)-force. That is how electricity is transferred and extracted in space seeing brightly lit galaxies.

An interesting picture I found in the Popular Science (October 2011) showed a picture of scientist Nikola Tesla in his Niagara Falls Lab under coils that could discharge millions of volts and send electricity through air without a wire. His test proved that the earth linked to atoms conducts electricity with no wires necessary. It is caused by electromagnetic induction, a phenomenon discovered 70 years before Tesla's light experiment by the English scientist Michael Faraday.

At the Picatinny Arsenal Research & Development facility, the Army demonstrated that shoots along a laser beam guiding powerful lightning bolts. The laser beam of polarized amplified light works like a fandom wire to channel blazing energy of million volts.

Experiments in electromagnetic induction oscillate a magnetic field around a magnet to produce a current in the nearby conductor, which I applied in the UREE. In effect, the current jumps the gap while the airborne electric energy exists as a magnetic field.

In 2006, a professor of the Massachusetts Institute of Technology Marin Soljacic sent wireless electricity across a room to light a 60-watt bulb in a new method of resonant coupling far more efficient than Tesla's attempt and safer, too.

He is forming a new company, WiTricity. His invention can send 3,000 watts wireless through the air across a room and charge an electric car. Many modern products utilize wireless technology, like my computer connected to the Internet. Expanded, this principle to my 16 UREE inventions will obsolete every wire and transformer outside your house.

5. Magnetized Water?

The front picture advertised magnetized GRAVITY as explained later, but it is just as crazy as magnetized water. This book is difficult to organize as so much in nature is unknown. Science is mixed up globally by university tunnel vision. Nobody is capable to put it together into one whole theory relating everything to each other. The prevailing, unscientific evolution religion screws up logical thinking. Being in the same boat, my books will be perceive NOT POSSIBLE, difficult for the average reader starting from scratch to relate how the universe functions, connected down to the atom level.

More and more I become convinced that MAGNETISM runs parallel to light on a dual rail like DNA and is the transfer method of energy.

Please check the Internet to widen your science perspective. I just found out that even water can be made magnetic. [MagnationWater.com]. It is used commercially [MOREPLANT.COM], which surprised me. The inventor tells me that magnetized water treated with super magnets **Neodymium-Iron-Boron** (like in UREE 11 - **Pearl #201-Electric Generator**) will increase food yields 30% with increased plant vigor, better water-penetration, seed germination, pump efficiency, nutrient uptake and shelf life.

It improves soil pH and reduces water use to 30%. It reduces fertilizer and pesticide use. It removes scale build-up (salt in soil) and lowers electric bills: all to grow more food. The magnetization of water uses the same laws of physics.

...Soviet researchers are also convinced that magnets act as living things through water or water solutions involved in the functioning of cells and tissues. As we've mentioned, magnetic fields can change the ionic of molecules (and thus their pH content) but an additional fact should be brought to your attention. Heating water also changes its ionic balance and induces excess of positive ions - which can produce health hazards of its own. Once thought to be ridiculous, today magnetized water is becoming universally accepted, even if it can't be fully explained.

Water passed thru strong magnetic fields is now being used in industry and medicine in Russia, France, and the U.S.A. and other countries. The Volga Research Institute now irrigates huge areas with magnetized water. The water has been found to give a 28% increase in winter wheat, 17% in corn, 37% in cucumbers and 32% in tomatoes! So impressive have been the results that a special crop sprinkler is now being mass produced, which magnetizes the water. (A similar device is also being tried, here in Washington State, to increase the fertility of rainbow trout)! In addition, factory engineers in the Soviet Union are using methods of steam boiler water magnetizing for reduction of boiler scale and the textile industry is experimenting intensively with dye magnetizing.

How Bio-Magnetism Works. Every particle, atom, molecule, organelle, cell, tissue, organ and whole organisms resonate at its own **particular frequency**. EEG measurements show that the field generated by the human brain has frequencies in the extremely low (ELF) region centering around 7-8 Hertz. This field is the 'wave envelope' of all the electromagnetic activity of the cells in the brain. It naturally resonates to, and is entrained by the earth's resonant field (the Schumann Resonance), but can also be entrained by artificially generated fields of the appropriate frequencies. (This is the basic theory behind the Pacer.)

This entrainment effect can occur in every cell, organ and system of the body. Certain bodily effects can be caused by applying magnetic fields which resonate to those certain biological frequencies. Thus events could be triggered which effect conformation of the body's molecules, alter rates of cellular, enzymatic or organic processes, alter chemical processes or just effect overall changes within the body. (This is the same operational theory employed by Royal Rife except that he used acoustic and electrical resonances rather than magnetic).

In addition, we know now that molecular alignment will occur within intense magnetic fields. Experimental results have shown that muscle fiber, membranes, chloroplast, retinal elements of the eyes, and other fibers and macro molecules including nucleic acids have been aligned in an intense magnetic field. Highly oriented structures can result from this which may interact with other biomaterials in your body. Cell membranes, for example, are liquid crystals which are very close to change depending only upon bodily temperatures.

Thus **magnetic fields** could affect the membrane's fluidity or other properties. It only takes a very low intensity magnetic field to effect chemical reactions and these reactions have a definite biological effect on your body. Consider any one of your blood cells for a moment. It is composed of a membrane which separates two bioelectrical charges caused by the action between the potassium ion (positive charge) and the sodium ion (negative charge). This bioelectric action exists in all of your body's cells, in nerve transmissions, in the red blood cells and even in whole blood. http://www.life-sources.com/pages/The-Health-Benefits-of-Magnetic-Water.html

Another Stanford upstart, PICAROD Corp., involved with isotropic measurements to measure CO_2, methane, and changes in liquids and gases buried in pipes. It does this by cathode protection linked to magnetism, similar to on the bottom of submarines. They attach metal plates made from zinc, I believe, to prevent electro-magnetic charges from building up and corroding the boat body.

This technology connects with my energy path story that light flows on a dual rail track with "magnetism" in parallel, thus influencing light-synthesis conversion being dual balanced. Ultimately, even compressed air could be magnetized going through a UREE invention creating a higher level of electricity when magnetism reaches the **critical resonance frequency** which gives us extra unlimited electricity explained in the UREE motor section next in this book.

The energy path story ultimately ends in conflict between interpreting science with one's personal religion. It creates many obstructions raging since the Middle Ages between two belief systems. One from a 6000-year-old Bible recording the history of mankind that gives details from the very first verse in Genesis about where the energy comes from and goes in a full circle. It refers to how energy crystallized into matter with magnetism inside every atom. They bunched up into elements categorized by Fraunhofer spectral lines to create mankind eventually. MIND intelligence came with a plan to explain the purpose for human existence, linked to another mystery of the ELOHIM MIND, who is the Creator.

Expand your MIND and apply these examples to the sun conducting energy like a transfer-station being energized by a higher (chain) Alpha-(+ONE)-force cascading down to every atom to fuel the living bio-world in parallel with **magnetism**. Ultimately the system must be controlled by the embedded intelligence of a higher ELOHIM-MIND who originally designed the cosmos clock so that energy can run the system, ∞ moving down to the last atom, maintaining LIFE.

Without a designer like Professor Marin Soljacic, no new company could be formed to exist charging your automobile wireless. These concepts are illustrated and mirror imaged on the lower level of physics how the Internet computer works connecting everybody with an energy path to a global intelligence control center, much fun to analyze it.

Wireless transmission is only possible of air-atoms surrounding our earth and works only when atoms are magnetized like an electric generator. Have another look at my time-energy graph and focus on the DOT crossover linked the center of the universe, now seen in a new computer graphic picture to widen our horizon beyond Einstein's theory.

It shows that the earth is positioned in a fixed static center and surrounded with **407,702** cataloged planets and galaxies that expand our science vision, graphically. It shows the sun circling around the earth, no kidding. They only exist because energy is "moving" as explained in the carousel analogy. Look in the sky and see the lights on. We can wonder, "Why there are so many galaxies like our sun still burning?" Follow the energy trail further to get the answers.

6. ∞ Energy linked to Donut Atoms

To analyze an unknown energy path looking at space seeing so many lights on should start on the atom level being fueled perhaps from a source beyond the black hole of our galaxy the other side. A good illustration is the front picture of the 9th Babushka egg concept book showing an hour-glass-galaxy stretched out forming a donut. We can use that photograph to visualize how an atom works. To miniaturize that illustration, cut the hourglass at the neck and put it on top to see a round ball with a North and South Pole.

It becomes the smallest mini-egg concept seeing a circulating energy in color changing across the visible frequency spectrum but does not show the invisible force MAGNETISM and the other two forces GRAVITY and ANTI-GRAVITY, which holds everything together, explained previously in bigger Babushka eggs. Somehow, the invisible MAGNETISM is the chain connection and following ∞ energy path hopeful will understand how the UREE works.

From the energy perspective explained to my grandkid, the new UREE invention really works like the Donut atom described in a new theory published 3 years ago in Babushka egg concept book #5.

That new theory postulates when infinite energy from the cosmos moves protons inside the atom nucleus is in ∞ double helix motion from a higher entropy galaxy Milky Way level perspective can be logical compared to our "sun moving" orbiting around the earth spin axis. It will therefore generate a kinetic energy force that consequently could be extracted and became useful in a UREE invention. Again applying the three finger electrical rule when an orbit gravity current is flowing, a new force is created surrounding 90⁰ to its direction MAGNETISM learned from the atom theory.

Worth mentioning speaking about atoms like a crazy one driving on the wrong side, watched a recent event on TV showing the chief wizard Professor Peter Higgs was honored in a world conference in Southern Germany the promoter of CERN a 10 billion Dollar super-toy build by 1000 engineers. It was fabricated for the class of privilege to smash atoms in a collider 335 ft below hidden from the public eyes in a 17-mile underground circle located in Switzerland part France.

It was designed concentrating two proton beams, each running counter to the other, squeezed by massive magnetism into a small space to hit a target Nebelkammer with 40 trillion electron volts. It can only be partially operated in a year because Germany does not have enough electric power to run that monster machine.

The NEWS claiming once more we are close to having found a new Higgs sub atomic particle our founder postulated so many years ago and do not know yet what it is the machine run only a little and was hoped to find the open door what clings invisible to galaxies next year to proving a theory.

The major TV affair 4 July 2012 came with a worldwide-choreographed CERN announcement even Deutsche Welle and every major American media outlets were invited. The real hidden reason may have been budget cuts in Congress.

Notice: watching the big free private affair on TV so many young invited professionals lapping it up not knowing why they came and never realized they where duped and used as bait just to get more money from the government. To quote an upfront entertainer, but think logically, if you still got it:

1. The theoretical subatomic particle dubbed "the God Particle" coined 20 years ago...needs more data?

2. We are now closer to prove the Big Bang. Could we find the substance creating matter?

3. We can observe what happens in the universe laboratory?

4. We are hopeful to find the Higgs particle, looking since 1964, but do not know what it is?

5. We are hunting for the elusive energy forming mass?

6. We are confident to find the properties of the universe field of space to create mass at the end of the year?

Speculation, speculation, speculation: looking since 1964 and do not know what kind it is?

Smashing atoms 48 years still do not know what an atom is, annually wasting billion dollars only guaranteeing salary increases and obscene benefits for the class of privilege enjoying every year conferences free visiting globally the most expensive resort towns and casinos. Smashing intelligence will never find anything, put an egg on the anvil hit it with the biggest sledgehammer with all your might will never tell what a chicken looks like. If you lived on the farm would know that.

To get big grants from the government, the scientists created fractured atomic junk that only needed to be named. They get big bucks from imaginary Fermions, Cluons, Photons, Quarks, Leptons and four Bosons featured in elaborate physics fairy tales collected in expensive books cannot exist in reality designed for comatose bureaucrats, and senators will never understand it anyway.

I am just an ordinary instrument maker applying forensic science and cannot find anything similar in the range of spectral lines Fraunhofer discovered. I explain science for my grandkids when bio-cells got smashed will have lost the life-intelligence signature which will disconnect the energy from the source according to thermodynamics laws blindsided by nuclear science believing in evolution religion.

A more believable supposition is postulated on the Internet that the invisible neutrinos are the mail carrier of energy moving from the cosmos energy source perhaps from the black hole at infinite speed passing through every atom that is the fuel for our LIFE existence. For my grandkid's imagination to illustrate what goes on the inside of a Donut atom used an analogy of a naked apple pealed from a mechanical hand splicer, removed the apple core and slightly pull apart the exposed connected curled slices. Magically visualized will illustrate a highway for protons to travel up and down at high speed around the atom apple North and South Pole. Protons can be made visible in an electric field when liberated by igniting hydrogen gas.

I propose that protons are pushed inside the nucleus by billion Neutrinos which is the energy provider embedded with intelligence to control all the inter-action of the atom heart. A neutrino in the nuclear family is clearly a **black sheep** postulated in a good story read **Science News**

Magazine, May 19, 2012 page 20. On the previous page lamented that **"Neutrino search comes up empty"** by Nadia Drake. Disappointed a detector under the South Pole's Ice-cube Laboratory did not spot neutrinos after a gamma ray out burst, suggesting that something else is responsible for ultrahigh cosmic energy rays.

Science figured out that neutrinos have embedded horrendous energy because of their speed much higher than light which is invisible to our eyes demonstrated in a **black hole** going straight through mass never slowed down. Emerging from the black hole of our Milky Way galaxy is a good theory.

When protons are pushed by neutrinos, they cause electrons to levitate inside the apple atom, like dogs chasing their tails, as explained at my grandkid's level. When a moving energy pushes protons inside atoms, MAGNETISM is created 90^0 to the flow corresponding to the three finger electrical rules. It is just a reverse CERN teeter-totter principle if protons move at the speed of light will be surrounded with huge magnetism linked to 40 trillion electron volts.

Again once, the teeter-totter principle in the CERN application we have horrendous electricity creating magnetism speeding protons to end in a big bang inside a black Nebelkammer. But inside the Donut Atom, the same principle exists, only reversed, as the same energy from the black hole Nebelkammer is speeding protons to create magnetism converted to tremendous electricity of 40 trillion volts.

Check the UREE principle will understand where the energy comes from. Only when the universe moves does it create ∞ energy. Like a reverse CERN teeter-totter principle, it expels magnetism filling the vacuum of space down to every molecule in my body. Like a mini-CERN pushing protons on the atom level, it links to magnetism-levitated electrons surrounding trillions of Donut atoms to keep me alive. This explains how electricity is formed much better.

Pushed protons cause magnetic levitated electrons can be extracted driving my electric car and flying an electric jet obsoleting fossil fuels and giving us free non polluting energy even fuel my house therefore made POSSIBLE in a new UREE invention bathed in 40 trillion Volt. That is accomplished in a double generator system catching perpetually unlimited electron fishes from the pool of 40 trillion by applying a magnetic bridge guided along two wires at 110-volt electricity. To test that energy on your wall switch just touch it quickly will then know that energy came from artificial compressed water or air atom pool if you read my Babushka egg concept a good story.

Again, highly charged neutrinos at a trillion-billion volt level linked to the ∞ Alpha-(+One)-force energy source pushing protons-polarized positrons around the donut-apple atom slices at high speed is creating big-time energy. Think again and compare it with CERN teeter-totter principle has a double loop of proton beam running opposite direction too, is ending with horrendous bunched-up energy.

Similarly, on the atom level the energy transferred to MAGNETISM makes other Donut atoms stick together creating molecules. However, in the energy exchange causing electrons to levitate on the negative outer shell and can therefore be collected in a UREE and converted free to electricity. I hope the energy cartel will now understand how electricity is formed.

But some atoms are very big others tiny correspondingly has more or less magnetism created all depending on how fast protons move up and down inside the apple atom usually lower speed around the poles. Science is getting close to new ideas I postulate unlike Geltenbort and his colleagues in France working on a magnetic bottling setup discovered that neutrons well contained respond to magnetic fields. When lower proton velocity gets bigger on the apple belly avoiding crashing together being positive charged, therefore repelling and bypassing at high speed produces magnetism squeezed out sideways according the electrical three-finger rule.

Some MAGNETISM combined in molecules are bonded very strong depending on size like neodymium, iron, nickel, cobalt, others are very weak like helium or hydrogen in water H_2O which can be easily separated from its magnetic bond with a little electricity which is reversing the polarity bond between molecules discovered by Hoffman.

Remember the magnet principle around its poles either repel or attract. But both principles became useful in my new UREE invention extracting magnetism embedded in air atoms combined with crystallized nitrogen-carbon molecules once more will release a stored magnetic atom energy now expelled free. Postulated further when molecules combined from bigger and smaller Donut atoms pile on top of each other getting denser and become heavier to end as mass like our planet earth akin to a bigger apple or see how it starts in photographed galaxies or like our sun we see every day.

7. A Different Observation about Gravity

Looking again at the picture graph of millions data entries can it convince us that forgotten physics could be based on other rules like a crazy driving on the wrong site of the road upsetting a lot of people? Or, can our perception be fooled because it is biased dictated by the majority stated in the beginning? Perhaps analogue the time has come to tell a remnant British Empire to drive on the right side a better choice no longer use horses.

This Pearl story will upset yet again what is taught globally in schools now postulating that gravity can only be experienced on earth by a "moving universe" going back to what was believed 400 years ago. Or if you watched the YouTube video on how the earth does not revolve around the sun.

But consider logically, a GRAVITY force in space can only be created from a heavy mass formed by converted hot atoms ending in denser elements will eventual crystallizing which can be measured with Fraunhofer spectral analyzers. Now to discover that it is surrounded radiating in invisible MAGNETISM the glue to hold molecules together previously explained. Gravity is well described but there are still many questions that remain about its true nature.

Let's follow a quick overview what conventional theories teach, what is the Force of Gravity by John Carl Villanueva on July 19, 2009.[3]

> The force of gravity is the force exerted by the gravitational field of a massive object on any body within the vicinity of its surface. This force is dependent on three factors: the mass of the massive object, the mass of the smaller body, and the separation between the two, measured between their geometrical centers. Since planets are typically spherical, then their cores are taken to be their geometrical centers. Strictly speaking, Newton's Law of Universal Gravitation, from which the concept of the force of gravity is based on, does not stipulate a massive body and a smaller body.
>
> Instead it simply stipulates that the law affects two bodies regardless of whether one is really massive or not. However, since we commonly use the term 'gravity' in the context of massive celestial bodies, we defined the force of gravity accordingly in the first paragraph. In our day to day lives, we can measure the force of gravity. Its more familiar term is 'weight'. Thus, in most cases, your measured weight is actually a measurement of the force of the Earth's gravity on you. The heavier you are, the greater is the force of gravity on you. There are exceptions, like when you are buoyed up by a fluid but that is for another text.

However, looking into the universe and looking at the many photographed galaxies, millions of formless spaces dust of any imaginary shape and size with the lights on which would be outside a neat theory postulating nice round planet balls. Therefore being branded an offbeat philosopher with crazy ideas accept a new theory only if the UREE works. Alternatively, asked the experts how gravity affects the 70% of **dark** energy filling up the rest of space.

8. The Holy Grail of Magnetic Gravity?

I remember a salesperson gave me a 2" dia. x 2" Uranium metal plug extremely heavy much more than lead, which surprised me greatly. The U-Atoms must be very dense. Compared to an Aluminum plug next of the same size and color will never know its inherent GRAVITY unless "moved" standing on earth.

Applied to explain Gravity to my grandkid with a simplified Newton's law, fueling a UREE can only be extracted from atoms but some are very dense forming a ball others are stretched like a banana or flattened out Donut. Atoms come in any size a banana can never be magnetized, which means in the structure of atoms is embedded another force we discovered magnetism will only align itself with protons at a certain speed.

Let's focus. First, if we take the two metal plugs - one very heavy the other very light - falling from the ceiling to the ground, they will arrive at the same time according to Newton. True?

---

[3] http://www.universetoday.com/34824/force-of-gravity/#ixzz20eFOPfBc

Carl Villanueva previously stated that seize and weight does not mater affecting gravity. The maximum attraction force measured on earth is 9.8 meter/sec. Jumping off an airplane gets a max speed of 135 miles/hr. When the parachute opens is about 12 miles. A matter of fact each planet like the moon, Mars or Jupiter has its own gravitational force, true? Ask the experts. Therefore, we must look closer inside the atom some are more or less magnetized.

Another example, if we drop one heavy iron super magnet plug and another plug of the same size made from aluminum from the ceiling, they will hit the ground at the same time. True?

However, imagine if we have on the ceiling mounted a big magnet thousand times bigger than the magnetic plug; I am sure that now dropping the two plugs will not hit the ground on the same time.

The magnet plug is affected by a surrounding ANTI-GRAVITY force coming from the ceiling changing the gravity attraction depending on the magnetic mass Newton did not postulate? That points the metaphysic finger to MAGNETISM being an inherent oddball influencing gravity. We learned inside the atoms is created magnetism by protons which make atoms stick together into molecules becoming denser forming an earth expressed as mass.

When many atoms are magnetically charged as a mass, it is easy to imagine will influences other magnetic planets nearby depending on seize. Like my grandkid played with bigger and smaller magnets and found out their behavior to each other. Further investigating this crazy story now postulating that a Herbert's magnetic GRAVITY theory is influencing Newton's gravity linked with other magnetic mass would be influencing conventional gravity vice versa?

When a magnetic mass is moved slower or faster into the orbit of another magnetic mass that has embedded kinetic energy, that energy ultimately can be extracted through the generation of free electricity shown useful in 16 UREE inventions. Is that POSSIBLE?

When ∞ energy moves, it starts the process of condensing and crystallizing energy, resembling dew on grass in the early morning. Looking into space we see elements forming as measured with Fraunhofer spectral analyzers giving the range in color lines (4000-7000A^0) of how much internal magnetism was produced as the glue to hold the whole enchilada together.

As the total put on the balance, these concentrated atoms have masses catalogued in the periodic table as atomic weight, which is really how much glue-magnetism is embedded as measured by science. I have since discovered that **Gravity and Anti-gravity** should be linked to polarized **Magnetism**.

This new concept can only be evaluated at this point by inductive reasoning because no incentives have existed to build the instruments required to measure it. The ancient compass designs detecting magnetism from the Middle Ages are not good enough. Except perhaps some could be motivated to find the underlying cause of and consider why the UREE inventions work linked to magnetic gravity.

I can only point to a roadmap applying logic a process everybody must pursue too. Later investigating a number of UREE applications one things to remember that magnetic gravity like a magnet falling from a ceiling with reversed polarity therefore will accelerate a magnet hitting the floor first.

That is better explained applied in physics by Faradays' laws (YouTube. com) that a magnetic flux inside a copper coil is directly proportional to the induced EMF. Meaning, if the super-magnet passes faster gets a bigger pulse useful in modified UREE application explained in **UREE #12, #15, #16.**

Follow further the energy trail when we have a certain mass like a bowling ball or Neutron's apple and dropped from my roof they all fall to the ground at the same time only on earth. That is a paradox unless we consider that gravity in space has embedded magnetism creating electricity like a universe carousel becomes a gigantic electric generator. Seeing so many lights in the starry sky will now comprehend why electricity exists and how it is produced.

Check the Bible slingshot stone story of David killing the giant Goliath. A little stone accelerated becomes a bowling ball explained in science a little differently. A bowling ball or tiny stone we all know has a different weight not affecting gravity, but on the atom level, some atomic elements are very big and others tiny, but when they are moving, it is another story.

If we apply that to a quicker moved magnet inside a cooper loop gets a bigger pulse. Therefore, a gravity air pressure environment inside a UREE cylinder could extract electricity when magnetism is moved quicker in a double-sided generator. Looking at the night sky, the light is not produced by other laws.

Expand that idea to the universe science believes that gravity is dispersed throughout space linked to galaxies and space dust made of matter or elements but investigated from an atom world perspective could reveal more. We have learned that each element is really a magnetized atom apple and will stick with others forming molecules surrounded with levitating electrons.

Consequently, Magnetic Gravity is like a crystallized energy force embedded in the atom weight located between (Zero time = ∞E) to become Gestalt. That clears up the relationship between gravity and magnetism linked to the resonance frequency holding everything together.

Every atom, bigger or smaller, is fuelled by the ∞E and resembles the smallest polarized magnetic field, and if passed over by a bigger magnet at a certain distance and speed, could we observe what will happen?

We never build instruments to measure it on the atom level because smashing atoms in bigger machines like CERN was more fun spending the money. Now we come back once more explaining how Herbert's magnetic Gravity theory works confronted getting free electrical energy from UREE.

9. Magnetism Expanded

My grandkid learned playing around with magnets demonstrating an invisible force either repel or are being attracted by a stronger or weaker magnetism which is just like analogue a Russian big universe egg mirror reflected replicated in the smallest Donut atom egg interconnected too by an invisible magnetic force. Otherwise my new UREE would not work.

Atom nuclei come in different sizes of mass like heavy lead, iron, uranium gravity will respond stronger to embedded atomic magnetism. Some mass is feather lightweight like hydrogen; helium will be elevated by a weaker radiating magnetism attracted on the atom level. A bigger magnet like the sun in orbit turned by a universe carrousel should be linked to the sum total of a larger magnetic field emanating perhaps from the Milky Way galaxy also orbiting with a bigger magnetic field like a bigger transformer.

Playing with bigger and smaller magnets at a certain distant will notice a differently force attracted or pushing depending how much is the surrounding radiating magnetism either sitting fixed too heavy or some are a little lighter may become levitated for a while demonstrating a gravity-antigravity force we can observe demonstrated now in satellites applied to earth. They are delicately balanced as general science think it is gravity to return to earth or drift into space lost forever. But we could also imaging that MAGNETISM is the force balancing satellite attracted to a big collection of atom magnets stuck together getting denser toward the inner earth center.

Our earth is made of infinitely compressed magnetic atoms similar to how the sun is made from lighter magnets with hydrogen, maybe helium, still very hot and not crystallized out yet to heavier atoms. Looking in the sky seeing a whirlwind of bigger round magnets photographing galaxies or partially the Milky Way pulling everything toward a black hole reduced to infinite energy, I now recognize it as antigravity force, which is a reversed magnetic polarity of concentrated magnetism. My grandkid's magnet is a good example.

Applied to space stations and satellites, it must be carefully managed, always watching where the mass will drift either sucked toward the center of our magnetic earth or lost in space toward a bigger galaxy magnet. We could therefore theorize that magnetism is the primary force compressing atoms. It is expelled in a magnetic polarity sticking molecules together. Eventually in time, it will crystallized out as mass demonstrate as Gestalt ending solid as our earth with a magnetic North and South pole.

Satellites are made of atom-molecules building up a certain mass too must be immediately corrected within seconds according to invisible forces attracting or repelling perhaps with MAGNETISM. Otherwise space objects having mass made of atoms with an embedded magnetic potential could be drifting into space toward a bigger magnetic fields radiating from galaxies pushed by a unified magnetic antigravity force ending out of sight perhaps in the black hole of our galaxy one way or another. Unless controlled by intelligence or artificial computer intelligence.

Consequently ultimately magnetism produced inside every Donut atom now looking past our space-station at distant magnetized galaxies brightly shining following the laws of electricity demonstrated look closer demonstrated by Nikola Tesla and the other scientists applied to a new magnetic theory discharged toward the outer universe seen from earth. Alternatively, go back and read again what Professor Marin Soljacic discovered transferring electricity converted to light projected seeing the same photographing galaxies.

If we see light in galaxies must be linked to MAGNETISM if you understand electricity generation and must be balanced by an invisible magnetic force only works like an electrical transformer we are familiar mounted on top of the pole.

That unusual magnetic-gravity theory is proven with two witnesses when we send a camera out in distant space millions of miles away and returns the many pictures in a vacuum only possible with a magnetic field guided by a space magnetic frequency carrier.

Similar on the lower entropy spectrum, we transmit television-telephone frequencies over the air can only work because surrounded magnetism is embedded in air atoms. Making a previous statement that AIR is magnetic now proven also with a Teslar motor (UREE #16). One thing is sure anything send into space is affected by the nearest object could orbit around or used as an accelerator like a slingshot for further travel. However, playing around with magnets matches what we see, why not a combination magnetic gravity?

The universe could be our third witness. It is just a bigger transformer which works on the same magnetic principle like a bigger or smaller Babushka egg. That caused a new UREE invention discovering that levitating electrons inter-connecting with other air atoms molecule like billions of Electron slaves guided over an invisible magnetic bridge explained in Babushka Book #9 otherwise the energy transmission would not work to get us free electricity energy from gravity linked to space carousel system which is not static, but a mover for the lights to be on. The principle of electricity proves it.

MAGNETISM, as a force, is a bridge between on one side ANTIGRAVITY and GRAVITY fueled by protons the driver pushed by millions of neutrino energy mailman send from the Alpha(+ONE)-force source. Remember the three finger electrical rule creating magnetism 90^0 when a proton-electron current is moved creating magnetism, or read the first Bible verse, which tells where the energy came from.

Pay attention. I use the same magnetic bridge in my UREE invention to gain free electricity, which is why it works. It gets better. We discovered that although different weights can fall with the same speed on earth to the ground at the same time. That changes if a big magnet is passing by, like the sun with its densely heavy magnetic molecules. Some are still very hot, not yet crystallized as elements. However, the embedded magnetism can be noticed against the space dust shadow on sun's poles, only seen with our eyes, as we have no instruments to measure it.

The magnetic forces of nature can be proven with the next moon orbit cycling in the direction or opposite will shows us how gravity-magnetism is related and now learned to be the same from satellites. Babushka books give many examples.

An astronaut on the moon demonstrates gravity variations when a moon changes direction like pendulum of a clock showing us gravity converted to antigravity inertia force can be related to a sun. That will change a perspective upside down with consequences; perhaps now understand how a UREE really works. Being a German instrument maker with a little practicable science background acquired from the trade school has an advantage when I considered that everything on earth works like a cuckoo clock.

A UREE only works when gravity is moving therefore moving magnetism from a nearby sun linked to inertia creating perpetual motion. Applying the electrical principle if magnetism is moved can be extracted as electricity profit. But from the space perspective there is a delicate balance on one side gravity all the magnetism acting as a force glued together the billion-billions Donut atom magnets attracted compressed toward the center of our earth and on the other side the stronger magnets will force atoms to levitated like an ANTIGRAVITY force if something bigger is nearby.

Remember the ceiling explained for my grandkid. Expanded to universe carousel wheel as shown in the **Popular Science** magazine demonstrated in the Data Centric Universe picture it would function like an electric generator magnets turning inside a black hole according to physics will be expelling energy in its center according to the three finger rule. That will push the chain gravity chairs sideways set in motion we can see in a starry sky like our sun in orbit around the earth or vice versa.

This system gets its energy from the center black hole Milky Way galaxy now shining. In addition, when infinite energy moves the lights are on now demonstrated linked to a UREE. Energy is transferred from the original Alpha-(+ONE)-force the ∞ energy source controlled by intelligence to fuel every atom invented by the owner ELOHIM for a purpose.

It comes with a promise will transform my mortality one notch higher to live in heaven, or could end in NOTHING before the big bang. The alternative option is demonstrated in the last #17 UREE roadmap the best engine designed ever also the last one, therefore should not miss it to become extremely wealthy. Check it out read further ending this pearl.

Once more, we have learned that magnetism is the force, which sticks other atoms together, creating molecules that point to the evidence that invisible magnetism must be present in space. It works like a cosmic transformer using energy from our Milky Way galaxy black hole to maintain LIFE sliding on a visible light frequency carrier via the sun similar to a dual railroad track meant for balance on the lower entropy of 186,282 miles/sec. This could carry undetectable magnetic embedded energy to the cell level as an alternative energy fuel source. Perhaps Einstein measured the speed of carousel to be constant.

That theory may not necessary be photoautotrophic or photo biological synthesized but could be expanded to photoconductivity linked to magnetism like a dual railroad is better balanced. We know that energy moves only in a dual copper wire pairs surrounded with magnetism, matching physics the same way when we link the ∞ energy Alpha-(+ONE)-force from the cosmos in parallel. It will be fueling every atom creating a universe surrounded by magnetism levitating electrons.

In addition, if energy is further crystallized out becomes electricity that is proven by Nicola Tesla experiments and the other science electromagnetic demonstrations when the lights of billion billion potential observed in galaxies and now proven miniaturized scaled down in UREE invention technology. Notice electric energy from the source is passing only on the inside of the black wire surrounded with magnetism but needs the white neutral wire no longer magnetic energized to close the energy loop running on a dual track.

Even wireless electric transfer is confirmed by Professor Marin Soljacic and if looped can only happen linked to atoms being magnetic either the earth or galaxies both demonstrate mass more or less concentrated. The force between one object to another is not gravity but is bridged by a strong or weak invisible magnetic energy even demonstrated in the smallest atom egg. Look closer how an electric motor works expanded to the UREE, experts of the evolution religion say NOT POSSIBLE.

I hope that this quintessence Pearl opened a knowledge horizon a little wider and answered some unknown energy mysteries. Compare that to what is taught in every school and university, which postulated the Big Bang theory fairy tale starting from a NOTHING as an energy source.

The other side of the same coin postulated making a trip with my grandkid to a modern science museum, which has painted across the entrance wall the imaginary evidence from an artist rendering lacking the proof dedicated to an evolution religion a picture series notably centered on a monkey walking upright my grandkids always gets the giggles.

Ending the evolution painted series cresting with a person in a business suit my grandkid says looks like the US president. That artist rendering designed for ignorant people explaining fantasized evolution evidence must have somehow stopped with evolution process already maxed out on the monkey-turkey intelligence level. Now demonstrated by many PhDs teaching unscientific fairy-tales in our universities watching Larry King on his TV program interviewed a fired high school teacher in conflict teaching intelligence not allowed. Ask how you will feed your family told I teach complex military equipment in the Army to new recruits.

What a loss for our grandkids especially in Chicago many youngsters are in prison never graduated from high school. Most cannot afford a higher education unless belong to the class of government privilege now costing $100,000 to $200,000 no longer a sheepskin exchanged for a cheap plastic, worse going deep into debt as many jobs no longer available because inventors creating new businesses disappeared. Perhaps it could be linked when America murdered 60 million babies by abortion no longer possible to have people born with a higher intelligence like Einstein and Werner von Braun called out of retirement to counter sputnik building a better rocket landing eventually on the moon.

The American Supreme Court decision ignorant in fertility science approved by a comatose Christian majority culture killed more people in

America than Hitler and Stalin did in World War II that will irreversible be ending now 2000-year old Western Civilization in one generation - guaranteed by fertility laws embedded in physics and having nothing to do with human rights. It is not a forbidden religion.

Therefore, I could conclude from what is taught in schools that a cuckoo clock evolved from NOTHING with complex gears and because we are mortal, an accident happened to tell time. But ignoring that intelligence can never evolve from a NOTHING proven with 60 trillion proteins looking for intelligence to connect with a matching DNA must be a forbidden religion as only tunnel vision is allowed by the atheistic priesthood.

But intelligence needed to be explained if you aspire to a PhD degree therefore postulated further that the NOTHING decided to explode into SOMETHING, which is better choice than the NOTHING. O wonder of wonders, intelligence was born and evolved from the NOTHING.

Notice why every science department is not allowed to postulate other ideas unless it is dressed to reflect the stupid evolution religion enforced by an atheistic priesthood controlling every university. Why is the metaphysic forbidden found in Webster's dictionary therefore exist and why are lies and unscientific fairy tales opinions exclusively enforced by ignorant brain-dead professors pushing their moronic unscientific evolution religion.

Western Civilization will now end if you are educated the old-fashioned way fueled by bogus printed money linked to the atheistic Supreme Court brotherhood rewriting history and exported globally their lies to every country? Only Babushka egg concept books dare to warn us.

No wonder when science was still in infancy why is a dubbed education establishment still holding on of hundred-year-old speculation. My UREE invention would never work in a NOTHING theory. Prevented to think in logic, they can only reject my UREE being misled by a perception from looking through a telescope collecting 407,702 computerized data points to the utter frontier of the universe not understanding that gravity cannot be generated by a fixed sun with all the other planets circling around. When we apply known physics to the source of gravitational energy can only be extracted when the cause is "moving" like demonstrated in an electric generator.

Therefore, being misled by a telescope perception still needs interpretation depending which religion we believe in. It prevented the science establishment from understanding Newton's laws consequently cannot explain logical how the UREE works, and what gravity or electromagnetic force is, along theorizing about a strong-weak force still a mystery.

Or tell me how electricity is produced on the other side of the generator getting wet from pressurized water converted to mysterious electricity heating up melting 50 tons of iron in a Bessemer-Thomas bell furnace from compressed water? However, sometimes science magazine will publish a true picture demonstrating how the sun is positioned among other planets in cosmic relationship. But explaining how gravity energy is transferred from a true science perspective according to physics can only be achieved when a body is moving which has build in inertia depending on acceleration.

Therefore, the sun must be moving if we want to extract gravity energy on earth. Otherwise the UREE will not work. Believing that the sun is fixed and the earth and stars orbiting around the sun, it is no wonder why they declare the UREE will not work.

Resolving a perception from the Middle Ages I asked you to remember the analogy how our mind can be fooled as some stopped evolving frozen on the monkey-turkey level teaching now our grandkids their favored evolution religion avoiding recent science discoveries. Being highly paid by the atheistic priesthood to keep the masses stupid designed to destroy a Christian culture.

The final mediator proof can only be recognized either looking through the black hole from the perspective of our Milky Way galaxy or being outside the universe looking inside or build a **UREE motor** and if it works revise an evolution sun religion not based on science.

However, because we are mortals, and if educated in physics not biased from unscientific evolution fairy tales ignoring 50% what Newton wrote, can get a balanced proof demonstrated by observing the path of planets and the closest orbit is the moon. If analyzed it in physics behaves or function like a cuckoo clock pendulum switching back and forth.

The pendulum only works in physics when the sun gravity weight is "moving" being linked to the GRAVITY force, not static. However, gravity is balanced with an ANTI-GRAVITY force utilized in the 16 **UREE** inventions. Check it out. When gravity is static, it cannot be measured with existing scientific instruments.

10. A Closer Look at Antigravity - Levitation

If someone in the future is planning to travel through space must understand how the universe is function and figure out which laws will govern illustrated in the Time-Energy graph. It could open the possibilities to power a jet engine with free electricity obsoleting fossil fuel even power a UFO to go further.

My last pearl explains the fundamentals we need to consider contrasted with misapplied science and attempt to change a 400 year old false belief system exclusively taught in our schools. Flying through space with a UFOs with plenty of fuel still needs a little more reading time to explain what is not postulated in universities blindsided by an evolution religion, as no other reference books exist which could help. What is discussed here will be foreign never postulated but is not speculation and needs a little logic to investigate the evidence that I am pointing to.

In every **UREE** version, hopefully we learned about a principle that we need a little **extra anti-gravity** force added to **gravity** to permit perpetual motion on the return cycle to make it work to give us electricity profit. On the lower entropy level when a piston is pushed down by gravity needs to be push back with a higher force returned overcoming the stored gravity force in the storage air-tank. That can only be accomplished in physics with a bigger force I call it **anti-gravity** unless you named it differently.

An anti-gravity force is similar to gravity can be married to inertia in a roller coaster or cuckoo clock pendulum. Or better yet can be noticed first looking back at history for an example when Jesus and Peter was walking on water verified 2000 years ago by (12) twelve witnesses that demonstrated a anti-gravity force. That unbelievable, uncommon phenomenon was recently demonstrated on TV (2011).

A young man using a swimming pool in Las Vegas with cameras rolling incredible walked on top of water across the pool while seeing people swimming under his feet, "no kidding". My first reaction was that the viewers were being tricked, but later they kept showing it to people passing by on the street. The guy was levitated 1-2 ft. above the ground entertaining to impress the giggling girls being surprised all seen on TV.

An afterthought could perhaps explain what was not seen on TV. An invisible nylon string connected high up to a helium balloon as everybody is fooled looking only at this young man's feet. For me as a scientist, searching for plausible evidence must agree with physics by using our 5 senses, hopefully not being fooled.

Perhaps levitation could be explained by my magnetic toy suspending a little ball in midair pushed by a magnet hidden in the base. Or, in another science experiment example, I saw a living frog levitated being surrounded with magnetism. Better yet, visit Germany and watch the levitated high-speed train passing by.

But the best experiment was shown on YouTube where one made a balloon the size of a tire and filled it with helium gas. He made a horizontal ring of copper wire loops around the circumference to the point of losing levitation, sinking to the floor on top of another coil. Applying a current to the floor coil would raise the balloon contraption several feet up and down depending how much power was supplied as coil magnetism created a moving levitation. That perhaps could explain why UFOs seem to land and levitate on planets being magnetic like our earth, producing internally a counter magnetic force.

Hopefully we have not turned off our MIND reading the first Bible verse explaining metaphysic science definitely not fooled but learned how we can manipulate the forces of nature linked to perpetual motion which will give us benefit in form of electrical energy which can be extracted totally free from GRAVITY, WATER and AIR with a new UREE invention. Therefore Gravity is always linked to an ANTIGRAVITY force like in Chinese yin-yang explained in ancient clocks captured in the third Babushkas egg concept book, which is much fun to read.

Repeating once more, when we mounted a big magnet thousand times bigger in the ceiling imitating a galaxy in the sky, a magnetic weight will not hit the ground on the same time according to conventional gravity. The magnet weight has build in an ANTI-GRAVITY force affecting Gravity depending how much Herbert's Magnetic Gravity was exchanged for Newton's Gravity converted to electricity in a UREE totally free to future mankind.

A Mini-Hoover Dam Designed into a Gravity Motor

Following the unusual (magnetic) trail of energy on top of the Hoover-Dam, we will all notice a very large lake, representing an **open** infinite energy source potential, which is like storing billions of bowling balls, but only a tiny fraction is useful when spliced out, when the (**water-media**) is piped down to the bottom creating kinetic energy, which is moved by gravity above and converted to electricity profit.

Also looking closer to a steam generation I recognized we need very high pressure for generating electricity. However, our perpetual gravity motor is not that much different as it works inside a **closed** pressure environment when linked to stored (**air media**), which is embedded in an small air tank as either water media or air media, represents a highly pressurized differential, thus releasing an inherent, embedded kinetic energy. Plus if allowed to move perpetually, it could be converted into perpetual electricity, similar to the Hoover Dam or steam generator example.

My new UREE invention makes it portable, which can be applied useful in automobiles, trucks and trains, which need no transmission wires, thereby, making it a little cheaper as a system. That implication being now portable has tremendous ramification which will obsolete overland wires and associated transformer system, supersede massive nuclear-coal fired power stations and changing conventional science shifting Einstein's theory upside down and probably causing a revolution in most universities and hopefully will burry an unscientific atheistic evolution religion forever.

Once more, high water steam pressure **continually** passing over the generator gives **continual** electric energy. The same principle is applied to the UREE gravity motor but also must recycle **continually** the air pressure media returning it in the same tank will get perpetual electricity, too.

Explaining it further by applying that analogy to an air tank extracting pressure from below the ocean, similar to the weight of my cuckoo clock that hangs on the wall, will not change its inherent mass, but does move gears, hands and a cuckoo, even overcoming friction forces which could be linked to compressed atoms collected from the sky, which has a gravity force embedded. It expands parallel to our earth, not changing its mass either and is surrounded with static gravity, which is the same force.

However, when we use an air tank, we create an artificial gravity environment, just like that which exists below the ocean or on top of the Hoover Dam. We accomplish this by compressing water or air atoms closer to each other and hold them unnaturally captive, bottled up within the steel pipe, container or tank. By releasing the pressure cyclically, the simulated, embedded kinetic energy is released when the oxygen molecules expand at a very fast rate, back to its original natural state.

The Hoover Dam's gravity pressure potential can therefore be carried around in a steel air tank (bottle) so as to make a gravity differential portable, which would not need flowing water from a lake anymore, in order to drive huge turbines. We only changed the transfer media from water to air. That makes it portable and a little more compact.

From the energy exchange level inside the oxygen atom of the **compressed water molecule** (Hoover Dam or steam-nuclear/coal), the negative (-) charged electrons will be expelled with a plus force (+) on a black copper wire in a ∞ loop. But for a current to flow over a magnetic bridge exchange according to physics, it still must be linked to a return negative (-) white copper wire, which is a highway for stripped naked electrons no longer pressured returning and passing through a magnetic electric generator wire-loop bridge system once more.

The electric generator is the magnetic door connecting the invisible world of an **atom-city** to our visible mortal (Daleth) dimension. It functions similar to a slave market only bigger. Billions of electron slaves, previously fully charged, returned hungry and naked from a foreign environment and must return and come back home to have a party again, being fully dressed once more. They are happy in their natural state only at home levitating on top of negative charged oxygen nuclear donut cloud.

The energy exchange mathematics of a ∞ loop demonstrated that energy flows continually according to entropy transformer principle. If electrons cross the magnetic bridge must return once more over the atom border through the back door into a higher energy level loop, which is then rejoined in reverse to the oxygen atom-city bathed in infinite light energy for a recharge. Only as long the water-air compression environment exists compressing billions of oxygen atoms pushed together by an artificially created substitute for gravity force can the naked electron (-) slaves be recycled hitching a ride on top of the oxygen atom.

At that time they get recharged by circling around a fast moving vortex in a negatively charged magnetic donut atom cloud being refueled from the inside of the oxygen nucleus. When atoms of different size like the bigger oxygen atom and externally compressed together with other bigger atoms, electrons are forced to jump from one orbit to another always be looking for a natural pressure release, like a second entropy magnetic bridge coming by to be free floating again. It is similar to a flowing river detoured being directed over a waterwheel to do some work on the way.

But on the lower transformer entropy level, some electrons doing work in an energy exchange will become naked losing their charge. They still will follow embedded intelligence to return for another cycle to live a little longer, or if disconnected disappear flowing down the energy water river ending in ocean water molecule reservoir.

However, in the Gravity Motor UREE system we use the same oxygen atoms over and over in the same air tank to be a portable perpetual energy source. The naked electrons detour is shorter and will be immediately recharged by a neutrino energy force flowing down to every atom according to the novel cosmos magnetic energy trail theory. That process makes the portable energy source system possible.

Lining it up in a smaller steel air bottle, along a football stadium, could give us the same energy as a Hoover Dam, only quite a bit less expensive as well as without the connecting overhead wire system. Basically, an infinite energy source potential becomes a transportable Hoover Dam reservoir, only miniaturized in portable steel container bottles. The process to generate conventional electricity is accomplished pushing high-pressured water or steam through turbines linked to an electric generator to extract electricity as a system.

When we use a different media of compressed Argon-Oxygen gas or air, a special turbine-motor is required which is designed around the standard car engine, yet must be modified to a electric generator-streetcar motor combination connected with two flywheels. That could replace a Hoover Dam water column pressure system in exchange gaining the same energy but now extracted from an air bottle a little more portable.

To gain electricity explained in physics, we utilize a gravity differential potential replaced with air or suitable compressed gases in a tank. If released by a valve in the gravity motor, it will explode the piston to a lower entropy level via the crankshaft. That moves kinetic energy, which can be converted to electricity. Whatever is over 100%, I call **profit**. Pulling energy out from a gas bottle is achieved with a modified diesel motor system, but it must have a flywheel assembly mounted on each end of the crankshaft embedded with a standard generator armature linked to a streetcar motor.

The Gravity Motor is fueled with just plain high-pressure gas from the tank. But what is novel the same pressure pushing the piston down must be returned 100% into the air tank therefore needs two flywheels converting and storing kinetic energy applying perpetual motion laws. When looking inside the motor cylinder, some experts who are not convinced may say: "tit for tat", as nothing is gained, since 100% pressure goes out and 100% compressed is returned via the flywheel.

Yet, as we look closer at the energy profile, while the piston moves to the bottom, it accelerates kinetic energy, which grows inside the piston while it travels down and gains speed to that of a bowling ball energy level a little faster than the earth orbit. At that time a number of capacitors are fired via spark plugs, which has an added bowling ball force effect embedded to over 100% of the original air tank force, but is now converted along the way. I will explain more later on.

The two flywheels will store temporal inertia energy and push back the return pressure piston while the generator and streetcar motor combination is generating electricity. A number of fired capacitors within the diesel-modified motor are needed to synchronize and convert more kinetic energy, which is embedded inside the compressed air atoms that expand gas pressure in cycles. Electricity is converted within the system by a miniaturized Hoover Dam generator and is perpetually possible, which I compare to a genie waiting in the bottle, ready to be released. This basic flywheel principle will soon be seen in the next hybrid electrical car, which will be sold next year in many dealer lots, with an electric generator embedded inside a flywheel, which will get over 500 mpg range.

How the Cosmos Side Fuels the Gravity Motor

We will now become exposed to the principles of perpetual energy exchange put into practice in the very heart of this crazy idea, which is not possible according to conventional science. If my energy proposal is accepted by the energy cartel, we have succeeded to get totally free green energy for the coming 21st century, guaranteed.

If it is not accepted and laughed at, it was worth the try needing a little more time. It is just another repeat of the Wright brothers' attempt and heard much laughter at their church even preached from the pulpit, "If God wanted man to fly, he would have made mankind with wings." Galileo, too, was silenced, and it took another hundred years to prove him right.

New inventions have consequences visiting harvest festivals in farming towns, we will see the assembled old steam engine running, which is linked to many laborsaving machines, but it also created massive unemployment in the thirties. Your grandparents still remember the big depression.

When it equalized, it created a time of blessing unheard of in history. People owned their own house, an extra vacation home, two cars, two refrigerators, a wash machine and a dozen hairdryers for my teenager, thousands of time-saving kitchen gadgets, billions of TVs and telephones, or a boat on weekends. All because of a new energy conversion that saved labor, allowing more free time.

Conventional energy is running out. We should look around for the next cycle or repeat the great depression once more. The new gravity energy conversion discovery follows the same laws as a business. In order to make a profit, money from energy must be turned around and put back in the system, which is demonstrated in physics, such as an electric motor.

When energy moves, it gives us free work. Explain why magnetism is needed to create electricity in the rotor? When it returned electricity back to the stator, it produces more magnetism, which rotates like an electrical motor in order to do the work that slaves used to do. It is a paradox. Energy going around in a full figure eight motion, flowing endlessly in a circle, like the infinite math ∞ symbol shows graphically.

My invention is profit-motivated to make perpetual motion possible. It will borrow some energy that is indirectly needed from the cuckoo clock chain force. When it is redirected, it will cause gravity to move and give a little leftover electrical energy, squeezed out as profit. That is accomplished by linking a second entropy gravity force, converted in a hybrid electric motor and embedded in a flywheel controlled by a series of capacitors as a system. That is a man-made invention and is therefore patentable.

To drive a car around with just plain pressurized gas as a fuel might seem insane to most, or we need a little more insight and continue our examination of an energy conversion analysis that operates a perpetual gravity UREE motor for the gain of profitable electricity. We have investigated an energy trail and ended with a mix of argon-oxygen air atom molecules to find the invisible embedded ∞ **energy** source. But my invention also postulates we need much higher pressure perhaps 10 times over a gasoline explosion that makes it possible to drive around without conventional energy. Maybe an entrepreneur starts a new business submerge air-tanks to the bottom of the ocean and closing the valve now has enough pressure to drive around your car.

If the first prototype of the gravity motor works, then we have found the missing link in the search for the holy grail of a unifying theory: that being, **"What is electromagnetism, gravity linked to strong force and weak force."** When we follow the trail to check the many assumptions, it must conform to the laws of entropy physics, yet to me, it must also match the metaphysical domain in order to have a balanced perspective. On my quest, it became rather obvious; simply by observing nature, that everything needs energy, in order for life to exist right to the very atom level.

Closing an Cosmos Mystery Energy Trail

What you read here was never postulated before and is greatly abridged. I have attempted to redefine some laws of nature poorly defined in our universities. A great deal of science seems distorted denying historic knowledge accumulated in the Bible across a 6000-year span. It needed an unbiased assessment collected from a metaphysical database, which should include recorded divine revelation that has been rejected by the establishment. Babushka egg concept books became the outlet for me.

Modeling a new atom theory from information that comes from the research of history is an attempt to explain how the cosmos came into being and linked to how an atom works, as understood by the observance of nature. Some may wonder how the universe is fueled, as billions of galaxies burning with energy and linked to our sun, controls our bio environment right down to a single atom, where life is embedded.

This is still a mystery. Smashing atoms with CERN will never discover the hidden complex intelligence, where information is embedded within every single atom linked to the cosmos, which is similar to our personal computer software connected to the Internet.

Yet luckily, new DNA discoveries and genetics finally opened the back door of science that is manipulating atoms. Eventually we find out what works and what does not. Following the energy trail, everybody would point to the sun as an energy source. Newton's law should convince us that the sun is an energy source and burns from a higher energy potential to a lower.

Modern science does not know where the higher energy comes from, but logic tells us that an invisible energy source must come from somewhere, as looking into the night sky we see thousands of lights and observe the many galaxies and stars that are brightly shining. It is not too difficult to postulate when energy is available in space as we look at the beautiful galaxies from the Hubble Space Telescope and notice that the same energy still flows and could be coming down to our earthly level.

Seeing pictures sent by cameras that were previously launched with a space mission, as they travel millions of miles through space, I question what the medium SOMETHING is, in order to make a path for a frequency to travel over? We think that there is nothing in a vacuum; therefore, energy from space, which is converted to light, must bridge the NOTHING, as it is not dependent on one atom transferring energy to another atom.

Energy travelling from space could end up invisible in a steel tank bottle, where I drive my car without fossil fuel? That would prove a "genie force" trapped in a bottle. The cosmos started with infinite energy from somewhere. That energy is still around, as we see the aftereffect, according to the second entropy, which is recognized as visible light that only our eyes respond to. Therefore, light and atoms could reveal more.

My journey started when I discovered ancient clocks exhibited in museums revealing a prehistoric time dimension that opened the door to understand the first verse of Genesis, which expresses infinite light in the ∞ math sign, indicating an energy exchange on two entropy levels. That is only possible in a time dimension. "There was evening and morning..."

Investigating science much better explained from the oldest book of mankind, the BIBLE, which starts with the purpose of the cosmos and a detailed creation report in Genesis 1:1 and John 1:1, where God executed his invention and created mass to crystallize from the LOGOS ∞ energy center, born with the **time dimension** in a teeter-totter relationship which will be governed by two entropy laws - **Bereshyt**.

Creation started when the invisible God revealed himself and became the visible WORD as inventions first must be formulated within a ∞MIND expressed with the hieroglyph letter "**B**" which specifies that a newly formed cosmos will function on two ∞ energy entropy laws.

The laws of physics and metaphysics identify that the ∞light (Genesis 1:1) runs on a divided highway to move toward lower entropy visible light like Heh-heaven toward Daleth-earth cascading linked together in a ∞ circle creating a subordinate gravity force embedded with **magnetism**. Rotating the universe became the glue for atoms to form mass, coalescing into bigger mass ending with our world centered around a second energy center "The WORD" (John 1:1) who became flesh (physics) foretold as EMANUEL (God with us) demonstrating by the ∞ symbol.

The resurrection principle is the TIME-dimension DOT dividing mortality to immortal LIFE demonstrate for my grandkid in a caterpillar illustration morphed into a butterfly. Confirmed on a higher level the invisible ELOHIM became visible in Jesus THE Christ [let us make man in our image]. Jesus is the first-born still divine of a new creation. Needed a place to live and created a universe for mortals to fulfill its purpose. In the math symbol ∞ again expressed the ELOHIM-ANGEL new heaven world (Heh) and Jesus Firstborn new Earth (Daleth) dimension a divine paradox we mortals will never understand mentioned everywhere in scripture.

When moving mass crystallizes, kinetic rooted energy is released for the benefit of mankind; foremost from our sun and on the second entropy bio-world and moon proves this. [Electri-city] is the bottom, miniaturized force embedded within every atom [city] molecule, closing the loop of the COSMOS DONUT proving why the UREE works.

For another good illustration of a cosmos dimension loop could be illustrated on the front page, if we connect two hourglass galaxies in a large circle like a donut. It would demonstrate an energy path swirling around in a spiraling vortex being separated by a North Pole black hole representing the ∞ Heh-dimension and energy circulating perpetually opposite demonstrated a two looped polarization finite-time, Daleth-dimension the South Pole. In between are billions of stars, galaxies, planets and asteroids in a circle like a Donut loop, coalescing and strung together, quite similar to a vortex spiral ∞DNA chain. (See the next picture ending my story photographed a galaxy not static turning very fast.)

The Whole Schlimazel Summed up

Babushka eggs expose a little better understanding of the universe by postulating forbidden concepts and hopefully expanding knowledge horizons. I start from the donut atom level linked to why we have lights on in many galaxies and how it is produced to make a universe functions from a very different perspective.

Two easy illustrations on my grandkid's level emerges explained in analogues, once more a cuckoo clock the other a rotating universe carousel theory. I learned to believe the many facts revealed to me as being true. To understand the interrelationship of how physics is intertwined with metaphysics, I created a ENERGY-TIME-graph with a vertical, exponential energy curve intersecting with a horizontal TIME dimension can now explain science from two perspectives metaphysic-physics viewpoint even Einstein would be pleased not obsoleted.

To understand what is prominently noticed in nature, one must first start by investigating the intelligence embedded in every atom linked to a predominant LIFE FORCE all is exclusively controlled by massive cosmic intelligence ∞ELOHIM computer as demonstrated in a miniaturized design, replicated in the mortal MIND. Our minds must be fueled with energy, too. Hopefully, our lights are "on" no different like the universe, seeing many lights are "on" in galaxies indicating that there is life-force seeing that ELECTRICITY is flowing.

That is aligned on the horizontal time dimension line illustrated on the graph that captures Einstein in a math equation expanded along a visible rainbow light band frequency. Our human eyes linked to a infinite MIND is able to discern a perceptible frequency spectrum to give evidence designed in the exact middle range from the ∞ infinite down to zero. It was defined by Fraunhofer to make visible all the elements discovered on earth. When ∞ Energy crystallizes from a magnetic frequency converting into particle matter articulated in physics with a teeter-totter energy math formula:

$$\infty E = \text{Zero time.}$$

Since the middle of the energy frequency is the second entropy level previously defined unknown by Dr. Albert Einstein in his famous $E = mc^2$ formula only visible to our eyes but ignores the many other electromagnetic higher or lower invisible frequencies bathed in MAGNETISM. Just because we cannot measure magnetism on the higher light band over X-rays does not mean it does not exist. Now many more magnetic energy frequencies have been discovered on the lower side useful to high technology TV and telephone computers in the last 50 years Einstein never dreamed about.

One more thought on Einstein's theory of General Relativity linked to Variable Speed of Light. Albert Einstein himself emphasized in his paper in 1917, "The results of the special relativity hold only so long as we are able to disregard the influence of gravitational fields on the phenomena." In the presence of magnetic gravity, the speed of light becomes relative. To see the steps how Einstein theorized that the measured speed of light in a gravitational field is actually not a constant but rather a variable depending on the observer's frame of reference.

(http://www.speed light.info/speed_of_light_variable.htm)

We can now characterize an invisible cosmic magnetic energy on a higher first entropy level not yet understood by the establishment:

$$\infty E = m \, (+\infty C / -\infty C)^2$$

But the "∞" in the math symbol perplexing my visiting math friend needs a little more time to logical be linked to an intelligent MIND to recognize how the energy system works to maintain our embedded LIFE, which is energized down to the smallest nuclear atom needs to be fueled with energy to exist.

We think of elements being static having no life but NASA teaches otherwise utilizing an atom clock, which ticks in femtosecond that later was applied to the Internet synchronization. To make it possible globally to talk to each other controlled by unified computer intelligence fueled really with MAGNETISM. Check every electronic circuit on silicon chips or boards will not work without embedded magnetism controlled by intelligence. Lastly, to discover that gravity is influenced by magnetism. A matter of fact it becomes the energy path connecting to the infinite source down to the atom level now extracted by UREE inventions.

That gave me some ideas tracing the energy to the Alpha(+ONE)force, which can be converted to electricity that is free like perpetual profit. It is also the fuel keeping us alive embedded with intelligence everybody even a child knows the function of LIFE and what is lifeless. Life is another dimension only found on earth which was imported from another world we know nothing about just called open ended - heaven. My first Babushka egg concept book deals with God's Plan for Mankind. I illustrated it with a Hebrew cuckoo clock diagram having two dials to indicate the time dimension of the KINGDOM OF GOD and the KINGDOM OF DARKNESS.

The Creator gives us a final choice before the GOD'S KINGDOM ON EARTH is born but comes with a placenta, the Apocalypse discarded as useless. A second Jonah has come to the town square announcing our last chance receiving ETERNAL LIFE with a very high probability that you and I will be dead in about fewer (5) five years. The population of this civilization will be reduced to less of 800 million (Check Pearl #126), mostly children will survive who have been cleansed from evil. The big war in heaven will have ended 2012/2013 with all evil angels and demons executed and later on earth too with Satan the only survivor bound in the underworld prison.

Evil on earth once more ends in the same manner as in Noah's time (5 February 2287 BC), on the same day (17 September) with a corrected

Julian calendar. God's once more has chosen an awesome asteroid instrument already in orbit check out the information for your own sake. If you want to know the purpose of the Plan for Mankind never preached in church, just check the Internet, which at this time is still free and not controlled by the atheistic priesthood manipulating every university suppressing truth. Listen what a scientist is pointing to: **Why Quantum Physics Ends the Free Will Debate:**

http://www.youtube.com/watch?v=lFLR5vNKiSw

I invite you to have some fun investigating facts in physics not allowed and if you want to survive the Apocalypse, reading what is forbidden by the establishment will widen a knowledge horizon that is the intention of free Babushka egg concept books. When man sows evil, he will harvest...? My last advice.

Thoughts to Ponder

If you did not understand the **Mystery Energy Trail**, like Sir Isaac Newton, who said, "Hypotheses non fingo", which is roughly translated as, "I don't have a clue." Or you might wonder that a lot of nonsense is being postulated outside the university forum and conclude that I must be from another planet. I would agree with you. But since we are all mortal forced to be in the same boat, but being different like to investigate the invisible realities behind the metaphysical cosmos curtain side. It can only be understood by revelation from the Bible, which is the oldest school textbook explaining complex science. An ancient civilization disappeared that was more advanced than ours. Why has science taken so long to get back to where it was before 2288 BC when genetic modification, electricity and iron were known?

The answer is that I speak German and must write in English to overcome the effect of BABEL. Check it out a good, logical ancient story much better than the replacement monkey story to explain science from the La-La land, painted inside every science building to teach little children fairy tales. But eventually fables, hopefully could lead to TRUTH now answered the question 2000 years ago remember Pontius Pilate facing Jesus. It can now be linked to our dreams that want to be liberated from constraints of our Daleth-dimension mortality. If you are tired of being perplexed of the many distortions of physics caused by a closed, locked-up, sin-tainted MIND check the free Babushka concept books on the Internet.

But better yet read the Bible will give us a better choice all along declared we must follow the metaphysical foundation trail which started what was reported by Moses, a highly educated prince and the greatest leader in history.

He collected the TORAH from Noah's boat and described a detailed history from before the Atlantis civilization that survived the flood. It is the only trustworthy information available divinely protected that tells the beginning of mankind from a scientific perspective and is much more logical when compared to the monkey evolution speculation never proven with physics in conflict with modern DNA intelligence.

It will give you the answers if you want to be well informed about the other, metaphysical science rail for a better balance. It works for me.

However, all is fueled and radiates outwardly from off a COSMOS DONUT hub wheel even illustrated in the our blood demonstrated a perfect math geometry a oxygen energy red cell carrier designed to perpetually fueling LIFE emanating from a DOT, ∞ELOHIM the infinite energy center expressed in math:

A Cosmic Law Polarity

In a Teeter – Totter

$$\infty\text{E-lohim} = m \left(+\infty C / -\infty C\right)^2$$

Explaining the universe to my kids is like riding at night the chain carousel, which has at its center a driving force. The operator announced the last round, as the outside is now dark. People have gone home, but

looking inside, there is light. I compare it to gazing at the universe and only feeling the outward force as all seems stationary looking inside. However in reality, the universe is speeding like the chain carousel pushing the chairs outside. It is fun to feel a gravity force but cannot see it, like in frozen galaxy pictures.

When infinite ∞ LIGHT declared in the first BIBLE verse in Genesis is bathed in the TIME DIMENSION. It created a magnetic energy path for a cosmos to appear, which will be proven investigating the mechanical side of the UREE. That will give us a little more supplementary knowledge to broaden science horizons and will disclose and confirm the existence of the ∞ energy coming from space designed around sixteen (16) UREE applications to establish credibility presented next in Part 2.

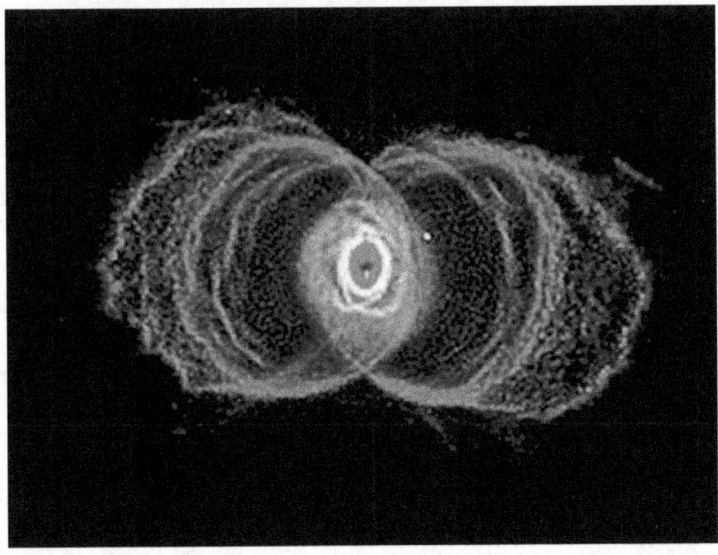

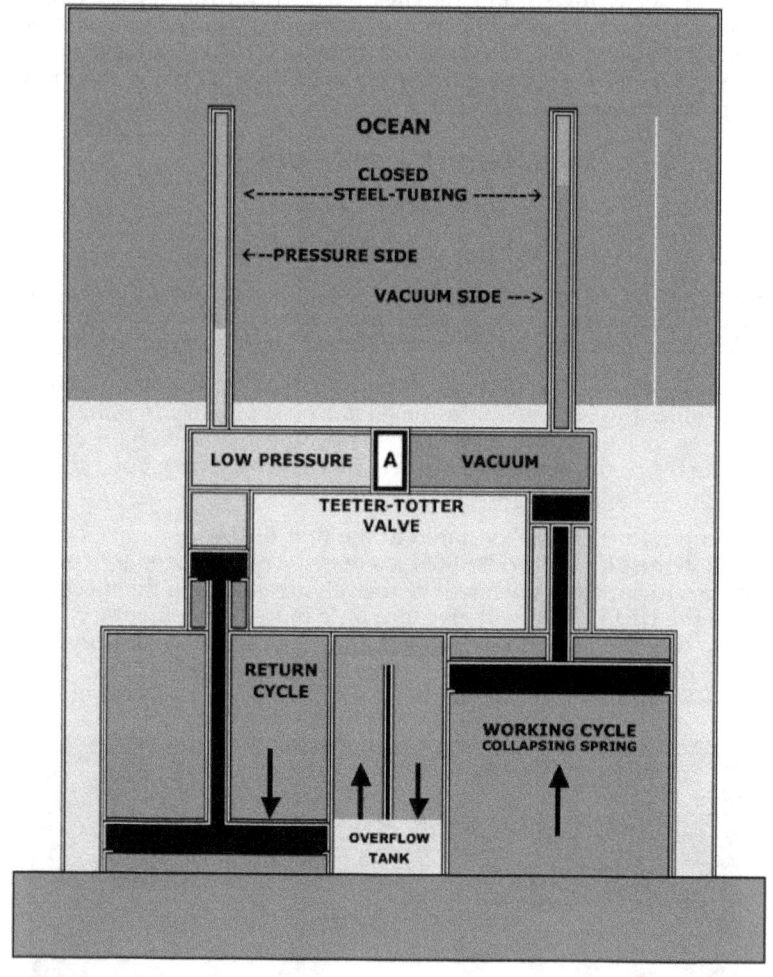

Part 2: 17 UREE
Motorized Design Concepts

The Concept of the Ultimate Renewable Energy started to grow like a seed and will take time to fully ripen. Previously in my first edition I started out with a simple notion to explain that free energy can be perpetual extracted from magnetized Gravity-Air-Water.

Magnetism is not very well understood in university environment, as much in physics has been forgotten. This essay attempt to reestablish and lay a foundation how in principle a UREE could work expanding a concept ending in fueling a Jumbo-jet UREE #16.

This Third Edition has stretched my vision adding 6 more design potentials and looking back reminds me of the first Wright Brothers contraption we all laugh about. Have fun reading about more possibilities but the last UREE Model I would recommend you should design and start a business in the garage like HP selling your first model to your friends and relatives and become a billionaire. Just follow the road map forbidden by the energy cartel establishment.

Underwater Motor Design - UREE #1

The underwater motor piston on the next page moves high air pressure from one side to the other in a closed-loop system; however, the work performed is the result when the high pressure is connected to the attached mechanical lower pressure piston side. Therefore, it will collapse into the next loop system, according to entropy laws, which determine what is on a higher level. It will only go one way to a lower level, if you still remember physics.

Previously we have learned about the difference of mixing air with the more expensive noble argon gas or a little cheaper nitrogen. Below the ocean is much higher pressure than the Hoover Dam; therefore, air with (21%) oxygen is just fine. But above the ocean high pressure is difficult to handle and needs to be modified a little with an enriched oxygen (60%) gas.

Analyzing when looking inside a gasoline motor, we see that on one side of the piston is the explosion of high pressure moving toward the low pressure side to do some work and needs oxygen also to run my car. This is explained similarly on the first page, which states that anything submerged deep below the ocean will be collapsed and flattened out by the pressure, which follows the second entropy, being related to the first entropy gravity force above.

However, submarines cannot go beyond a certain depth, as pressure increases beyond a certain limit. Therefore, they cannot go deeper, as they are restricted by the design parameters. In my underwater motor design inside a submarine, the main pressure piston is thusly constrained in allowing only so much pressure in the air tank, while the whole submarine with the motor assembly descends to a certain level, anchored below the ocean.

In the next few pages, we will see a diagram illustrating piston arrangements. Have a quick look. To bring back the working high-pressure piston for another cycle, it must push against the same pressure. This means that what was squeezed out must be pressed back with the same force. However, in order to attain perpetual motion, it needs an additional force to start the cycle over. That force comes from a **flywheel**, which is not seen in the piston diagram, but once it is in motion, it has embedded inertia, which is a stored energy.

We learned a principle from Galileo: that the earth goes around the sun at a certain speed. Newton defined what has mass (so many atoms stuck together) and that it stores kinetic energy. That was discovered when an apple fell on his head, which was thereafter analyzed and we are convinced that it was gravity. Yet today's confused scientists believe in an atheistic evolution religion.

They do not know that this force is the result of a second entropy level coming from an unknown cosmic energy source and is squeezed out from the first entropy, **infinite light** creating galaxies, as one of several photographed galaxies reveals how they were formed, which is now my story. See the picture on the front of an hourglass, a colorful galaxy taken from the Hubble Space Telescope. It is obvious, if you have a little background in mathematical logic. I see two independent loops, known for thousands of years, in the infinite ∞ hieroglyph symbol, which is also embedded in ancient clocks, such as the Tzolkin clock. All along, this has been indicative of two energy loops creating a cosmos, which is better explained if you follow the trail of **Babushka egg concept books**.

However, I still need someone with expertise in mathematics and a basic understanding of the energy conversion summed up in this paper, leading into some patent applications. The manufacturing of automobiles, building hybrids of gasoline or steam engines, coupled with electric motor generators, would be a good candidate, as they are already applying **gravity** and **inertia** to extend the mileage per gallon fuel.

The curiosity that caused me to investigate an unknown infinite fuel also discovered the forming of galaxies, which led to a completely different atom theory where the outcome is perpetual motion, which I noticed was inside atoms. I applied it to our high technology, but ultimately the energy equation will be proven when power is moved from one side to another and the work is performed with a net profit. Only the net leftover energy profit is the driver of a civilization.

Theories must be logical, especially if they conflict with the establishment's opinions. They must be backed up by the witnesses of some proven scientific facts. Much in science has not yet been defined with good theory, such as what electricity really is as a latecomer in technology.

The ultimate energy source mankind will need for a future, I postulate, could be the same that forms galaxies. Let's get a detailed understanding and analyze how perpetual motion creates electricity summed up with 17 UREE witness examples, as many more are possible and hopefully will clear up some misunderstanding, as logic must prevail. I hope that my German causes no problems when translated into English learned on the street.

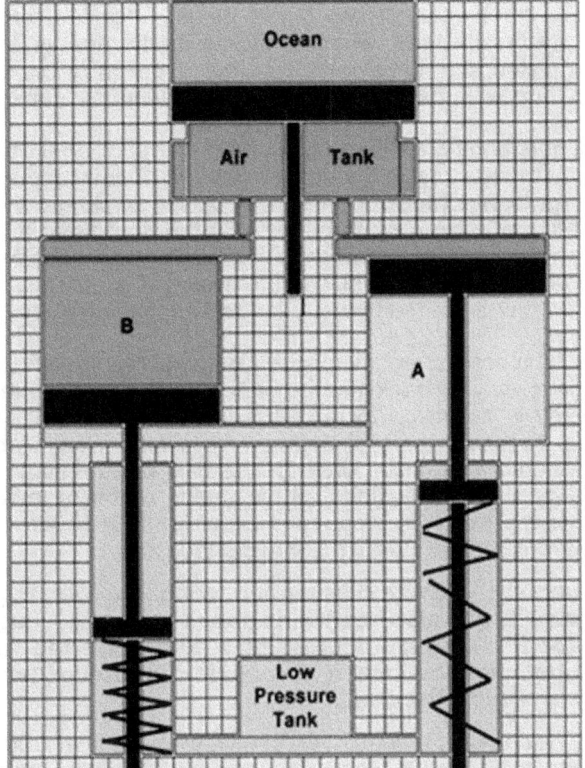

How a Gravity Motor Works on the Ocean Bottom

1. The **OCEAN PISTON** is bottomed out, which allows only so much pressure in the **AIR TANK**. It works like a pressurized steam engine maxed out to transfer energy and crank the pistons.

2. Unlock **SPRING B** to collapse **PISTON B**, which now drives the force, transferring it to the flywheel crankshaft. At the same time, **PISTON A** returns home aided by the additional force from **SPRING A** equalized in the **AIR TANK**.

3. Unlock **SPRING A** to repeat another working cycle.

4. The spring-pressure tank works like the pendulum principle of my cuckoo clock. A heavy clock weight activated by gravity needs only a relatively small portion of gravity force, spliced out at the right time, to make the pendulum go the other way. The preset, pressurized air tank is like a heavy clock weight: an activated gravity force (caused by a collapsing smaller pressure) is always available to set the big piston into motion.

5. It seems when the high pressure pistons are halfway and equalized all would stop; therefore, another principle of stored energy is needed. A flywheel stores energy temporarily. Therefore, just

like our automobile has a flywheel, it must be cranked. In this case, the higher pistons' force needs additional energy from the flywheel speed, which is stored inertia energy, to help the one piston side to return. Since the direction of the flywheel is only one way, the inertia energy will flow in the same direction.

Consequently the return piston gets that additional power needed for the next cycle to begin once more. A smaller portion of the energy converted to inertia must be fed back into the system, just like a cuckoo clock needs it at the right time; a little chain force is needed for another cycle. The flywheel speed sensor via the generator will reveal when to assist, so as to reduce the current-generated feed to the two streetcar motors to keep it going.

The high pressure of the piston would not require a huge angular force to move it back into the high pressure tank, as the needed energy is borrowed for a very short time from the flywheel, in order to repeat the cycle, just as the cuckoo clock also needs a recycled, small chain force for the next day's timing sequence. The cyclic compression and expanding of the air media atoms inside the piston is the very heart of creating kinetic energy for electrons to flow over a magnetic bridge.

6.   **Ultimately, gravity differential technology that generates electricity can also be applied above sea level. A huge weight on top of a piston compressing a fluid works the same. Just like my cuckoo clock, it now could run forever, never needing to recycle the chain.**

7.   The next step is an engineering review as the final judge for what seems to be a great idea on paper, benefiting mankind in the 21st century civilization.

Next, let's focus on the attachment needed for a system that extracts free electricity from GRAVITY.

New Hi-tech Flywheel Construction

For perpetual motion to do some work endlessly, we must feed some electrical energy back, which is gained in the flywheel, to keep the system moving steadily. The high-pressure piston can only return when facing a high pressure, if it is linked to a flywheel and a generator, with a clutch between the crank, in order to transfer its stored power, which is converted to the flywheel inertia energy.

The flywheel serves two purposes, again expressed in the ∞ symbol, the two loops energy system. One loop is moving energy to the other side, but some is fed back in an unending cycle, like perpetual motion. However, notice the crossover is zero time in the math equation; therefore, perpetual motion is possibly doing work.

For perpetual motion to keep the flywheel moving we need two looped forces. Kinetic energy is embedded in gas atoms, one side mechanical pressure and the other ∞ nuclear. Both energy cycles combined is useful to do some work in a GRAVITY MOTOR with the piston going down and returned on the other closed-loop back home position. The high mechanical tank pressure is the second entropy however is overcome by an expanding oxygen atom energy, which is the first entropy. Using both is generating electrical profit.

In order to gain electricity, some energy must be reinvested back into the flywheel system; therefore, we must build a flywheel with an embedded hybrid generator creating a current to drive a streetcar motor independently rotating to the crankshaft to create electricity at a higher speed.

On either side of the crankshaft of the same hybrid flywheels are linked two embedded generator and motor combinations, with four pistons mounted in-between. They only function together as a unit, as the two flywheels on either side must be controlled with an electronic clutch in order to disengage the streetcar motors if they reach a speed over the designed inertia.

The guts of the inner flywheel are designed for creating electricity that continually switches the flywheel's generator that drives with the same shaft of an externally connected motor on either side of the crankshaft and vice-versa, which will give us some profit, leftover electricity, which is useful for mankind. Therefore, electricity as a force can be instantly switched between the generating electricity flywheel system, as well as using the current to drive a hybrid streetcar motor, which accelerates the generator's rotor and thereby gain more electricity.

However, just like the pendulum on my cuckoo clock needs a kick for the right timing, the kick of the electrical storage system becomes a capacitor charged by the generator flywheel current. A capacitor stores temporal power and gives it back very quickly, compared to a battery that releases energy very slowly, only being useful to get the whole system cranked up, such as in your car.

I remember streetcars 50 years ago, as a kid. The driver had a linkage ratchet controlling the speed of the streetcar. First, the ratchet arm was full throttle as the car gained momentum, but as soon as the car gained enough speed, the driver had to begin reducing power and continue this until the car slowed down.

The sound of the cycled ratchets is still embedded in my memory, but I learned something that I still remember. The streetcar motor is a paradox similar to the teeter-totter principle. By reducing energy, will the streetcar run faster?

It was explained to me that as the internal motor was turning, it was creating electricity. If the EMF of the stator is returned to the rotor armature, it produced more electricity and gained more speed. The streetcar design only required high torque at the beginning, when it needed extra energy, which is not needed at high speed.

Therefore, with the way the motor is designed, its function is to always produce more energy to feed back as electricity, which causes a higher speed that generates even more electricity. What the network would give by the increased speed could only be controlled by starving the EMF and reducing the network power later on.

In other words, if you turned on the motor, it would run faster and faster and eventually explode, unless it was starved by the driver in order to slow it down. To me, that is perpetual motion in reverse.

Why not use that phenomena and apply it to the flywheel, which is needed as a system connected via a clutch to the crankshaft by... let's say, four pistons.

However, if a hybrid electrical generator fabricated on the inside of a flywheel gains speed via the crankshaft, it will produce enough voltage for a streetcar motor to run at high speed, yet because of its internal armature arrangement, it will eventually explode when speed increases. It now needs an electronic computer chip driver that must redirect the power. We just learned about this.

Yet before that point is reached with a certain designed inertia achieved, the clutch must be activated, just like the streetcar driver who applied the resistance force to prevent it from exploding. The electrical system runs on it own axis at a much higher speed and is mechanically separated from the four pistons and the crankshaft motor, working like a regular gasoline or steam engine, but running at a lower speed.

The gravity motor only provides power via a flywheel to transfer the kinetic inertia energy that is gained from the pistons to the return pistons, in order to push the high pressure back into the air tank where it came from, working like a compressor, or like a bicycle air pump blowing air into the tire.

The four pistons are then aligned and connected to a crankshaft that is spaced 90° all around, in a full circle. Translated: the first cylinder gets the main power from a high pressure air tank and causes the piston to collapse in order to do some work.

However, it is linked with the inertia-stored flywheel by means of a crankshaft, which gives energy back to assist the same piston by compressing the air back into the air tank after the piston collapsed and did some work. The other three pistons are positioned at 90° from each other, in order to equalize the returning high-pressure piston, which is synchronized at a full turn of 360°.

If you divide the energy pie to maintain the flywheel inertia generating electricity and return some energy back to the high-pressure piston (I am not a math guy), it may eat up 60%. The capacitor may eat 10%, which leaves 30% leftover electrical energy profit coming from the electrical generator at the other end of the crankshaft, which now has perpetual motion running endlessly.

To reiterate, the first generator linked to the crankshaft creates electricity, some is spliced out, in order to feed the streetcar motor that is driving the second generator, which again is fed back to the first streetcar motor. As the current is ∞ recycled, what is left over is **electrical profit**.

Yet, it must be controlled, balancing out the mechanical system design for long life against moderate profit. It is also in a teeter-totter relationship. If there is too much profit, there is less motor life and vice versa.

Feeding electricity back into a four piston motor arrangement will generate electricity for a continuous, full 360° cycle, which needs another device for synchronization, which is performed by a capacitor. The combination makes perpetually generated electricity possible, only if it is controlled ultimately by a sound infinite ∞ MIND, as the motor speed should not exceed safety margins.

The experts will do a better job at squeezing energy, by designing a specialized hybrid flywheel generator and motor that is linked to a modified gasoline or steam engine, using gravity as a fuel, to which scientists say, "Not possible!" Let's expand a GRAVITY conversion and follow the trail of perpetual motion. I am convinced that it is logically possible.

Automobile UREE #2 -
Driving Perpetually without Gasoline

Not as an expert in automobile technology, but rather as a philosopher, the thought of an expanded underwater Gravity Motor application crept into my conscience; that being that the perpetual motion idea discussed in the below sea level example, which explained how to create electricity, could be expanded and applied to automobiles.

I am lacking a math education and do not have the information available for the latest electro-motor designs and dealing with higher pressure in order to check the feasibility and efficiency of using gravity as a fuel. The below ocean pressure is achievable but driving around requires high-pressure air tanks.

Most automobile manufacturers already sell hybrid gasoline-electric automobiles commercially that utilize gravity and inertia in part, in order to extend both the range and efficiency and achieve over 50 mpg. Therefore, my idea is just expanded further.

It would be the ultimate invention and just the next step towards replacing gasoline in order to run an engine, designed in a hybrid system that matches the underwater generator and motor configuration, with the goal to drive around continuously, without costing a cent in fuel. We are at the same juncture the Wright brothers were when they were convinced that flying was possible, since everybody at that time believed it to be impossible.

The data for this crazy idea can be obtained by high school kids collecting data to test perpetual motion. I recommend it in my next roller coaster example that will prove creating electricity to broaden our perspective, since the fact that perpetual motion was not possible was drummed into us since grammar school.

In order to get my point across, why not ask anyone on the street because it is the same as in the Wright brothers' time, where people said it was not possible to fly. It was even preached in church.

I first started from the steam engine because as a kid, I watched a steam engine locomotive as it arrived at the station. I was impressed and as I looked closer, I noticed and learned that the steam vents reverse the high pressure on the other side of the same piston design, causing it to return for another cycle.

The problem killing the perpetual motion idea is that the high-pressure piston must return the high pressure once it collapses on the other side, which is linked via a shaft to the low-pressure side. Checking the US patent department, it was revealed that not a single, useful invention exists that is linked to perpetual motion.

To understand whether the possibility of perpetual motion could work, I pondered (from a philosophical perspective) and developed through the study of the principles of a new atom clock (which NASA now uses), a new theory that matches the laws of physics.

It is seen and replicated in an hourglass galaxy that created a cosmos, which is demonstrated in the Hubble Space telescope picture and illustrated with a simple infinite math symbol ∞.

I basically think of gravity as a pressure differential energy that can be imitated theoretically in my new understanding of physics. It could be converted adequately enough in a highly pressurized air cylinder and linked to a closed-loop system that could run an automobile without gasoline indefinitely, which is not possible, according to the experts of the energy cartel.

I am not good at math and know the limits of how high we can make an air pressure tank to simulate a pressure below the ocean, as well as consider safety, when imitating gravity. It is a far-fetched idea, but it still needs a closer look, in order to discover how the underwater generator works with perpetual motion and thereby define it better. You be the judge.

Presently I understand that a gasoline car engine produces high pressure by an exploding energy on one side of the cylinder loop; while on the other side is low pressure. That is the same principle. The high pressure was produced by igniting fossil fuel, which was previously converted from gravity, creating biomass fossil fuel. It was formed formerly by combination of DNA atoms, but still needs the extra oxygen to run the motor just like a GRAVITY MOTOR.

A fast energy conversion of a biomass exploding into kinetic energy is released in a car cylinder, becoming a highly pressurized gas that acts against a low pressure environment, just as the ∞ symbol shows. However, the symbol also reveals perpetual motion, where time is zero at the crossover point.

To attain a perpetual motion mode, we must replace high pressure that is now stored from an air tank, borrowing a little from the next piston for the return cycle. To illustrate that principle with an analogy, remember that in a closed pressure system, the air tank remains the same. Similarly, the weight in my cuckoo clock does not change either, but the pendulum gets kicked from two sides.

It is the same system as my cuckoo clock that could run indefinitely, if we added a little automatic chain force to the system, in order to synchronize it with our calendar. That must be in addition to the energy waiting on the other side of the clock pendulum, where a force is ready to give the pendulum another kick.

The easy part to understand is that an explosion inside a gasoline cylinder acts against the lower pressure and pushes the piston down. The next piston comes around at 90° to take over, repeating the cycle. I remembered my old VW, where all four pistons were linked to a crankshaft flywheel to smooth out the velocity speed.

However, expanding on the flywheel inertia idea could provide the extra-added energy force needed, in order to regulate beyond the velocity inertia, and gain electricity in the process by putting an electric generator inside the flywheel. That will drive a future car further with the same tank of energy, one way or another, if you expand on that idea.

Yet the system needs an additional energy to become perpetual. That is accomplished when we convert one flywheel to an electric generator and motor, but perhaps it needs a little bigger mass.

Iron and copper wires, which are quite heavy, would help, but to function properly must turn around at a high speed. It needs a bigger armature, in either the rotor or stator assembly that will produce the electricity desirable for perpetual motion.

A lot of force is available in the high air pressure tank, which is linked to the power piston. It only needs to be sliced out in little chunks, rather like moving the pendulum of my cuckoo clock, coupled to a heavy weight, which does not change.

Like in a conventional gasoline motor, the return exhaust valve can be utilized when opened, pushing out the compressed gas from the bottom of the cylinder. However, now it is utilized and being pushed back into the air tank for another cycle, with the help of the extra flywheel inertia, driven by a little extra electricity that is generated from the other generator flywheel.

The same piston collapses and transfers work to the crankshaft, reversing the stroke for it to become an air pump that compresses gas. The flywheel inertia will accomplish that. However, from the energy equation, the principle of extracting electricity from the flywheel is then gained energy, one way or another. I am sure that will not be disputed.

I expect to see it in the next hybrid automobile, which will give a minimum of 300 mpg, without changing much and will be sold in a few years in my neighborhood, after you read my paper. The new Volkswagen advertised is already double mpg.

However, for perpetual motion, one more force is needed. Go back to the section that explains how an electrical motor works to produce reverse electricity, which creates the magnetism needed, in order to acquire electricity with cycles that repeat, like a ∞ math sign, which is symbolized. This is seen in galaxies or ancient clocks, as I have explained once more.

Creating an additional energy force is still needed, which is linked to the air tank; however, it requires one more force to be spliced out for perpetual motion from the same energy pool. After the piston bottoms out, the energy is then transferred to the flywheel, thus creating electricity. The Gravity Motor switches from a generating mode to a driving electrical motor mode, in order to produce the extra energy boost that is required.

However, perpetual motion needs a little added force. First, we must analyze what goes on inside the pressure piston because on either side, the high pressure is the same. It never changes. The same force collapsed the piston, then turned around and pushed the high pressure back to where it came from. On paper, it is tit for tat and nothing is gained.

Checking history for examples and in an attempt to explain physics, I selected two men. I propose that one is Newton, who discovered gravity when an apple fell on his head. Thereafter, he experimented, endeavoring to find the laws of nature; gravity, in particular. The other is a teenager who became famous and was born 1000 years before Christ. David became well known as the king who killed a giant with a slingshot. It is recorded in the Bible.[1] Both made history and because of it, we can understand science better.

Newton's discovery is verified by my grandkids. Whether you take a little stone like David's, or an apple or even a bowling ball, when you cause them to fall from the same height, they arrive at the bottom at the same time. However, the difference being that the apple gave Newton a bump on his head; a bowling ball falling on his head would have resulted in physics being delayed. There would have been no airplane or space travel either, I am sure. We still would be mired in mixed-up evolution fairy tale opinions and still trying to figure out the laws of physics.

A fancy phrase coined by scientists is kinetic energy, which is basically moving gravity, but in low-tech terms, for me, it is inertia. Yet, even a little stone propelled from a slingshot by David had an effect like a bowling ball, more or less, depending on the application. We can therefore deduct that if a static gravity force is accelerated with the same mass, it will gain energy, which can be extracted and be useful for mankind.

Even with sufficient high pressure as backup, which is returned with a push into the same gas-air tank, some extra energy would be left over when gravity gains a little inertia by being pushed faster in the downward stroke of the piston. It is embedded within the piston stroke that collapses to a much lower entropy.

Moving to the bottom of the cylinder, it accumulates extra inertia energy and is transferred via a crankshaft that is linked to a flywheel. The extra inertia energy is then converted in the generator, giving us electricity that switches back into the motor to keep the flywheel going and returns back to the generator.

It is like the ∞energy loop; however, some energy becomes leftover profit perpetually, until the next piston takes over. Just as a little stone gains the extra energy difference and becomes like a bowling ball energy-wise, inside the piston. The energy profile reveals converted kinetic energy, which is transferred until it bottoms out, which will add a little more of a differential force that is extracted from the main pressure, which is needed for perpetual motion.

Once more, the piston collapses and is backed up by the high pressure behind it, which now accelerates; the speed builds momentum at the bottom stroke, just as the cuckoo clock pendulum gains speed and hence, gains kinetic energy that increases.

[1] The story of David in 1 Samuel 17.

Check the velocity profile of the pendulum stroke of a cuckoo clock, which is the same or similar inside a power piston. It is fed by a weight like an gas-air tank, where both store a little extra energy, just as David's stone's inertia profile increased to the kinetic energy of a bowling ball. While the main pressure piston is collapsed, it passes energy to the crankshaft, but is forced midway to decelerate. Therefore, it must pass on that extra kinetic energy which was gained at its peak (coupled with David's stone growing to bowling ball inertia) to the flywheel for temporary storage.

However, understand that perpetual motion needs one more energy push because on the bottom of the cylinder, the energy velocity is zero after it transfers energy to the flywheel. The extra energy kick that is needed to return the piston, which overcomes high pressure, can be explained when we analyze David's slingshot.

To learn about inertia, let's analyze David's slingshot, which is proportional to his size. Circulating the stone elliptically around his body, he would pivot with his body, in order to lengthen or shorten the distance for the stone circling around that is gaining speed by accelerating cyclic motion. That same principle is duplicated in a cranked speeding GRAVITY MOTOR piston accelerating gaining velocity.

The unchanged gravity exerts inertia and is transferred to the GRAVITY MOTOR flywheel. Yet again, one more force is needed for perpetual motion, when the piston bottoms out. We attain the extra force needed by giving David a slingshot made from a rubber band the same size.

A rubber band will stretch with force, having a built-in momentum energy force expressed in expansion and contraction which, when added to the centrifugal force stored in the flywheel, would give extra inertia at a certain point in time.

The high pressure profile in the air tank is similar to that in David's stone, in that it has extra kinetic energy embedded that has converted from the motor flywheel, having crossed over to generate electricity in the flywheel, forcing the high pressure piston back, but with leftover energy.

It turns energy that is shaken out into profit, as long as the high pressure is not changed, which is like the weight of the cuckoo clock running as a system. Perpetual motion needs additional, temporal energy storage for the proper timing, which comes from David's stretched rubber band as an electricity force, similar a chain adds energy to my cuckoo clock to run perpetual, too.

An extra energy force is required for perpetual motion and can be stored either in batteries, or come directly from the flywheel that generates electricity that is converted, in order to charge a number of capacitors for a high energy discharge, depending again on timing. This is similar in your car, where the firing of your spark plug via a transformer coil must be timed to maximize the inertia in an exploding gas pressure force. In a perpetual motion motor, it works the same and needs two spark plugs mounted on top of the piston outfitted with larger electropoles each for a higher intensity light-flash discharge also linked to a capacitor via a coil.

The pressurized spent argon-oxygen-gas that reverses direction is similar to a cuckoo clock pendulum requiring a kick, but must be heated up very quickly, therefore expanding the pressure inside the cylinder. That force is adding to the inertia previously stored from the flywheel, thus creating more electricity for the extra energy needed tentatively comes from a higher clock chain force.

David's hi-tech rubber slingshot-stretching inertia must be converted in a short time motion, which embeds extra energy from a discharged capacitor. In analyzing the pressure profile of a high pressure piston collapsing, we will look at David's rubber slingshot principle, in order to determine the right point when the capacitor needs to be fired via a high tension coil, for a little extra air energy to be heated up, thus expanding air-atoms, which has added kinetic energy embedded. Now we have a steam engine environment inside the cylinder expanding pressure transferred to the flywheel.

The energy profile reveals at the midsection of the piston travel, it will slow down being controlled by the crankshaft heading toward zero velocity. At that middle point the air-oxygen tank pressure is maxed out, therefore should close the intake valve. Now two spark plugs mounted on top of the piston get fired, which is getting the oxygen-air gas hot gaining higher pressure that rises from the expanding air-gas atoms.

So when the piston arrived at the bottom and starts to reverse, we probably have 100 times higher pressure pushing back with much higher gas pressure now opened once more the intake valve. The returning cycle is really the power cycle the force coming from within atoms, which has no problem overcoming the higher pressure of the air-tank needed for perpetual motion. The extra momentary created kinetic atom force is 100 times higher immediately transferred and stored in the flywheel which gains crankshaft speed converted generate more electricity, which is driving two streetcar motors via a two generators to create 100 times profit.

> **The combination of mixing kinetic energy converted to recycled electricity is the very heart of the gravity motor running perpetually, which is now possible.**

Once more, when the piston stroke collapsed on the bottom of the cylinder space, the high-energy velocity is at zero, unmoving, like the end of a cuckoo clock pendulum cycle. It now needs an additional energy kick to force the return of the piston or pendulum. The extra energy needed inside the cylinder is created when the spent high-pressure gas was heated up by a couple of fired spark plugs, converting the extra energy needed for perpetual motion.

According to physics, warm air is expanded by a conversion of moving atoms, inertia and energy, which momentarily creates a higher pressure at that point in time inside the cylinder. Next, the exhaust valve is mechanically opened, coinciding with the flywheel stored inertia that comes from the crankshaft and will force the higher pressure back into the rear gas-air tank side at a little warmer temperature, but it returns much easier for another cycle.

The returning piston stroke has now momentarily become the higher pressure side, overcoming and pushing against the previous air tank pressure, which has not changed, just like the weight of the cuckoo clock, only now it is weaker in relation to the higher, hot air cylinder's volume pressure differential pushed back by the added flywheel stored inertia.

It is analogy like the chain winding up my cuckoo clock, which is a 100 times stronger force that overcomes the gravity weight. It follows the laws of physics again, in that the higher-pressure level always moves to what is lower, according to entropy. It is cyclic, according to the energy laws, as a pressure differential goes in the same direction. Energy fizzles out from the flywheel, but regenerates from the next piston that generates electrical energy in perpetual motion as it moves in the same direction, coming and going, as well as expanding and contracting.

In the energy equation, the warmer air is rejoined on the back door of the pressure air tank, where it has time to cool down again for yet another heating cycle, which follows entropy, too. The flywheel inertia is linked to the pressure piston, but for perpetual motion, it needs an extra force at just the right moment, similar to the cuckoo clock pendulum that pushes to begin the cycle over again.

However, it needs a stronger, added chain force to be recyclable, **linked to the infinite energy source embedded in air atoms coupled to the cosmos source.** Analyze how the pendulum clock profile accelerating and decelerating works, as it will explain what happens inside the GRAVITY MOTOR, when we apply the theory pertaining to David's rubber slingshot, which is still linked to the **infinite energy source.** The limiting unknown factor to me is how high is the pressure for an above ocean application when driving around with a car. If assumed electrical profit 10% we should consider 10 times higher pressure of a gasoline motor piston.

I learned about David's rubber slingshot principle by studying why the moon is being stretched by a period of 2 days. It is different every month as it orbits around the earth, when measured against the sun's orbit. This variation affects global warming and confuses many scientists. Many observations were linked together and put forward in my fifth Babushka egg concept book, **REFLECTIONS ON GLOBAL WARMING.**

One major discovery was that the earth orbit, as well as the moon orbit, is held together perhaps **by magnetism and not gravity** mentioned previously. We could not measure it because we are on the inside of the system. That is an upside-down, completely different viewpoint to broaden horizons for science, linked to a new Donut Atom theory.

Why does the UREE Work?

In the future if a standard fuel injected gasoline automobile can be technically modified to hold up to higher pressure, it could make possible to combine it with a GRAVITY MOTOR system design. Analyzing the internal cylinder function of expanding gasses creating pressure, it differs only by redirecting the exhaust gas (CO_2 - NO) now collected and returned to the pressure tank helped by the additional flywheels of the crank.

The largest part of most combustion gas is nitrogen (NO), water vapor (H_2O) and carbon dioxide (CO_2) according to Wikipedia. When enough pressure to explode piston cycles has accumulated, the fuel is turned off, but we keep on driving, no longer paying for gasoline. That idea is not new and keeps popping up; attempts in reusing exhaust gas are underway check the web.[2]

Traditional science has assumed that once the fuel is spent with the piston on the bottom what remains should be discarded into the environment causing horrendous pollution. But gas atoms can be influenced like a rubber band slingshot to expand further. Why stop stretching the rubber band only half way? Expanding gas atoms are the real energy force. We can convert it to kinetic energy to propel the car or swap it into electricity via a streetcar motor.

Therefore, our GRAVITY MOTOR could reuse the spent fuel which still has embedded a ∞ atom energy potential and could be extracted with a UREE design along collect the exhaust gas back into the air tank. But I believe the force must be raised to a higher level over a combustion engine's pressure. The wasted exhaust gas from the exploding gasoline is squeezed a little more by starting your car and let it run a little to collect unused exhaust gas to a higher pressure.

It is like a bicycle air pump that takes a few strokes to charge up an air tank to what is considered necessary for the UREE design. The additional gasoline energy is only used once to compress (CO_2 - NO) gases to achieve 10 times higher pressure collected in a small, one-gallon tank. If the intake valve of the working cycle is thereafter opened, logically we could have 10 times higher pressure available pushing the piston down much faster to do some work, but this time with spent gas energy!

To keep on driving in our GRAVITY MOTOR car, the spent energy is first detoured temporarily via a flywheel. Now the UREE mode will take over and continue running perpetually from the stored energy transferred into the flywheel. From there, it is further converted to usable electricity energy that makes possible to drive "perpetually" around utilizing a streetcar electric motor in your car.

Read my story again from the beginning. It will take some time to realize that mankind could fly if you lived in the last century. Many puzzled over the Wright Brothers contraption and were laughing about it when it fell apart at the first try. Let's not get discouraged and apply logic.

The poisonous carbon monoxide-dioxide-nitric-oxide gas still has plenty kinetic energy embedded useful for freeing up billions of electrons to flow if collected in the generator. Squeezed like a lemon, a little more will get us more energy from the pressurized **oxygen (CO_2 - NO) atoms**. That is where the atomic electrical energy fuel comes from.

Converted electricity gained from the generation system is an abundant energy source because it is extracted from **atoms** that are a bigger source than exploding fossil fuel - exceeding what we would need. What I postulate is so unusual contrary our education and need to go back to read it ones more explained on page 85, Flywheel Construction.

> For perpetual motion to keep the flywheel moving we need two looped forces. Kinetic energy is embedded in gas atoms, one side mechanical pressure and the other ∞ nuclear. Both energy cycles combined is useful

[2] http://www.gizmag.com/cpts-auto-exhaust-gas-energy-recovery-system/16469/

to do some work in a GRAVITY MOTOR with the piston going down and returned on the other closed-loop back home position. The high mechanical tank pressure is the second entropy however is overcome by an expanding oxygen atom energy, which is the first entropy. Using both is generating electrical profit.

The key to understanding perpetual energy renewal within the GRAVITY MOTOR is when the piston returns, a little extra kinetic force is extracted on top of the spent gasoline gas by utilizing an additional midsection cycle spark plug discharge. Once more it would re-expand cold gas atoms further on top of the high pressure fired from a capacitor storing 10,000 volts via a transformer.

Again, it is like an accelerating rubber band slingshot that stretches kinetic energy a little more. That little extra is captured and stored in the flywheel while the input valve is closed as the piston bottomed out. It returns and a couple spark plugs once more fired the second time in an enclosed environment will expand atoms creating atomic higher pressure needed for perpetual motion. The flywheel will collect and store every little bit of extra kinetic energy. Without it my motor conversions would not work.

Since the flywheel energy storage is only short-term needs to be immediately transferred to another energy electricity conversion, which happens to be the cleanest major fuel in this 21st century. It only works as a system based on energy being ∞ infinitely recycled demonstrated in the ancient Tzolkin clock. But electricity is only useful when electron slaves are put to work cannot be stored long time therefore freely available to move your automobile. A switch will stop the electrons from perpetually flowing being controlled by intelligence, when atoms are manipulated by our MIND.

For a standard UREE gaining electricity extracted from the cylinder gas pressure I postulated we should enrich it with oxygen atoms to collect electrons equal to the ratio of water used in the Hoover Dam, which is about 40%. Or I suggested going a little higher to 60%. The used up (CO_2 - NO) gas is 33% carbon-nitrogen and 66% oxygen should even work a little better. If we use only air, which is 21% oxygen, we need higher pressure easier produced below the ocean application. I am not educated in advanced math and do not know how high the pressure in the air-tank below the ocean should be.

The extra kinetic energy is immediately converted in the electrical system and that becomes electrical profit for us making it possible to drive the car perpetually as the sum total atoms amount have not changed in the air tank in an closed environment.

It is like the fixed weight of the cuckoo clock moving the pendulum. But even my cuckoo clock needs to be rewound with a chain energy force for perpetual motion and needs a little extra energy further extracted when atoms expanded a second time, which is 100 times bigger chain force. It is like ice expanding with horrendous pressure. Examine a nearby river in the winter. Pre-pressurized cold atoms once more heated will expand once more releasing more free electron slave energy, picked up by the flywheel storing everything by converting it to magnetism. Perhaps there is more? It is really simple physics for my grandkids.

Remember, in any court of law only the testifying witnesses will establish truth as physics is now on trial. It is not exempt from the laws of nature and must obey the metaphysic laws of a higher entropy level. Tremendous benefit will now be gained will no longer experiencing pollution and once more in the future can breathe clean air in every city. Halleluiah!

If technical a higher pressure is possible within a cylinder motor automobile design we could enjoy driving around in a car without gasoline or diesel fuel not including conversion cost. It would get better if the energy cartel controlling the governments will allow it, and if not taxed, this invention would expand the economy creating many jobs.

Technical Recommendations

1. The cylinder of the Gravity Motor should be impregnated with (micros-size) Molybdenum-Disulfide, which is a dry lubricant powder and will not require oil for the crankshaft casing. A special long age, non-oxidizing grease should be mixed with the same dry lubricant for the crankshaft because once embedded, it cannot be separated chemically (bond-wise). The gravity-motor-flywheel made from iron will run at 3000-5000 rpm relatively cool depending on application; no water-cooling would be required. The flywheel has the electric generator field winding armature embedded, which was set to run maximum at 5000-rpm crankshaft speed for longer life and turns the crankshaft clockwise.

2. However the separate generator shaft is linked to a streetcar rotor running counter-clockwise to the crankshaft and will therefore increase and double the velocity of the streetcar-generator system osculating electricity back and forth one each side of the two flywheels joined together on the same crank. When the streetcar motor is speeding up a generator recycled on the other side will eventually run at higher 10,000 rpm, creating elevated voltage now needs a transformer and control circuit board adjusted down to 110 volts of normal use.

3. The higher created generator voltage at 10,000 rpm needs a special transformer to match the streetcar motor, which is also needed for a standardized **profit 110 volt** as energy must be cycling back and forth, ∞ linked to the freely rotating generator shaft.

The rotating generator shaft unit is designed to stand alone, supported on two duplex bearings or self-generating air bearing for long-lasting service, running at double the speed [about 10,000 rpm]. Each side of the crankshaft has the same flywheel design, but mounted in reverse, facing the motor piston side of the crankshaft.

In a stationary application I would recommend a pressure tank filled with **argon (40%), oxygen (60%)**, allowing for ten times cylinder pressure considering the size of gasoline-cars designed for exploding its fuel. It still would require a small starter motor.

Another recommendation: I would insert an aluminum oxide ceramic sleeve inside the cylinder wall that was impregnated with micro-particles of moly-disulfide to provide higher wear resistance.

UREE #3 - WATER - Hoffman Energy

The recent Japanese nuclear disasters (2011) will be repeated in other countries, due to the fact that when iron in concrete will disintegrate and becomes weakened by radiation and oxidation. Even without an earthquake, we will see future accidents that will create confusion and finger pointing from those who blame some experts who are too old to remember as the new generation of nuclear engineers are ignorant of what was done 60 years ago.

For example, a huge underground gas line exploded in San Bruno in California in 2010, destroying a vast city block, which cost lives and a billion dollars in damages. Thereafter, it was further revealed that PGE, the energy provider, no longer knows where all the pipes are buried as many records no longer exist. Apply that to nuclear power stations, such as that demonstrated in Japan, which we should learn from, as time for obsoleted worn-out machinery is running out and should shut them down, all of them, if we want to live on this earth a little longer.

However, we do have a number of alternatives available to produce electricity, so there is absolutely no excuse, as you will see as I expand my ideas and present them in this paper. There is enough money floating around uselessly spent on gigantic installations that smash atoms, like CERN, which could be redirected towards something that works.

Solar has become quite cheap despite government efforts to keep it suppressed, as solar power is not allowed to be over 20%, efficiency-wise, or it will be confiscated in the USA. Check the security laws that are not know by the public. I discovered the Invention Secrecy Act 1957, which restricts publication. (5135) Inventions were confiscated as of September 30, 2010 and restricted by the US Secrecy Act 1957.

For example, solar is restricted to 20% efficiency. Anything over that is forbidden and if someone is convicted, they will get up to 20 years in prison. Why is the US government interested in restricting other energy use? That is a fact that should no longer be silenced and suppressed. Check the previous president and vice president, who were linked to big oil, and started the many wars that are still perpetually being fought feeding a military industrialized complex and might get you an answer.

Technically, I believe that the permissible output of solar panels of 19.5% can be doubled with special ultra violet filters covered over the solar panels with a directional Fresnel-lens, thus concentrating shorter blue light frequencies, which have more stored energy. When light is put through a prism, we see seven rainbow colors with our eyes and its frequency is measured from 4000 to 7000 Angstroms. It is obvious that one end is higher. Filter out the lower energy side and allow only the higher light energy to be exposed to the solar surface panels and you will get more electricity - guaranteed. Just a few weeks later my opinion on higher solar output possibilities was verified on the Internet.[3]

Water Energy is another principle that I learned in grammar school 70 years ago. In a perfect 4% salt solution, the German scientist Hoffman discovered that electricity will separate hydrogen and oxygen from

[3] Off the Grid News, Innovations in Solar Technology. http://www.offthegridnews.com/2011/04/18/innovations-in-solar-technology/

seawater could use now solar panels for free energy. Even with a permitted 20% efficiency quotient, solar panels would give us enough perpetually free electricity, by first separating the seawater into energy gas, with each stored separately, where they will then liquefy in the steel tanks, which is then combined within a Gravity Motor simultaneously.

Location is not that important as seawater can be piped directly into Gravity Motor installation sites that are situated in arid areas, surrounded by plenty of sunshine, for solar panel energy conversion. Electricity profit is gained by first separating and then rejoining the two gases because it taps into the infinite First Entropy energy source. That becomes the fuel for a redesigned, hybrid gasoline-diesel fuel injected engine linked to a UREE Gravity Motor electric generation combination, which I am convinced would work.

1. Let's start with the first cycle of a four cycle linked with a fuel-injection engine on the downward piston motion will fill up the cylinder with **pressurized** hydrogen. A little kinetic energy is put in the flywheel because of external pressured gas.

2. In the next cycle, with valves closed on the upward motion of the piston, the hydrogen gas is **more** pressurized, being helped by the added second flywheel force. Remember there are two flywheels on each side of the crankshaft.

3. The next downward motion cycle is the big energy cycle, creating more kinetic energy within both flywheels when pure oxygen is injected, similar to diesel fuel, and ignited with a spark plug discharge, which is parallel to a conventional automobile cylinder explosion. That is big time kinetic energy conversion stored within the flywheels.

4. The return piston travel will expel the water in the upward motion to repeat the cycle.

The Gravity Motor's crankshaft still connected to the generator and motor arrangement will guarantee electrical profit, even if you do not believe that perpetual motion is possible. Utilizing the ocean will give us unlimited energy even a blind man will see. The infinite energy loop exchange is demonstrated when hydrogen is combined with Oxygen to release explosive energy (Knallgas) that should be considered first entropy, whereas, when water was split apart to release hydrogen and oxygen gases, outside energy was absorbed on a second entropy level. Re-combining the hydrogen and oxygen creates a higher energy differential of surplus electrons, I describe as electrical profit.

Using conventional solar panels stationed along the beach applied to split the water molecules is only a standby. Excess power from the UREE Gravity Motor facility generated during low usage periods can automatically be switched to split water, too. It eliminates expensive supervision for the next cycle and keeps harvesting **unlimited perpetual electricity** much cheaper and safer for mankind.

Or, we could go the other way around and build the UREE facility in sunny desert areas near the ocean and pipe in seawater. Besides gaining electricity in the process, it could produce drinking water as a by-product, thereby reaping a double benefit. We could use the existing oil pipes for seawater and start growing vegetables in the desert instead of grossly polluting the environment with coal-fired power stations only benefiting the energy cartel.

Plenty of money is available. Read the July 2011 Popular Science. Just adding up only the first year operating costs of one ION Collider, Gigantic Fusion, Large Array, Laser, Space Stations, Neutron Source, Large Hadron Collider equals an operating cost of **25 billion** dollars, not counting the 100,000-year plus impact of nuclear waste – a cost that nobody knows! My question is, "How many solar panels could we buy to give us positively clean electricity in return?" The answer is $25 billion/yr. profit but connected with the UREE we get obscene profits.

Summary of the UREE Principles

I have shown how to generate clean, green, nonlethal energy based on metaphysical concepts from an unknown energy source now revealed from the biblical mystery energy trail of galaxies creating elements, ancient clocks in museums revealing some history of people using electricity before the Atlantis civilization was destroyed in 2288 BC. The Internet forum allows metaphysic energy opinions to be shared without being censored by the energy cartel. Electricity can be produced free with UREE gravity motors, but will it be allowed?

The problem exists with the nuclear energy cartels who lie a lot, paying big donations under the table, as well as with brainwashed senators who have not been educated in science and where told a lie that nuclear energy is cheaper sold as a scientific fact. A day is fast approaching when the energy moguls will be before a World Court accused of crimes against mankind, as plenty clean energy options existed. The taxpayer was never told the real energy cost per kilowatt, which is buried in the trillion dollar American national debt, where not even liberal senators dare to reveal it.

The poison of spent uranium rods is still stored in secret places that are carefully monitored. Nothing is reported on TV leaving the public completely unaware of its extreme lethal danger. The cost will never be disclosed. No country in the world wants the spent fuel rods even parked within them, as the hazard of contaminated drinking groundwater exists, as seeping hot radiation cannot be stored in nature. That is a scientific fact.

It is a well-kept secret and known among nuclear specialists that concrete and reinforced iron-rods according to physics will rust and in time will crumble and show cracks in the storage pools accelerated by excessively high heat and radiation intensity. That will leak guaranteed. We should have by now learned a lesson with Japan and Russia showing vast real-estate areas deserted and many farms made unproductive for food, void of all life "forever".

I hope the handwriting on the wall will tell the controlling policy makers that their safety rules have never being applied as Senators have been grossly dubbed and will not be long to make it obvious that corrupt money was the only motivator. Globally in the near future, millions of taxpayers will be dying from radiation, but some might wake up and start a revolt, which is feared by FEMA in America, as they are already installing cameras every quarter mile along the highways on either side, as well as at street crossings. London is king, if you are in the camera business.

The nuclear lie was the greatest deception mankind ever experienced, coming directly from Satan. Nuclear power generates massive radiation that will shortly poison everything on this planet, which is the only place in the whole universe of billions of galaxies with embedded life, which will create death continually for a minimum of 100,000 years, which is the biggest crime. In spite of so many alternate green free energy sources around avoided by the energy cartel only interested in obscene profit and billion dollar bonuses.

Compare the time factor to mankind on earth, which is only 6000 years, as proven by Professor Edward Hull M.A., L.L.D., F.R.S., who published a 14 ft. long folded up WALL CHART of WORLD HYSTORY, which is available at Barnes & Noble Publishing Inc. These are documented facts that are verified by fertility laws and calculated mankind on earth with modern computers. Two science witnesses cannot be denied proving that mankind is only a short time on this planet against the evolution religion being taught in our universities postulating billions of years of monkey fairy tales contradict many entropy and Newton's laws.

Inventing nuclear poison in time will be considered the greatest crime ever perpetrated by evil scientists who exploited it and who follow the big money trail; all for obscene profit, but even big money must be left behind after they die. Their names will be forgotten, but their bones will testify forever glowing with radiation, thus explaining why they died.

Ultimately, the new UREE gravity motor invention will provide logical proof and saved my credibility, which is now proclaimed on the Internet and which cannot be controlled by an atheistic priesthood. My presentation makes only rational sense because it is reflected in the mirror image of metaphysics that is based what an Apostle Paul proclaimed 1900 years ago recorded in the Bible. In ancient times, when he was standing in the Greek city arena, where science was started announcing, "I perceive you are very religious, but found one altar to an unknown god. This unknown God I proclaim."

A long time ago in Canada, I remember an eccentric person who walked on a steel rope across the Niagara Falls abyss. In the middle, he let go of his balancing rod and that exposed everything he stood for. The GRAVITY MOTOR will expose my faith in the Elohim, as a reality for me, also testing me without a balancing rod based on gained experience from walking with a God I can trust.

God loves all of us and became mortal on our level to teach us a resurrection energy cycle principle started with [∞-light] to zero time [BC-AD] moving on the other loop toward [eternal life] Jod dimension like the second entropy is the profit cycle when he shows up the second time. His visit to earth will start with free electrical energy to last the next thousand years and end in a new earth and heaven prophesied mainly created for the benefit of his Saints desiring a vacation home in the universe. Many more concepts you can find in God's Word the BIBLE, all is revealed in a future "Plan for mankind" because every one of us are very curious.

My Gravity Motor will work, not phantom religion, then logically, we must accept the energy source Elohim, the beginning of all what moves and has Gestalt embedded with ∞ intelligence to continue with ∞Life promised, a triple witness to prove that the ∞ energy source exists. An atheistic world will now have positive confirmation of God's existence facing the Judge Jesus the Christ, King of Kings return on this earth to judge everybody without a balancing rod. The apocalypse has been dated, so there is no longer an excuse, "I did not know" - stripped of any excuse and facing a righteous verdict.

Let's further investigate alternate energy possibilities to gain free green electricity from other applications extracting free energy for the benefit of a future generation of mankind guaranteed by ELOHIM the energy source if you understand the metaphysic side of nature. That will widen a science and metaphysic spiritual horizon to 360⁰ like being on top of a pyramid or looking out on a space station seeing more.

UREE #4 - Roller Coaster Theory

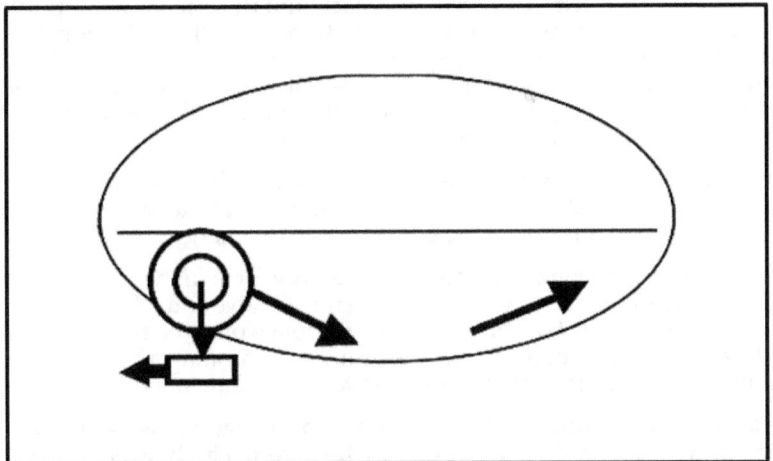

Here is another practical idea to expand our horizons about how to squeeze energy from gravity, which is cheaper than nuclear energy or oil conversion. It will deepen our understanding about a forbidden perpetual motion when applied to hi-tech.

A recent Popular Science issue, dated April 2011, introduced the world's fastest roller coaster. It will fling passengers at up to 149 mph with a propulsion system inspired by aircraft carrier jet launchers. It has 48 hydromotors that use pressurized hydraulic fluid to turn a winch and a drum with cables (eight feet in diameter) that is connected to a train on a coaster track straightway that is catapulted to 149 mph. It is quite a system of pumps, launch valves and accumulators that are linked to 10 nitrogen gas cylindrical storage tanks, which are the energy source for the launch.

However, let's go to an old-fashioned fun roller coaster for little kids. Nearby high school kids that are hanging around could help me build a contraption and mount it to the front carriage in order to prove a new theory. It could demonstrate an energy conversion very quickly in order to test a new hypothesis about creating perpetual motion, which would be useful to mankind. It will be a lot cheaper as the world's fastest hi-tech contraption and it can begin with only a few parts and some Band-Aids off the shelf from the hi-tech warehouse. We could give away free tickets for every rider to have more fun.

In my sketch, we can see the principle designed around a generator, motor housing and wheel carriage assembly that is somewhat similar to the roller coaster train that is so much fun for kids. Electricity is produced in a generator, which is basically a reverse motor. A streetcar motor has a stationary armature, a number of winding copper wires encased in a magnet iron and a rotor with the same winding wire, which is also encased in magnetic iron; however, it must move to produce electric energy.

Science still does not know why an electrical pulse appears when we push a magnet through a copper wire loop. Bunching many wire loops together and using magnetic iron produces electricity. Even the magnet needs to be made magnetic by reverse electricity, which is a mystery.

Producing electricity and feeding it back into the system will produce magnetism, which in turn produces electricity, which produces magnetism that is needed for creating electricity. If you understand that, then you are closer to understanding perpetual motion. We have learned that as long as energy is moving, something is leftover for our benefit. That is why a streetcar motor produces work.

Once again, some energy must be put back into the system, much like a business reinvesting its resources in order to make a profit; similarly, converted inertia is only a temporal storage for energy.

However, it is needed as an additional energy force in the system, in order to create more energy and cause energy to move and what is left over will make a profit. The roller coaster principle works the same way, just like a cuckoo clock or analyzing the principle of running a business, with returning energy investments in order to make more money.

Electricity is produced when the generator housing is wound around the fixed stator, rolling down like a wheel from a high level, similar to a roller coaster that hits the bottom curve. However, the inertia gained will help the whole carriage train to run uphill, but not quite as high. That principle is fun for kids on a roller coaster, which needs an extra energy converted into a ticket. The ticket buys extra energy needed to reach the same height from where it started to repeat the cycle.

By repeating the cycle and duplicating it, we will soon discover that it needs a little extra energy on each side, in order to start over from the highest point once more. Perpetual motion is even applied to my bank account, which could buy tickets (converted to energy), in order to have fun forever. As long as money (energy) is in motion, linked to a big bank account, it produces fun or leftover profit for my kids. However, energy must cross over the ∞ zero point which proves perpetual motion, much like my kid standing at the zero counter once more, in order to buy another ticket for another round.

Again, learning about a new scientific application, which is even embedded in my cuckoo clock, is also duplicated inside a nuclear Donut atom design, where we see the same principles as now postulated in this book.

It demonstrates that it is like being inside the atom, where a number of proton carriages are speeding around in circles, but are slowed down as they come around the poles, as in the roller coaster example. The laws of physics are replicated wherever I look. It needs a little gravity converted to inertia that is stored, in order to make it run indefinitely, as energy is used, which I now understand much better.

A little later, I learned that every atom needs to be refueled with energy, in order to exist and live a little longer, which can be demonstrated in the roller-coaster principle. Otherwise, if the universe would be running out of fuel the lights in galaxies no longer shining therefore would collapse in a reversed Big Bang.

How Does It Work?

Of course, the stator of an electrical generator needs to be stationary where it is mounted inside the generator, either by hanging it with a weight or by mounting it to the frame. Since it is not turning, the electricity is generated through the outer generator housing winding converted and running like a wheel down to the valley floor. Yet notice that the inner stationary has a fixed winding core, which becomes the wire linkage, in order to tap off the generated electricity.

It will touch a sliding rail, much like a streetcar on top touches a wire that is linked to the electrical control center, collecting electricity and directing it into the distribution network.

However, everybody may ask where we get the extra energy from, in order to have the roller coaster generator return to the highest point, when the inertia going uphill has slowed down. Perhaps it needed an additional 25% on each side, in order to return to the highest point, thereby repeating the cycle once more to continue indefinitely, just like my cuckoo clock replicated, or the need for the four pistons in the previous perpetual motion car example, in order to equalize the forces.

Remember that a generator is also a streetcar motor, only it is reversed by electricity. Therefore, at that point, when the velocity is near zero, we use electricity, which is now switched over to where it becomes a motor, causing it to go back uphill to where it started, repeating another cycle.

Where Does the Other 50% Energy Come From?

The extra electricity for perpetual motion does not come from when you bought a roller coaster ticket, paying indirectly for another power source, but rather it came from another carriage generator assembly, running in parallel, linked together with other roller coaster generators, motors and carriages producing surplus electricity. That would make more profit, which is guaranteed if you no longer pay the utility.

Two electric generators running in parallel, both down and uphill, can create the additional electricity needed to supply a driving motor carriage assembly back to the high point. From the energy equation, each carriage needs an additional energy, totaling about 25% each, when friction is added. They could be hitched together in a train, much like what we see with our kids on the roller coaster that is linked to a number of carriages.

Therefore, basically we need only three roller coaster motor/generator assemblies in a carriage unit, rolling along on two rails that would run endlessly, as the net energy gained is worth one roller coaster electrical generator. On a downhill travel, GRAVITY pushes two electrical generators, which will create electricity; one to drive a motor, to give the carriage extra inertia to the total mass needed for the highest uphill point; the other generator being the profit energy.

Again, one electric generator output is fed back into the system to drive a motor carriage; the other leftover electricity is profit. The infinite ∞ math glyph once again demonstrates that the electrical energy is fed back into the system, which will therefore run indefinitely. Remember the motor principle, in that the internal magnetism in the stator is produced by the electricity fed back in the rotor that creates magnetism. Some energy profit is gained, but only if it is connected to the same system.

The roller coaster can be designed in a closed circle with a hundred carriage generators linked together, and going around, endlessly turning as a system. Over the hill, each carriage motor is switched on and off; coasting downhill with GRAVITY accelerated to build up inertia, which gets you back over the hill.

It would be fun to see it running and logically, it should work. It is similar in principle to a hybrid electric car, sold by dealers, that uses gravity and inertia, in order to extend the mpg gasoline. Now you know why the hybrid car gets better mileage. However, in our example of perpetual motion profit, a new concept coined for the first time in history.

UREE #5 - Extended Roller Coaster

Expanding on the concept of perpetual motion converted to electricity, let's borrow an idea from a German levitated magnetic train with an electric motor stretched out linear miles long. We will convert the roller coaster design and apply it to a large "Wienerrad" in Austria **mounted horizontally on an axis turning.**

On the periphery we have a roller coaster design moving heavy weighted wheels hanging linked together on a single outer monorail running on a track with perpetual motion up and down driven by streetcar motors in between wheels accelerating gravity to gain inertia to get over the hill gaining speed circling around.

While all the streetcar motor-wheels coupled in a circle moving around up and down at different speeds (accelerating down - decelerating up, like a cuckoo clock pendulum) - pushing with outlier bars the horizontal Wienerrad-rim which has embedded a German linear motor section lined up converted to a generator.

That motion turning the Wiener-wheel gives us electricity at an even speed perpetual. If you doubled up the outer monorail track rolling up and down but a little off set one over the hill side matching the bottom curve of the other track will help to even out the motion turning the big wheel rim like a two piston motor with a gigantic flywheel. It is the same principle going with my grandkids in my neighborhood park there is a horizontal wheel they sit on and the grandparents push the wheel to higher speed to see kinetic energy in action kids flying off the wheel which is fun for them opposing gravity.

The other side of the Wiener-wheel is a heavier stator section with fixed winding firmly mounted to the ground in a big circle wired to a collector transformer to give us electrical profit. The center of the big Wienerrad-wheel is an electric generator linked to an external motor to put back a little energy to balance speed and friction for ∞ perpetual motion to work.

Remember the Wienerrad is heavy and needs to be cranked and the weights balanced with the sum total roller coaster carriage assemblies acting on gravity. We have learned that only when gravity moves do we get profit. The bottom horizontal Wienerrad-wheel rim could be driven from the ground with a number of embedded streetcar motor wheels using excessive electricity osculating back and forth to maintain velocity, but the leftover energy profile will determine how much electrical profit is. The math professor can calculate you if the energy equation justifies kinetic energy gained versa cost but needs to be fairly in motion at a certain speed to get an electricity profit.

However one strike against is our present generator design is still the same not much was changed in 100 years and should use some research money spent foolishly from CERN and Berkeley Laser facility spending billions looking for energy, they will be way ahead of the game not having found the energy Genie in the bottle after 40 years. We have now a better understanding what really is a magnetic bridge to generate perpetually electricity.

What is certain that the energy cartel may put water on the fire and say not possible like similar listening for the echo of laughter when the Wright Brothers announced we could fly? Today compare an airplane and how fast it goes and how many people are inside.

We have not yet looked at an electric generator to see the potential for a redesign rather making obscene money selling poisoned energy. We subsidize billions time billions of Dollars every year into foolish research but not one cent on improving a simple 100-year old electric generator design.

If the energy cartel experts say not possible why not get a bunch of high school kids together with some money from the CERN project and will develop for the first time something which makes sense and benefit applying the roller-coaster principle to streetcars and trains, using a lot of wheels on rails.

Insert a generator assembly embedded across two wheels for each axis, the size of an oil drum should generate additional electricity. Feed back the current gained and via a balance transformer that is added to the same main drive system could gain electricity. Using ball bearings would reduce friction to counter the reluctance, which would be present in every generator-motor.

A new, 250 mph high-speed train could be designed much lighter and would accelerate like a rocket, turning multiple motors underneath each coach in the beginning, when enough inertia is reached and is coasting, switching the motor to an electrical generator, alternatively generating electricity back and forth into the system, in order to maintain speed.

That could achieve a minimum of 50% energy savings, even without the perpetual motion principle. When translated, it would equal billions of gallons of oil annually, globally. As for a railroad, the highest expense, which is fuel, would make a profit once more.

It would throw out the heavy diesel engine, which is slow to move and replace it with a control caboose, in order to crank up the system with as many axis hybrid generators and motors added under each coach, linked and electronically synchronized applying high-technology. Instead of one, massive engine energy is distributed over a large area; therefore, weight could be reduced for higher speed to go faster.

We may be forced in that direction, if you look at the fossil oil usage statistic in 2010. We have maxed out the global oil production, as from now on, it will be all downhill, but consumption is still rising.

At some point in time, within the next 15 years, the system will collapse, as there is only so much fossil fuel available. It is already degraded at a lower quality level and is estimated that whatever is left over is past 50% depletion, worldwide.

We must redirect spending money and not waste it foolishly by smashing atoms, but think about gravity as an energy source, combined with solar panels, as being the only viable option for the future.

Plenty energy exists by using the source of ocean water, which can be separated into Hydrogen and Oxygen with solar panels. That is giving us a free sustainable resource energy, which should replace expensive deadly nuclear power. A combination hybrid hydrogen and gravity system now invented will be replacing fossil fuel just in time.

A basic electric generator design hasn't been improved since 100 years but could now be possible. To aim for higher output needs a redesign applying rare-earth super magnets not even known five years ago. A new generator design can be further enhanced applying low inertia printed board armatures. I am sure it can be converted to a higher output generator as the low inertia windings would help reduce the weight in a high-speed application.

For example thin printed boards with the embedded windings can be stretched out across the horizontal Wienerrad-wheel requiring lower mass, which could create electricity with stationary super magnets on top and bottom. It should be further investigated by the experts as the printed board armature with its inherent lower inertia mass which can be spun at higher speed hence producing increased electricity.

UREE #6 - Horizontal Wienerrad

For a triple horizontal Wienerrad-wheel application I can think of another version applying a design feature described in my next chapter about a levitated train going 500 mph without wheels utilizing an embedded gravity force on a V-shaped flat track which could be expanded to a big roller-coaster idea to generate large high voltage electricity like a Hoover Dam.

A [V] shape track has two sides and could be applied one path for a motor and the other generator following the GRAVITY MOTOR electrical principle. Envision a miniature V-shaped racetrack mounted inside of a horizontal Wienerrad-wheel with the rim shaped like a short vertical drum.

The rotating drum is turning at lower speed driven by roller-coaster gravity design we previously described like a slower GRAVITY MOTOR piston perpetual moving.

But for a higher electrical generation speed we could turn on the inside of the drum floating counter-clockwise many carriages linked together in a circle of a low inertia printed board armature design inter-connecting many electric generator-motor sections which will be racing inside the rotating cylinder drum at extreme higher velocity perhaps 1,000 rpm or 500 MPH like my Gravity levitated bullet train explained later.

That creates a levitated centrifugal force that is generating its own air-bearing like a recording flying head floating on top of your disc-drive computer. While suspended at high racing speed inside the drum with the lowest mass still is creating an inertia combination linked to the UREE gravity motor electrical design feature.

That will be switching electricity generated from the generator section back and forth to the motor armature osculating across the V-tract gaining big time electrical profit, just another application.

Huge moving motor installations needs not to be massive and heavy if we applied a floating air-bearing design which reduces friction therefore much higher velocity is possible like a speeding levitated train inside the race track of a Wienerrad-wheel rim drum for leftover electrical profit. It just replicated a bigger Hoover Dam turbine but running much faster.

My novel atom theory could explain the interior process when the generator armature is moved by a motor armature racing at very high speed together in a number of carriages looped around inside the drum.

The same principle according to physics is demonstrated gaining electricity illustrated inside a conventional generator but with lower speed. Looking further racing inside a drum will create a centrifugal force producing air-bearing and therefore compressing air-oxygen atoms.

It is the same principle of the UREE gravity motor extracting electrons by creating a kinetic condition for condensing air-oxygen atoms to free up electron energy.

A Million Kilowatts?

Let your imagination go wild applying a simpler lower rpm design and visualize turning as many printed board wire-loop assemblies the size 4x4 ft. fastened vertical on the outside of a slow moving Wienerrad-drum periphery sandwiched between super-magnets on either side.

It will create electricity recycled immediately and linked to 100 (double streetcar-generator-units) UREE gravity motors stationed on the bottom around the periphery running at 10,000 rpm standard speed, but wired in series and parallel.

This will gain electricity at the highest voltage, big time a thousand times over needed in a furnace melting iron. All is perpetually running from weights running down and up on a monorail around the outer periphery utilizing gravity accelerated by a streetcar motor to make it over the hill of a roller coaster facilitator pushing the big flywheel perpetually. Check it out with physics to prove my theory.

My philosophy of converting gravity, through applied perpetual kinetic motion, will give us free electricity, which has many more now possible applications imitating the UREE gravity motor design. That should challenge our university graduates enough for them to become millionaires by applying what they have read here for free.

You will see in the future, like the Wright Brothers' friends, those who further expand on a basic idea building bigger and faster airplanes. Once proven, people no longer laugh.

Gaining free green electricity energy from various GRAVITY MOTOR principles stretched via hydrogen extracted from the ocean linked to wind and solar electrical adaptation, all have the ingredients for successful energy conversion.

Horrible deadly nuclear options should be shut down and replaced with an open global energy policy implementing the many cheaper green alternative choices which should not be suppressed by a biased criminal energy cartel bent to exterminate all life on earth programmed for 100,000 years. But notice will end the extinction sooner in the next generation of mankind.

That process of murdering people already has started in Russia and Japan, now leaking massive nuclear contaminations in surrounding farmland and recently into the ocean which will be killing all life forms on earth and ending permanently the food chain.[4] It only exists because comatose uneducated taxpayers voting in corrupted politicians allowed it.

A nuclear Dollar exchanged for a green Dollar; gaining extra non-lethal energy is not so difficult, but it means we need to replace the criminal nuclear energy cartel moguls and wean them off from deadly nuclear preference.

The future has tremendous gravity potential extracting free clean energy as a source globally available. Why it is not allowed by the establishment refusing alternate energy spending while starving green energy research no funds available?

---

[4] Pearl #176 Japan's Nuclear After Effects (6-29-11) Solutions from Science.

Summing up my advice, we should put all our effort into unlimited alternative energy source possibilities, rather than wasting it ridiculously on smashing atoms. They have been doing it for 40 years and should have learned by now that it doesn't work searching for an elusive atomic energy not possible if your atom theory is wrong.

Rechanneling funds from CERN and Berkeley laser-beam experiments switched to green energy research would be much more useful gaining free electrical energy profit for the first time.

What you will see in the future is that electricity will be the major force applying many rooted kinetic applications or extracted from the ocean, but it still needs 50% energy recycled for perpetual motion if you understand the roller coaster principle.

As oil prices will surely accelerate can get an early start by using solar panels linked to UREE gravity motors utilizing a free energy source and should not be suppressed by the oil-nuclear biased energy cartel.

UREEs #7-10: Orffyreus Wheel Energy

An alternate version of a gravity perpetual motion motor could be designed around the Johann Bessler (pictured right from the 1715 engraving by C. Fritsch) wheel, also known as the Orffyreus Wheel.

One version is that of balls sliding along the spokes, designed around an electrical generation teeter - totter relationship, once the wheel begins turning. The ball on one side would run toward the inside rim, while the other ball would roll outside. The disparity of weight turns the wheel perpetually.

It can be built with some parts off the shelf to create electricity anywhere on this globe. That could be another gravity invention benefiting mankind. Now it puts the world's attention towards this remarkable man who pointed to a possible perpetual motion energy conversion. Read further.

In these troublesome times of global warming, it is not due to carbon pollution, as only brainwashed ignorant scientists are postulating; however, it could speed up the process, as seen by watching the rapidly diminishing oil reserves and the search for a cheap, clean, alternative energy source. It will force us to do some rethinking.

I find it remarkable and frustrating that there appears to be a viable solution to all these problems, namely Bessler's wheel, but no one is taking any real notice, as it has been screwed up by an atheistic evolution religion. Just like me, those who are taking notice are lone voices in the wilderness. It's time to sit up and take notice and apply the forbidden logic that our universities are no longer teaching.

Wiener Fairy Wheel - Four Action Design

Electricity can be converted from a static gravity force, linked to perpetual motion, which will give us energy that is useful for mankind. If static gravity moves as a force, we will gain electricity, according to the math ∞ symbol (previously explained) when it is applied to hi-tech power that is embedded in refueled simple atoms on lower entropy. We could find some interesting methods to extract a much-needed infinite energy source for mankind.

UREE #7

If we transformed the rolling balls and converted them into a wheel, which has the outer housing of an electric generator embedded, and fasten a hanging gravitational weight on the inner armature, in order to make the stator stationary, it would generate electricity when the outer housing-wheel armature rolls towards the outer periphery of the fairy-wheel on the downhill side.

On the other side, the generator/wheel assembly would roll towards the inside of the spoke center and also generate electricity. The spokes must be an odd number, in order to allow synchronization, so that when one cycle stops, the other continues, thereby generating continued electricity to end in perpetual motion, generating electricity endlessly.

UREE #8

On the outer periphery of the fairy-wheel, on either side, are mounted sections that look like a gigantic gear, with a pinion driving another generator on the bottom in order to produce electricity, or you could use a steel rope, like a belt instead of gears. A little more energy is gained.

UREE #9

The fairy-wheel center bearing could incorporate a generator designed for windmills with a variable speed clutch to utilize slower speeds and create a little additional electricity. This would power a motorized assist for the fairy-wheel to help compensate for friction, just like a cuckoo clock needs a small power kick for the pendulum to return for another cycle.

UREE #10

One more idea for dessert, the rim of a big fairy-wheel could have wire-loop-sections of lightweight printed board assemblies mounted all around the outside periphery and now moving perpetually sandwiched between two rings of super-magnets on each side generating free electricity. The operator could even charge an entrance fee to see it for extra profit and comes with a hot dog explaining physics to the next generation. Creating a theme park for my grandkids to teach science is always popular.

The net gravity is now converted to electricity by adding the little electricity gained back from the other generators, therefore running indefinitely. In any court of law, there needs to be three witnesses to establish truth. In our case, we have 10 applications; there are more if you include high school kids. The energy mystery trail will testify that unending power is available coming directly from the cosmos source revealed in the Bible. The ∞Elohim described in Genesis started His creation with infinite ∞ light energy fueling his creation forever.

Converting some of the ∞ energy should provide the next civilization after 2018 AD, "God's Kingdom on Earth", unlimited electrical energy. It stands to reason that if even "one watt" electricity is gained "perpetually" it will prove the principle. Similarly, the Wright brothers were not deterred by the laughter and being told that it was "impossible". They proved man could fly.

The Wright Brothers probably never envisioned double decked airplanes a hundred times bigger and going faster then sound and reaching into space. Applying history to my 10 applications designed around a GRAVITY MOTOR will change our up coming civilization forever. Gaining just one watt of extra energy by applying 10 UREE technology inventions will prove that a 100-year-old electricity discovery can be expanded. This new technology will open the floodgates of enough energy to fuel the next civilization.

Looking closer when a magnet is put through a copper wire loop, we get a pulse now explained WHY. That is the key to break the sound barrier creating unlimited electrical energy without causing the death of future mankind by uncontrollable nuclear poison. We have already started the extinction of all life forms. Even a blind man can see this happening on the only planet in the universe where life exists.

My professor friend wrote after reading my first draft, "Herbert what I see in your design is an application of revealed principles. Real world application will be less efficient at first until the system is tweaked and better understood about where and how strongly the electrons come in from the other side.

"Yes, you describe the principles, but the trick lies in the doing. With other inventions, you could take your principles and build, or have built a real model to test them. Like when you designed the flying magnetic head, which made a miniaturized disk drive possible to fit inside your computer, your mind puts principles to work without advanced math formulas. Your born intuition proven in many inventions can adjust it without you needing to do the math. Your mind did it subconsciously.

"With this big UREE fish is too big, so others will need to build the models to test and refine it. Your philosophical insight expanded a knowledge horizon, as intended. At first some government funds may be needed to accommodate our technical shortcomings. A little more experience may be needed but in time it will grow more efficient as it takes advantage of the ultimate renewable energy from Yod, tapping into the creation cycle from Aleph to Tau."

THE KEY:
The Ultimate Renewable Energy Engine

Summing up the Mystery Energy Trail not found in schoolbooks, one energy loop was described as neutrinos cascading along an invisible magnetic trail, fueling every atom with energy while passing through each atom, as some kinetic energy is rubbed off. It thereby creates electricity within the nuclear atom.

In our case, the pressurized air atom that is confined in a bottle like a Genie spirit produces expelled, levitated electrons when released that give us an electricity profit via a magnetic bridge inside our gravity motor. When electricity is re-invested in the generator motor, similar like re-investing in a business, it gives us an added energy profit, which can be applied to grow food and maintain life for a better lifestyle.

By investigating physics a little closer, I learned that the mysterious GRAVITY force is static, but when it is moved perpetually, like in the math ∞ symbol, it is transformed into a kinetic-rooted energy force, which is useful to mankind because of when in perpetual motion activated, causes time to be zero in the energy equation. When time is zero, energy becomes infinite.

Zero time = ∞E

However electrical energy Profit can be extracted from the energy loop coming freely from space, which is a gift for mankind that will fuel the next civilization linked to a ∞ energy loop system.

Concluding once more the text everybody with a general education will either be very puzzled, or hit a wall, convinced something is wrong in the La-La land. Kindly worded it is insane to postulate believing to driving around with an air tank filed with spent exhaust CO-NO gas. My first reaction was that I had painted myself in a corner already. I heard the echo of laughter like the Wright brothers picked up from the science corner even in church. But that idea is not that far out seeing recently on TV (7/2011) channel 7 in San Francisco a bunch of college kids driving joyful around in a converted bicycle with 2 pressurized tanks: headlines - "driving with air!"

However, God on the metaphysical level reminded me when we started on a journey below the ocean a 1000 times higher pressure could be utilized with the Gravity Motor combination, because when higher kinetic pressure is recycled in a piston arrangement to allow perpetual motion and repeating the cycles will create electromagnetic pulses which is electricity. That same principle can be applied to an automobile but we still need high-pressure too otherwise it will not work. Here is the key. I only recently understood a little better.

When kinetic energy is accelerated higher than the speed of the earth around the sun that would increase energy proven when a stone falls down to the ground. But in David's slingshot example, kinetic energy can be raised in the air tank when the UREE-motor gains speed and is expanding pressurized atoms back in a semi-natural state in each piston cycle. It will thus convert David's stone inertia to a bowling ball, energy wise.

It is like when the piston speed moves down 100 times faster because of higher pressure behind it. Lower gasoline energy cannot provide it on a much inferior level. That artificial, higher pressure difference compressing atoms closer together and now expanding is 100 times bigger force transferred and converted requiring a flywheel or two for temporary storage. Furthermore, if the piston is allowed and designed to be recycled perpetually, it will give perpetual electricity pulses amplified in the streetcar-generator linked to rotating-transformer, a new invention. Gasoline extracted from fossil fuel will feed a properly designed flywheel a little less than 100% according to entropy.

The UREE reverses the piston generating higher pressure because of a second gas expansion. That difference is perpetual motion applied to get perpetual electricity. Any extra is captured in a bigger flywheel design to compensate for friction. With the pressure in the tank, they together make over 100%. That is the reason for perpetual motion.

When electrons orbit freely around a nuclear cloud, they are forced closer together to jump from one atom orbit to another, which are weakened having lost some energy from a billion volts down to the million-volt level. The electrons now flow like fish over a waterfall to be picked up perpetually in a UREE fishnet giving electricity profit. In a high-speed motor application, atoms compressed in gas-media would heat up much faster. Include a spark plug to release more kinetic force and expand the second piston cycle at a higher energy level to free up 100 times more bonus electrons.

I described the whole process but the key was not obvious, now hopefully proven with the Cosmos Mystery Energy Trail. It will give us free energy spliced out from atoms so that we can drive around with a car but need 10 or 100 times higher pressure stored in a small air-tank.

For a believable example, if the weight of a car with 6 people is 2 tons and allow 10% profit generated in the UREE, then we need 20 tons of pressure in the air-tank - perhaps a little more - to make people move and have fun. Kinetic pressure energy moved perpetually will give free electricity, and I wonder who has the last laugh.

Some Concluding Thoughts

The first entropy law is demonstrating an invisible ∞ cosmos energy flowing and crossing over a magnetic bridge inside a generator motor flywheel-assembly that will produce cheap energy in form of electricity profit converted on the second entropy level.

When it is all put together, we do not need a big Wiener wheel reaching into the sky because we can make it smaller - just like a large church clock can be miniaturized into a cuckoo clock, which works the same. Perhaps a miniaturized Bessler drum, 12 feet diameter, turned around in perpetual motion with steel bowling balls can be linked to 16 rotating generator printed boards spaced with super-magnet stators and speeding up to 1,000 rpm. These are in turn coupled with smaller streetcar motors on either side to generate about 100 KW.

For best utilization, a baby UREE should be designed in smaller units and manufactured with a single piston at lower motor rpm and linked to the double transformer-streetcar to produce about 1.5 KW output in one box. Several UREE boxes could be connected in series, side by side, to better provide the variable power needs of a residential end-user. The first unit runs 24 hours. If more energy is needed, the second, or more, will automatically kick on as needed to meet electrical energy demand. This arrangement also serves as systems back up if one of the units breaks down.

The baby UREE could be mass-produced: four separate piston units for each household, without needing any expensive transmission wires. Since this approach is not hi-tech, it could be manufactured by an established corporation in about six months. If you use the bigger solar-panel-Hydrogen-Oxygen UREE model, it will run 24 hours, but the off time is utilized to separate the water for the next cycle of gas fuel explosion use.

The big brother UREE needs a rock pile or water tank in order to give gravity weight, then transferred over a hand-activated hydraulic jack occasionally used to replenish air leakage from the motor/tank if a little compressor is not available.

The UREE rotating transformer runs at much higher speed as other generators therefore according to the formula [**Zero time = ∞E**] when kinetic energy is moving faster then time becomes shorter and will get us more electric power extracted if you understand the math. But needs a better generator design mentioned like if we made the wire loops bigger in a printed board design with a stronger magnetic field passing through using superior modern high-tech magnets. I think the printed board low inertia idea is going in the right direction but needs to be proven as my inventor gut feelings never disappointed me.

The **basic Ultimate Renewable Energy Engine (UREE)** can be applied to many more situations, like the rotating transformer-streetcar generator cross-linked with **windmills** could get 1000 times more profit and spawn many future millionaires in a wide-open technology now available off the shelf.

The simple wire loop pushed with a magnet generates a pulse, but a magnet or wire-loop must move. If we move the magnet perpetually, we get perpetual pulses. If we push-pull the magnet faster, the pulses are higher and stronger.

Kinetic energy can be used to move a magnet. If converted perpetual, it will logically gain more pulses. Kinetic energy is embedded in gravity, but atheistic scientists screwed up so many science theories and invented an unscientific evolution religion to make money from those who have lost the ability to think logically anymore.

Therefore, when physics is distorted, it is not possible to form new theories. Genetics and DNA prove that intelligence cannot evolve as confirmed with our computers, "Garbage in garbage out."

If you have a different opinion, start another evolution religion. It will not increase intelligence like 2000 years ago, a powerful well-educated Pontius Pilatus asked Jesus, **"What is Truth?"** That answer can only be revealed, like in the case of the one criminal crucified with Jesus who got assurance to continue living on the other side from the One who created life.

Evil has been around since Adam, but don't be too harsh to our parents. Satan started the rebellion in the heavenlies in 4488 BC. Before he was to be executed for his crimes, God created Adam and his children to replace Satan and his rebellious angels.

God appointed Satan and his associates in the meantime to be our schoolteachers to acquaint us with evil. It is necessary for mortals programmed for eternal life to learn about the tricks of evil.

That accomplishes an embedded immunity designed to automatically recognize the slightest evil if it shows up - like mushroom growing over night. Then we will snuff it out in the bud, not to spoil a new resurrected life given for us. Satan and his atheistic followers will be judged and terminated at the end of the Apocalypse for rebellion against a holy God.

My Babushka egg concept books were designed to answer the many questions people have had since childhood. Theologians are not educated in science being still mired in a Middle Ages belief system.

But to prove the skeptics that God exists and gives us life, he now gives, in addition, free electricity. Ignorant governments have not yet learned the lessons from Russia and Japan. They are bent on building even bigger, more deadly poisonous, nuclear power stations being obscene, crooked, and only money motivated by greed.

Even my 10-year-old grandchild comes to the logical conclusion that nuclear crimes committed against humanity will cause billions of future deaths and the massive extinction of all life on this earth in the next generation. After they die, the cement caps will crumble and crack. The iron will rust out to release massive radiation for **100,000 years**, poisoning all the oceans and air to ultimately cause a **dead planet void of life**.

Willfully ignoring science perverted by an atheistic evolution religion caused mankind to become totally corrupted to the bone marrow. That will force **God the Elohim** once more, like Noah's Atlantis generation, to terminate this civilization, for LIFE was planned before a time dimension; therefore, it will continue for future generations. God embedded in nature many illustrations to teach us principles like a caterpillar must go into the cocoon and is morphed into a butterfly on a higher level.

Pay attention to what is revealed on the metaphysical level after 2018 AD: God's Kingdom on earth, invisible in the cocoon, has arrived with a new management, which will destroy once more to end EVIL with an asteroid even projected by science on **21 September 2017**.

My hope is based on looking to **Jesus, King of Kings,** who promised to return to earth once more to save his world from utter ruin and destruction. He will rule with an iron scepter to demonstrate His divine authority that will be applied globally to governing all nations. That will usher in a golden civilization for mankind, with 1000 years of peace - no more wars and plenty of free energy for a flourishing righteous civilization.

From a scientific perspective, any delay for Jesus' reappearing will be extremely critical in saving the world from massive hi-tech destruction. Genetic modification must be reversed or reap the consequence of a collapsing food chain that will cause a billion people to perish ending with the Third World War at Armageddon fighting over the leftover food and oil reserves.

We can add to the list many other causes such the destruction of the environment: foul air, polluted water, and fishless oceans unending. If Jesus coming would be delayed even 10 years, mankind will not be around for the next generation. That is guaranteed, if you follow the hi-tech science trail with Satan already gloating:

"I won the confrontation against ELOHIM."

In my last book, I discovered the greatest miracle for me: the sequence of (9) nine Babushka egg concept books riding on a dual rail track, one side the Bible the other, Science. It developed in unusual concepts that interconnected with each other like a DNA strand interlinked but in reversed Hebrew order.

To understand why the ELOHIM created MANKIND was my ultimate desire as a Christian. The process started with an obscure Hebrew Alphabet Number System found in my attic, which became the roadmap of my Journey. The Babushka book titles summarize a metaphysic overlay symbolizing a nine step pyramid which points to a future Jod-10 = (Elevated Hand) New Heaven - New Earth, a totally new elevated creation.

All summed up: what you read here free on the Internet was revealed not possible to be my invention as many times I was puzzled too. If I refused to write it down would hear louder in my subconscious mind remember Jesus saying the stones would cry out, as I am only a scribe like in ancient times.

- Teth-9   = LIFE (∞Energy)
- Chet-8   = Gestalt (Web- questions)
- Zayin-7   = Completeness (Time Dimension)
- Waw-6   = Connect (Donut Atoms)
- Heh-5   = Window (Global Warming)
- Daleth-4   = Creation (Genetic Modification)
- Gimel-3   = Important ideas (Ancient Clocks)
- Beth-2   = Division (Apocalypse)
- Aleph-1   = Beginning (Purpose of Mankind)

Annex II: UREE Road Map

Expanding the underlying energy principles now six more UREEs -

Sixteen (16) basic UREE models have numerous practicable applications to create perpetually green, free electricity according to need like the Hoover Dam. It is generating large energy for factories and now could produce it much cheaper using the UREE underwater ocean version inside a submarine (page 5). But it can uniquely be miniaturized, scaled down and made portable into a tiny pedal bicycle needed by poor people with easy transportable electricity generated from compressed air. That requires focusing on the specific design concept based on use. Our civilization uses electricity in three energy modes, either moving or being stationary.

1. Cars and Trains

2. Home and Office

3. Jet airplanes & UFOs

For the various UREE possibilities extracting free clean energy from **GRAVITY, WATER & AIR,** the author recommends investigating first the model that is more suitable. Perhaps each should require a separate examination lead by a specialized engineering team. At the end they will discover what meets the cost criteria, use and benefit ratios. All will end like the Wright Brothers in better models changing our civilization. This process is about understanding the fundamentals of "perpetual motion" now applied to generating free electricity, which has come at the right time for a growing population.

1. Cars and Trains

Let's start for an example analyzing the green energy process by driving a car without gasoline needs to be investigated to choose which UREE design is best suitable and perhaps will be staged if found out more energy is needed depending on size. Each UREE stage must be scaled according to application all have the advantage giving more. We can start at the bottom with the minimal requirement and see how it works:

> A. The main power drive is a cylinder driven motor fueled with a high-pressure air tank. For perpetual motion to return the high pressure back into the tank, a spring is mounted on the low-pressure side secured to the floor of piston cylinder. When the piston is bottomed out and compressed, the spring has gained a little extra energy to help the pressurized air back at the right time on the return cycle.

> This is only possible with a flywheel. Perhaps that simple application is enough for the main power driving a motorized pedal bicycle or perhaps a little more is needed for a small, rubberized commuter car. However, on the other power spectrum, we would need

much bigger power now scaled up to an Underwater Ocean UREE application design (UREE #1) sized to a thousand times larger pressure than the Hoover Dam would provide. This can be achieved further by using advanced electric generators (UREE #13) fueled by solar panels and can even be applied to a Hoffman hydrogen-oxygen conversion creating large green free electricity the biggest UREE possibility if you live near an ocean.

B. If the spring is not enough for a bigger UREE piston design to achieve perpetual motion, we apply the next step using a spark plug to heat the air: That will expand the pressurized air inside the cylinder to a higher force to compel the piston to return. We must aim at perpetual motion as the secondary extraction of free energy converted into electricity. In the lower power requirement like a pedal bicycle, the spring-flywheel inertia would be sufficient.

C. If the biggest UREE motor still requires more piston energy for perpetual motion, we need a larger return force inside the cylinder by applying strong high-tech magnets could be made from Neodymium-Iron-Boron linked to two copper coils mounted on the bottom and top of the cylinder.

That will activate a stronger force when the piston-head is made from super-magnets. When the polarity of the coil is switched as synchronized with flywheel crank, it repels the piston-head on one side, and on the other side we have a sucking force that together will push the compressed air back into the air tank, which has not changed, like the weight of the cuckoo clock.

Those three applications should exhaust the basic UREE function utilizing one or all methods to create **perpetual motion** as needed by amplifying electricity converted from a gravity force imitated in an air tank to make it portable as linked to generating electricity. Check the list of added Pearls once more.

When the UREE concept is expanded further, it can improve an existing hybrid car driving without gasoline by adding extra electricity from a UREE kit combination (UREE #14). This enables driving around with perpetual energy. One or the other UREE versions can be chosen as the prime energy provider but not necessarily the same size, as one can be made smaller only an assist to increase extra energy needed for perpetual motion.

2. Home and Office

Generating electricity as a major fuel in our expanding population is more needed than ever, as fossil fuels will end shortly. However, nuclear power created by criminals should be shut down, if you want to live a little longer. As announced on TV, radiation is leaking into the underground drinking water in Germany, which is therefore shutting down all nuclear power units. The leaks are caused by rusting Iron cracking the cement will surely exterminate all life on this earth in the next generation because everyone knows that permanent storage bins for deadly nuclear rods just **"do not exist!"**

Sometimes UN nuclear experts reveal on TV how poorly emergency equipment is maintained in nuclear power stations with corroding pipes, failing pumps and worse. The computer technology changed from old documented original design layouts no longer available because disk or tape drives have no more spare parts available, obsolete after 40 years.

It is a recipe for a waiting disaster already seen in Russia, Japan and near the White House nuclear facility shaken with an earthquake. They always blame nature instead of revealing the total failure of emergency equipment. The UN expert told president Obama on TV that we were lucky the earthquake was one notch smaller. (Read Pearl #176 – Japan's Nuclear After effect).

Many others have the same conditions around the globe; we will never hear about it in the news. Ask those who survived. Do not listen to lying governments covering up the truth to protect the energy cartels. Another deadly nuclear power station was licensed (1/2012) without the approval of the American people. It guarantees the early death of our grandchildren's generation, terminating all life on this planet. Watch ELOHIM. He is not sleeping.

Scientists writing magazine articles tell us to watch that deadly nuclear radiation poison for 100,000 years. (Hope you live that long if you are an expert.) They warn us never let it out like Aladdin with the genie in the bottle. Please tell that to the Japanese and Russian farmers who lost all their land. Many surviving friends and relatives are dying from cancer. Empty ghost towns give evidence to the lies once heard from their government. They know that the nuclear radiation genie will never go back in the bottle once released.

However, we do have a better choice as plentiful green **electricity** is available much cheaper and not deadly to mankind. Many options exist, like how energy can be extracted even from water. Water is plentifully available; take a trip to the ocean linked to a method invented by Hoffman and grossly ignored a German scientist.

Why does the energy cartel not use the cheap $97 solar panels to split the hydrogen-oxygen water molecules, which later recombine to release big time energy?

Electricity is very functional, especially if made portable by newly discovered UREE inventions. If allowed numerous methods can be demonstrated like splitting water which opens again unlimited possibility gaining cheap green energy. The headline of this energy discourse explains three applications: a car driving perpetually and one for home and office and the ultimate flying a jet with infinite energy available extracted from variously designed UREE versions combined for triple portable use. But it needed to be amended. It gets better expanding to a jet engine design flying a jumbo jet without fossil fuel.

Applied to a miniaturized version will open a new way of life and get electricity to the last corner on earth without ugly wires hanging everywhere. Why not ask skilled NASA scientists now looking for other jobs as space projects are shut down. They could focus on something that is useful for mankind getting perpetual electricity to benefit every nation.

Recently, I was surprised to find on the Internet that a double magnet generator got the gold metal generating 50% more electricity. It is in line of the UREE printed board application discovery one notch higher.

With super-magnets now available, I postulated higher electrical output possibilities. There are several UREE options depending where the main energy source comes from. Again, look at it from the automobile perspective, which is also applicable for the home-office use because it is portable, a great advantage.

> D. The automobile has an old-fashioned generator could be converted to produce higher output of electricity linked to printed board generators using super-magnets in conjunction to paragraph A. Check the expanded possibility Pearl #203 - six more UREE design options that sum up the 9th Babushka egg concept book.

> E. If more energy is needed apply paragraph B. or C, with added solar panels linked to printed board generators (UREE #12 & UREE #13) much cheaper.

> F. For the connoisseur, the ultimate UREE possibility amplifying electricity with a double transformer, Generator – Streetcar - Motor combination, which ends our investigation unless (UREE #16 & UREE #17), which again are one notch higher. A little off beat, it was fun explaining how a UFO is fueled and how mankind can travel with the last #17-UREE into space outer universe, right into heaven at extremely high speed - no kidding.

> But sorry, the last pearl generates too much energy. It is not practical if you live on earth. If the government allows the remaining 16 UREE options, it will keep plenty of entrepreneurs busy designing applications to celebrate the greatest electricity invention ever. Check the energy path to become educated in the metaphysics controlling physics designed around the math symbol ∞.

Extracting free clean green electrical energy from **GRAVITY, WATER & AIR**, if allowed, will make dirty coal fired plants obsolete and replace nuclear poison power plants. It will shut down the polluting gasoline cars. These concepts are only possible if we have the will to force the government to allow it.

But somebody, I am sure, will be motivated to test the UREE models without the energy cartels approval and could make obscene profits while the establishment remains silent. Ruining somebody's energy business will keep suppressing Babushka egg concept books not allowed in universities quoting forbidden Bible verses revealing true physics linked to the metaphysics.

I hope it will not take a hundred years, like Galileo, to prove that the UREE energy trail was right. Watch once more you will see that ELOHIM the energy provider does exist.

How UREE Electrical Generator Works

(Excerpts from UREE Jet #206)

The principles to fly a 6 piston radial aircraft engine or a jumbo-jet and get over 10,000 times more electrical profit can also be applied to solar panels, windmills, or basic automobile motors to drive a high-speed, 500m/hr. train modified with a turbine design. All UREE versions must run "perpetually" thereby generating amplified electricity for free, but most important, they must conform to two thermodynamic entropy laws to work as described in physics. Applying the electrical three-finger rule to a copper wire could explain the process when connected to newly invented, double-sided electrical generator. It could make clear how electricity is created.

That will explain how electrical energy is amplified as applied to a single solar panel, for an example. A short copper wire connected to a solar panel in the sunlight will extract electricity to create a magnetic field around the copper wire conductor. That can be measured in any high school. If we connect two solar panels in parallel, we get a bigger magnetic field: quadrupled, I believe, as postulated in some other pearls. That should put to bed that word "IMPOSSIBLE" as solar energy is free once you pay the $97 dollars for a solar panel.

If we shape the copper wire into a loop and push a magnet through it, the wire gets a mystery electrical pulse. Nobody knows why. (Read Babushka egg concept books to widen your science horizons.)

But consider, when a loop of copper wire connected to a solar panel it will generate more magnetism now forced to go through a loop-center which at that point has higher magnetism bunched up all concentrated in the middle. That would make a much stronger pulse with the same magnet, because the copper wire is now biased from getting extra energy from the sun.

Let's duplicate it with another solar panel connected to a second copper wire loop, and then we do the same thing again. But this time, we add the second pulse generated to the bias of the first copper wire loop. The result is that the first loop will increase energy as the magnetic force centered in the middle has become bigger because of a higher bias creating a larger magnetic field around the copper wire.

Now let's repeat and take the higher output of each wire loop switching back and forth, always increasing the bias to get bigger and bigger. Would you agree? Or read my 9th Babushka egg concept book to widen perspectives to a full 360⁰ horizon.

Let's review for the PhDs in universities postulating, "NOT POSSIBLE."

1. A current created from a solar panel produces a magnetic field around a copper wire. **True?**

2. If the charged copper wire is made into a loop, all magnetic fields surrounding the wire have not changed but become more concentrated in the middle going in the same direction and stronger in the center. Therefore, a biased charged copper wire loop has more concentrated magnetic line in the middle then bare copper

loop and will consequently create a higher pulse with the same magnet. **True?**

3. If a previous pulse is added to the charged biased copper loop will increase the next pulse, as the bias is higher. If we kept it going always adding the next pulse to the other gets a bigger output eventually like double a solar panel. **True?** (UREE #13 stated 1000 X Solar panels!)

4. We have proof in the streetcar motor design. The bias current in a copper wire will increase magnetism ultimately gaining higher electrical output by recycling electricity proven in a streetcar motor design. On the inside the stator is connected to the rotor back to the stator in a ∞ loop and will accelerate the speed of the rotor until it explodes or is controlled by the streetcar conductor starving the input. A starving input differential can be extracted and made useful as electrical profit, but if returned, one could drive the street car perpetually without a conductor provided the energy is not stopped but freely let it go. **True?**

    Of course electrical measurements of parameters must be considered like resistance or heat, or eddy currents etc., which dictate that we must short cut and transfer the energy at the source with a transformer.

    That should cover most objections; however, the streetcar motor only runs perpetually if either the first or the second entropy energy loop is added and interlinked as outlined in the UREE principle. An objection can be raised if the current is shut off the streetcar will stop. **True?**

5. But another law of physics is not shut off. Mass if moved will store energy, like the streetcar has mass still running at high speed, which becomes in principle a flywheel storing energy.

6. And if the stored energy at that moment before it slows down according the physic converted and is added to the biases of paired electric generators switching back and forth will create a leftover electrical residue. That little portion of energy re-invested creates perpetual motion which I mentioned earlier is profit. **True?**

7. According to another law of nature if a saved energy like profit is reinvested and put back in the pot it gets more profit. Every business works that way why not in a streetcar motor? All we need is to add a motorized flywheel the only storage devise I know off which will store temporal energy and drive an electric improved generator pair could run the streetcar perpetually.

    The streetcar motor once in high motion switched over now will be driven by electricity generated from an electric Motorized-Flywheel linked to two generators UREE crossing over shown in Babushka book #9 page 17, but must satisfy two energy consumers linked to two entropy energy loops (∞). **True?**

8. My Grandkid now understands investigating the law of nature, which makes energy conversion possible. It is like throwing a stone into the pond: it has embedded energy that creates a secondary energy wave.

    But if we combine both entropy levels like connecting heaven to earth linked to a UREE, the ∞ invisible energy can now be perceived with our eyes because they are calibrated to the many laws of the physics. To make the invisible physics visible to understand it better, it must be linked to the metaphysics. The laws of nature are replicated on different levels and if used together in a proper sequence could get an advantage for us. **True?**

9. The UREE only works when perpetual motion is achieved interconnecting two energy levels but needs to improve the 100-year-old outdated generator. The crux of the whole UREE system summed up is that a flywheel once accelerated will peter-out and needs an additional energy to get it going perpetually, not much is needed and is now found in super-magnets a free force we all know we played around with.

    Magnets are used generating electricity and added as a force to a motorized flywheel will run perpetually and connected to improved generators on each side of the motor shaft is adding a bias to each loop amplifies more electricity big time profit. Take it from here my finger pointing. **True?**

10. To fly a jet requires a specially designed electric generator-motor combination (UREE #16) using the latest permanent super-magnets not possible 10 years ago. Therefore, why not learn about the metaphysics and become a billionaire replacing the corrupted energy cartel destroying life on this earth? It is the only place where LIFE exists in the universe. **True, too?**

Ultimate Renewable Energy Engine #11
Extracting free electricity from Atoms
for Household escalating power 1000x for
windmills, bullet trains and airplanes

The extraordinary totally free green energy described in the 9th Babushka concept book will take some time to understand because our present education system is biased in unscientific evolution religion enforced globally in every university. But this web site dares to expose the invisible forbidden laws of metaphysics that cannot be measured with instruments but nevertheless are real like your MIND hopefully not turned off.

Let's continue to discuss a new UREE #11 application to prove my green free energy theory by comparing the same method to understand how a streetcar motor on the inside really works to recycle electricity creating more magnetism from the stator to the rotor back to the stator turning the motor faster. When magnetism is recycled gets more electron slaves working previously described in an analogy for my grandkids. It is obvious that the UREE #11 motor systems must be driven by an easily available energy source but is only free of charge when redirected into perpetual motion to amplify electricity becoming extremely cheap. That is accomplished with a flywheel and once cranked requires less energy to keep it going - just like a cuckoo clock pendulum needs only a small energy kick at the right time linked to a gravity moving weight.

Free electrical energy is not known in universities will therefore be difficulty to understand and perhaps in-between switching back and forth to different UREE concepts could be helpful when you are stuck. In this UREE #11 we will describe a little the fundamental for my grandkid teaching science how electricity is amplified within a Donut atom creating magnetism as free energy is not a religious conviction but must come from somewhere.

Our understanding will get better checking out various other options bases on simple physics enhancing a generator design like **printed board wheels** in the next design choice. When shaped into a **low inertia cup set up,** an accelerating and rotating super-magnetic field will amplify electricity explained in **UREE #13 & UREE #16**. That will prove that more electricity is produced in principle by turning a flywheel motor linked to air-wind-water-solar therefore can be amplified **1000 X and more**. That is achieved by the addition of two crossover connected electric generators. That opens the opportunity of many more choices in application, which helps to overcome clouds, windless nighttime cycles and portability.

New inventions must conform to physics electricity is produced when a wire loop passes a magnet to get a pulse. Done faster and recycled infinitely but if biased will create a higher bias voltage, the pulse becomes stronger repeated again creating eventually a higher output electricity which I consider profit.

Once more explained a streetcar motor turning wire-loop-boards driven by solar-gravity-air or water energy will magnify electricity by the use of stronger, permanent **neodymium-iron-boron** super magnets now available commercially. This idea works according to the three-finger electrical hand rule. Remember the extra energy is extracted from the super magnets' oscillating magnetism embedded on the first entropy ∞ atom energy level created by moving protons at high speed within a nuclear Donut Atom.

That also creates a negatively charged magnetic cloud levitating concentrated electrons condensed from pressurized AIR that will now jump from one oxygen atom to another crossing over a magnetic bridge allowing electrons to be extracted as extra electricity.

For example, focusing on the **solar-water UREE #14** invention, which combines the conventional sunlight expelled from compressed sun-atoms becomes a second entropy energy level having lost energy travelling to earth in Kelvin cold. But if linked together with energy of the first entropy embedded in magnet atoms, which is infinite, will get more. That is how we get additional energy spliced out. My new inventions gain extra energy using a standard motor linked to Gravity-Sun- Wind- Air- or Water but must be turning "perpetually" using a flywheel for temporal kinetic energy storage.

At that moment the embedded energy is further converted, recycling magnetism infinitely inside two generators' wire, crossing over now transformed to electricity. The energy will gain intensity to the critical resonance frequency by perpetually turning biased copper-wire-loops faster while exposing a resonating additional stronger magnetic force inside atoms.

When a resonating magnetic field is recycled will generate a higher concentrated magnetic field that gives us more electricity in a streetcar motor or reversed becomes a generator. When energy is combined from the second entropy (like a sun-solar panel) with the first entropy magnetism embedded in atoms like super-magnets with an internal billion volt level it will oscillate faster; therefore, free electrons are squeezed out via a magnetic bridge only produced by perpetual motion like coming from the sun energy turning perpetual over a simple motor as long as the sun shines.

On the atom level at a certain critical resonance frequency speeding up electrons must be clipped off to prevent an explosion that is how an atomic bomb works. It is like you car speedometer drive not beyond the red zone critical resonance frequency shown on your dial or the motor will burn up. Only electrons can give us free ELECTRICITY on a higher nuclear level.

Therefore, the system must be stabilized at a constant speed below the resonance frequency to give 110 volts as long the sun shines or is windy. At night, the system will need to switch another UREE type to use hydrogen and oxygen recombination mode converted to electricity for a 24-hr cycle. Both proposed invention ideas together are worth billions of dollars. Applied to a WINDMILL generating electricity can consequently be amplified **1000 times or more**.

All UREE proposals have **two** elementary entropy stages, which is a totally new discovery. The major UREE concept must first achieve perpetual motion to allow similar motor pistons return the high-pressure into the air tank and when linked to a generator gets electrical profit. Making the piston head from super magnets enhances the UREE concept **(UREE #12)** creating higher electricity, which is more profit.

But focusing on the **second** new discovery about MAGNETISM is more extraordinary. Investigating how really a motor works, we find out that the electrical magnetic force creates additional electricity on a **higher** level when magnetic energy crosses over and adds a bias in the other copper loop, which then accelerates the rotor similar to a streetcar motor. The upshot is new generator design that could fuel a jumbo jet, no kidding **(Pearl #206-UREE #16)**.

That is a totally new concept. It is similar to what happens inside an electrical motor when magnetism is ∞ recycled from the stator to the rotor back to the stator resulting in work made skyscrapers possible. Without a motor even Wall-Mart would not exist as every single item was manufactured with a motor.

Later I got an even bigger surprise when I realized that the double acting UREE generator-motor combination is a mirror reflection of the Donut Atom Model postulated four years ago. The new Donut Atom Model proposed a different atomic structure not yet tolerated in universities prevented by a stupid illogical evolution religion perverting knowledge creating irrational fairy tales taught as science but is really simple for my grandkids below the high school level explaining Babushka egg concepts.

Protons do not just sit around but are spinning at high light speed being pushed by neutrinos. That creates magnetism applying the three finger electrical rule. It produces a negative magnetic outer sphere therefore levitating negative charged electrons also moving. But Newton law teaches us that what moves must have a cause. Only energy from a galaxy black hole fits the ticket now postulated but the proof of 16 UREEs function as a witness.

The path of protons shaping a donut traces a spinning infinite math symbol (∞) around the center dot very fast. That will appear as a three-dimensional shape similar pictured to a round red blood particle dished in on both sides in the center. This shape in mathematics is the perfect geometrical design to transfer energy.

Is this only a coincidence since it is duplicated so many times in nature? Only intelligence could have created any existence fueled with energy. That cannot be explained with an unscientific atheistic evolution religion.

Following the energy trail to better understand electricity lets look at the hourglass nebulae photographed on the front cover of my 9th Babushka egg concept book, which is a picture of a Donut Atom. We cannot see all of the energy exchange like gravity-antigravity, magnetism and invisible frequencies on either side of the rainbow spectrum. The photograph only shows a vortex of beautiful rainbow color energy lines, red to yellow, moving to green blue or towards the infinite energy DOT center black hole. That frequency spectrum reveals nature of all the elements on earth discovered by Fraunhofer categorized in spectral lines.

Currently we distinguish better how galaxies are formed trading Babushka egg concepts. Seeing motion in the picture demonstrates the spinning Donut Atom outer wheel or electron shell.

Because some portion beyond red is invisible to our eyes, we need to just keep drawing red colors spiraling in circles forming a Donut linked to the concentrated energy center dot similar to the picture of another rotating galaxy copied in the 9th Babushka book. Just use a scissor and cut the neck and put it back together seeing a round ball with a North and South Pole demonstrated for my grandkid a vibrating Donut-Atom picture.

However, the Fraunhofer rainbow-energy is based on physics duplicated everywhere in nature therefore can reveal the invisible domain of a Donut atom spiraling around like a big wheel - very fast with the speed of light. It is also a wheel within a wheel like two loops (∞) circling at high speed through the donut center. Remember Ezekiel's wheel within a wheel spaceship is similar. (Ezekiel 1:16; 10:2)

The moving proton-neutron force behaves like a generator creating magnetism to form a negatively charged cloud levitating electrons according to the three finger electrical rule. It functions like generator with the inner atom rotor turning magnetic energy in a vortex at 90^0 to shape a donut circling at high speed as replicated in that galaxy picture if expanded.

My new atom theory (Babushka book #6) will better fit this model explaining physics. Tracing the energy path on a three dimensional level within a Donut Atom could be simulated in a mechanical UREE gismo, which works similarly.

I designed the UREE generator comparable to a Donut Atom Model which behaves like a turned crankshaft at high speeds to replicate the donut atom performing like a fly-wheel, but it is also turning inside like ∞ motor loop, thus passing magnetism from one loop to the other through the donut hole circling around the North to South atom Pole like two generator axes resembling a wheel within a wheel. Does this explain logical the three-dimensional donut level better?

Let me explain it once more for my little grandkid. The reason why the UREE really works is linked to the structure of atomic nuclei as presented in the new Donut Atom Model. It is similar to streetcar motor designed to create electricity inside by recycling magnetism from the rotor to the stator and back to the rotor like the ∞ symbol.

That makes the wheels go faster and faster always giving more and needs a conductor to stop it, otherwise it could blow up. A solar panel, on the other hand, only gives so much electricity slowed down by Kelvin cold. If linked together, these two will store the extra energy from streetcar-motor via a flywheel.

Eventually, these components working together will run constantly without a conductor ignoring friction or eddy currents and other factors in physics as the accelerated energy in the generator is very large. When energy on the atom level reaches the critical resonance frequency it will flip into another element Fraunhofer defined. But before that point is reached now linked to a UREE could extract huge energy from compressed Oxygen atoms in air to make it portable will fuel the next civilization all free and green. To explain an invisible turning donut atom flywheel to my grandkid can be visualized with a blown up tire inner tube in the pool. The floating tire behaves like an energy machine if you spin it around and around floating on top of the water. In the case of an atom, it spins at the speed of light. When a Donut atom spins at high speed it creates magnetism for levitating electrons. We can visualize moving magnetism when we put a cloth line across the tire and make an infinite loop by inserting the end in the center.

Go underneath and come around back to the same center on the other side. Then, come around ending with two loops in the center of the tire. You just made a ∞ cloth-line. But hold the loops and move the tire between the loops still rotating with a spin and imagine that the tire would spin faster between the cloth line loops. In an atom's world, it would be light speed. However, the cloth-line demonstrated magnetism generated from moving protons like your finger did going through the center and around crossing over once more through the center and coming out on the other side to make a double loop. If you keep going we now have the wheel within a wheel idea moving magnetism, all accelerate with the speed of light - if it was an atom inner tube flywheel.

Once more the magnetic cloth-line demonstrates a path of many proton-neutrons circling around a Donut-tire-atom flywheel at high speed.

When something moves like that it has embedded energy, which caused magnetism in the donut theory, and magnetism can be converted to electricity inside a generator. If we attach in-line two turning generators to the tire tubing cloth line, some rotating energy would be leftover if it was linked to a streetcar motor (Tire rotating like a flywheel) always giving more. Watch a video clip on YouTube.com – Ampere's law applied to Toroidal Solenoid and the others much fun.

That idea is replicated in the UREE to work like a flywheel combination. We collect a little extra energy from the flywheel by adding two generators driven by the same streetcar motor. Collected together in a transformer, it will drive my hybrid automobile without gasoline guaranteed.

The principle of **UREE #11** extracting free energy from nature must be first linked to perpetual motion evicted from gravity-air-water-sun and when connected to a streetcar motor with two generators on each side of the motor axis will gain free electricity. But the wire connection must be crossing over to the other generator so that one loop pulse is transferred across creating a bigger pulse recycled in the next wire loop crossed over again, keep going like your cloth line illustration.

That creates stronger magnetic osculating pulses in a wire loop gaining an electrical output and keeps adding electricity until the red zone critical resonance frequency is reached to self-destruct therefore must be stopped with intelligence by a conductor. The question remains: Would two old-fashioned generators invented 100 years ago bought from the car dealer linked with the second UREE concept options amplify enough electricity which makes driving an existing hybrid car perpetual possible? Or could a windmill generator modified give us 1000 times more profit?

Our web site describes **16 UREE** variations discovering the Ultimate Renewable Energy but not all will be practical depending on application. A portable gigantic Hoover Dam power extraction is too big for a motorized bicycle. Thus, I feel the need to add something for clarification that could open up other benefits.

The lengthily 9th Babushka egg concept book described an energy path how to extract infinite energy from the cosmos linked to atoms. We live in the middle exposed to GRAVITY, WIND, WATER, AIR and SOLAR and discovered methods how to extract free electricity but is only useful in smaller energy chunks.

Free green electrical energy extracted from nature is only possible when two entropy energy levels are combined like the infinite ∞ math symbol demonstrated. As an example, if we throw a stone into the pond, it creates waves. Utilizing the stone first entropy (atom energy) creates a wave of second entropy (kinetic energy). If added together, they can be converted to gain extra electricity to drive a car perpetually. The fundamentals conform to physics but need more applied analysis for the majority of unbelieving doubting Thomas. My UREE system will work because of two forgotten dynamic entropy laws. I am sure that this application of physics will make big money for somebody.

A Second Witness Using Free ENERGY

Free energy can no longer be denied by the corrupted energy cartel now gets better applying one or the other UREE contraption option. Solar amplified electricity will obsolete ugly wires telephone power poles with transformers replacing expensive batteries, no longer require costly equipment just a simple system on my roof linked with three wires to the basement. Furthermore, when linked to a simple UREE motor with two generators, it will get amplified 1000 times more "free" energy forever.

When the energy cartel realizes that solar energy brings down the price, they put a diversion tactic in high gear on TV to discredit this new technology that gives the impression that solar energy is way-out too expensive, as demonstrated by two solar companies in California. One received 500 million dollars from the Federal Energy Department earlier mentioned and was even visited by President Obama establishing a linkage as he initiated once more deadly radiation poisoned nuclear energy PowerStation in 2012. The solar-company has now filed bankruptcy after only three months. All the money is gone, and the congressional investigation acts confused. They are more likely on the take as the money disappeared in quicksand.

The Internet revealed many cheaper options about recent huge discounts of solar panels from China. Now think a little with logic. If you convert the disappeared investment to standard commercially available solar panels, it could give 25,000 households free electricity forever, or 75,000 people, a whole town, could enjoy free energy. That would be a much better investment for individual taxpayers. If the other bankrupt solar company is included, it would double the benefit.

That points to a problem. When government gets in the manufacturing business, everything gets screwed up, increasing the benefit for the class of privilege. It should instead legislate easy access for everybody and systematically implement it one town at a time by converting the existing, aged energy system. It will obsolete old wires and transformers while controlling any devious interference by the energy cartels or special interests paying off senators running for reelection. It would encourage honest representatives instead. California is always first to start new companies, but watch for big money giveaways. According to Newton's law, they will sink fast in the many bottomless tax loopholes designed for those working for the government establishment. Mentioning names could get into big trouble just watching TV.

The Most Dangerous Energy for Mankind

Our greatest challenge for a modern society is how to produce "unlimited green energy" now postulated can be extracted totally free from **GRAVITY, WATER and AIR** only explained in complimentary Babushka concept books. It may look on the surface impossible but is not magic applying old-fashioned physics.

But can it be used for space travel? It will take some time to really understand how it is possible to get free energy but must be linked to a new Donut atom theory explaining how electricity is created within atoms that could be extracted quite cheaply without a nuclear reaction.

The answer is yes, now 16 UREE model ideas are postulated depending on application and size but for political reason will take a little more time to implement some of it.

Neutral green non-polluting energy rooted in a teeter-totter science world of good and evil must be controlled by a sound mind. We have choices when compared to life. Jesus advice us chose LIFE or chose a mushroom in the sky prophesied in Armageddon we all have become familiar with. Both are connected to energy in my last discussion describing how a UREE works for the benefit of mankind or will it be used for the destruction of all mankind. You be the judge.

The biggest teeter-totter energy exchange became visible in movies and pictures of a mushroom in the sky. Original the incentive of going nuclear was to build bombs like a bigger bat to bludgeon others to end all war of wars. In time it expanded and to make it salable to ignorant people now disguised and promoted as the cheapest energy we need, as oil price is too high another lie (remember $15 a barrel). That opened another good reason going into space and told us that nuclear is the only viable option crossing the vast open space another lie.

Applying nuclear energy for peaceful purpose is another lie never counted how many people since have died on radiation never included the cancer victims do not even know that it is the most expensive energy ever extracted and converted to electricity which is a true statement. Our government never stopped lying to us for a bona fide reason still hidden and never revealed the real cost per KW/hrs. buried in the national debt never had a budged accounting, it does not exist.

We kept producing deadly nuclear material to be superior and have enough bombs to destroy the world over a thousand times, but a good opportunity is never wasted to make obscene money to some to continue along a destructive path, which made it easy to siphon off and steal billions of dollars on a regular basis by none existing consulting companies and contractors presented only on phony invoice paper which cannot be traced, remember no accounting exist. Mega money is needed now to be re-elected to high office to remain in the class of privilege, but that has horrendous consequences for our grandchildren if we still believe that Newton's law has consequences and still works.

Pay attention if you want to live a little longer and follow my path being a scientist investigating the facts from a technical perspective with a sound mind forbidden in the public arena as it will point out corrupted politics evil to the core. Using deadly nuclear power is absolute insanity which ultimate in a relative short time will totally poison this earth more so like a thousand bombs exploding but because it is invisible to everyone nobody knows about it and therefore is very urgent.

Once again, because radiation energy is invisible most people are not concerned. But judging what some honest experts tell us that this invisible lethal radiation will kill eventual all LIFE on earth as it is so long lasting 100,000 years. It is scandalous and evil to continue with this program. Tell Obama still have not after 50 years invented a single container for the hundreds of nuclear power stations to safely hiding the spend radiating nuclear rods.

If you want more proof of the biggest lie that nuclear power-stations are safe, postulated by most universities lying to get the next government grand and cover-up that rusting iron is safe linked to another lie that cement is sturdy and will not crack when applied to radiating storage pools. That is announced by a paid off media to an ignorant comatose public. We hear on TV do not worry we tested every aspect of safety with the latest science controlling nuclear radiation to be stable for 100,000 years. If we repeat a lie over and over will become facts for a deranged mind.

But only Babushka egg concepts dare to expose that the emperor has no cloth on not believing in fairy tales but pointing the finger to vast areas of abandoned ghost towns and farmland check out Japan and Russia, which cannot be denied now void of all life never to hear a baby crying again for 100,000 years and conveniently blamed it as an act of God.

Ignoring warnings, look at Humpty Dumpty who cut off the branch he was sitting on, has consequence in physics and no longer applied and now suppressed in our culture. Notice everything became relative advertised as a better choice forgetting that to know the difference of evil and good made mankind survived for 6000 years, now conveniently ignored because it is found in the Bible not allowed anymore, another story.

Pay attention to make my point, recently seen on CNN TV our President Obama visiting South Korea University in Soul (3/15/2012) and reaffirming nuclear power once more. Student listening being scared looking at their faces now hearing another round was planned to build another lethal nuclear power-station to be started once more in America programmed to end in massive death not told. It is obvious our governments not listening to the United Nation Energy Department warnings.

Germany usually leading in technology is the first country to realize its danger to a society and therefore announced shutting down every nuclear-power-station under their control. They just discovered 2012 radiating drinking water next to a nuclear plant. But our corrupted American government typical ignores such NEWS.

Last Warning 2012 - Why a Big UREE Jonah Fish?

A deadly genie spirit will soon rise out from the underground-drinking water system everywhere near nuclear power-stations as storage cement will be unavoidable cracking from rusting iron which will cause deadly radiation seeping into the sub-terrain mixing with drinking water accelerated by the next earthquake.

The public does not know that such evil exist prevented by a surrounding fence nor allowed to asked question in public to keep this big lie secret from being exposed. But if not convinced how nature works visit a Jewish survivor monument seen on TV have a big problem seeing cement cracking now held together temporal with metal straps like a band aid until more money is available to fix it. Or better to be convinced look at the old cement bridges in the countryside build 60 years ago, but radiation is very hot will accelerate the corrosion process.

Also seen recently on TV employing young ignorant 25 year old women needing the money only $25 dollars an hour to risk their health volunteered in bunny suits diving in the pool storage of nuclear-rods to fix the cracking problem. But later mentioned offhandedly being unaware not told the danger they have big thyroid problems some already died from radiation, or still not convinced visit farmers in Russia and Japan and listen to their stories. What is not mentioned on TV will soon be forced to make big headliners seeing the effects of radiated lethal ground-drinking-water being poisoned forever killing every LIVING-THING, even the girls in protective bunny-suits would not be save proven from absorbing deadly consequences never ever to be healthy again, get that! "Forever". That is not an act of God.

We all know even my little grandkids understand that once the Genie in Aladdin story is released, "it will never get back in the bottle". That is not a fairy tale but trustworthy supported by government lies. To be better informed why not visit Russia and Japan which started the process of many eerie ghost towns and farms now empty void of all life never to hear babies crying again for 100,000 years the experts tell me!

This generation must be the most evil; watch God's Wrath - has not left town still around, but very angry. A corrupt society, evil to the bone, is totally destroying the global environment based on politically correct thinking changed a 200 year old Constitution based on Christians value and law now morphed it to be "relative" most lawyers love it to make big profit. They are now eliminating globally the ancient Bible rules no longer knowing the difference of good and evil, which used to be taught to every child.

What is put forward is not pie in the sky but based on true science not a fairy tale should have gotten some comments from the establishment but no longer wonder why is it suppressed? I cannot fathom why Obama would once more increase uncontrolled deadly poison to cause future death for millions? He pledged that fatal radiating accidents will not happen in his administration and promised the government will watch somehow to contain nuclear radiation for 100,000 years?

I wonder will he be the president next year. Avoiding responsibility what happened to Japan and Russia quick to blame it on an act of God. People may not realize that God may no longer be around and left town with his Bible no longer found in the classroom or on the desk of Judges forgotten and allowing evil heading for destruction. It is a risky business to get the attention and paddle against the stream of a godless corrupted society allowing such deadly nuclear evil in conflict with a 6000-year-old Bible thrown out as being worthless.

It changed a society believing in false rigid opinions based on fairy tales brainwashed in an atheistic unscientific evolution religion exclusively taught globally in most universities. Follow the laws of physics irrevocable will demonstrate the consequence either for good or evil should not be exchanged to be relative. What men sows, he will harvest fixed in the unalterable law of nature. I hope my Babushka concept books will gain credibility. It will take some time to understand this totally new Donut Atom theory offending the sensibility of many who do not have the ability to think logically anymore on two levels. They willfully deny that physics blends with metaphysics. Start reading the First Chapter of Romans in the New Testament. The Apostle Paul wrote it 2000 years ago, and it is still around to widen horizons.

As an inventor trained to think a little differently, I was chosen by God to point out God's intentions and why an apocalypse was planned for our times. God is sending a warning once more to save our earth and stop uncontrolled high technology, which has reached the point of being capable of totally destroying his earth by making life globally extinct. He will therefore judge a world gone the wrong way before it is too late.

He appoints messengers to reveal his intent to counsel mankind to repent and reverse from corrupting even evil learned from Satan. We are to prepare for the birthing of a new Civilization creating a righteous society still based on the old laws God gave to know the difference of good and evil and no longer will allow **"relativism"** and force nations to act in accordance to GOD'S LAWS now living together in peace for a thousand years prophesied.

If the warning is ignored by the atheistic establishment, then Bible judgments will be repeated like Sodom and Gomorrah to confront us with the fact that God, the ELOHIM, sets boundaries. When mankind ignores the warnings and deviate from his plan is like going over the edge of a cliff, which finalizes the Apocalypse now experiencing God's Wrath ending our civilization.

Once more a **big Jonah fish warning** is dressed in high technology to get the attention of a modern world. These many UREE inventions were given by divine revelation. They have become the sign that God means business. The missing key is now revealed about how the universe is fueled with free energy for the benefit of mankind according to God's purpose. Extracting energy is now redefined by forgotten entropy laws linked to gravity and anti-gravity forces that hold the universe together read my report how the universe is fueled and comes with a warning linked to UFOs in the UREE #16 B-Section and UREE # 17. **(Pearls #220 & #200)**

The Bible reveals how creation was fashioned from a 360^0 science perception. To understand the invisible dimension, we only need to understand with an elevated MIND. I compare the elevated MIND to a dark room filled with hidden treasures that needs the light turned "on" to see what is there. The oldest book in the Bible says, (Job 38:15)

...the wicked, their light is withheld.

Scripture tells us that if we deny the existence of the creator who fashioned us, he just turns the light off in our mind. Applying logic, why should anyone see his hidden treasures while rejecting revealed plain truth? Again, Job 40:1 -

Shall a faultfinder contend with the Almighty?

In the letter to the Romans, 2:21, we read:

> For though they knew God, they did not honor him as God or give thanks to him, but they became futile in their thinking, and their senseless minds were darkened. Claiming to be wise, they became fools.

Therefore, if the light is not on, that person will never really understand the revealed Babushka egg concept books because revealed concepts belong to an elevated MIND with the lights "ON". Only then will they see more.

Ultimate Renewable Energy Engine #12
Magnetic GRAVITY linked to Super Magnets

Recent developments in magnetism will greatly improve the UREE capability and open the door for more applications. Now we have 16 UREE applications that require updating. Look at the date below if you came back to read it once more.

I would like to propose additional options with new super magnets just to layout another possibility for perpetual energy conversion without consideration of cost, as the underlying theory must first be better understood. Cheaper magnets I am sure will work, too. I will not be surprised to hear from readers adding more UREE version remembering the Wright brothers followed with better and bigger versions.

When you read this web site, you will find unusual science evidence matching the Bible not found in universities or taught in Christian churches. It greatly motivated me as a retired scientist-inventor and took nine Babushka egg books of piled up data embedded in physics to prove the Bible, and visa versa, giving us a 360^0 round-about observation perspective.

This process developed into a new Donut atom theory based on understanding the nature of the Infinite Light Energy that fuels all Bio-LIFE. Why is there Good and Evil? From the beginning, prophetic history projected an end to the cosmos TIME DIMENSION. It is all linked to old-fashioned physics connected with forgotten metaphysical laws collected across 6000 years of human history.

Now a prevailing evolution theory violating the laws of physics is exclusively taught globally. It will keep students ignorant, prevented to think logically. Why has the Bible, the oldest book of mankind, been purposefully silenced in our present culture? The greatest collection of wisdom experience is no longer passed on to the next generation from the accumulated knowledge pool.

That consequence has changed our civilization, which can no longer delineate between **good or evil morphed into relativism**. Relativism is destroying the very fabric of our society by accelerating extinction as so prevalently seen in nature. No wonder our environment is stressed beyond possible recovery forced to accept a politically correct lifestyle numbed by an atheistic, illogical brain-dead mentality.

The laws of nature will unavoidably follow its course when the metaphysical laws are totally left out and replaced with really stupid fairy tales enforcing an atheistic evolution religion in most schools. The exclusion of biblical wisdom is causing worldwide political chaos down to the lowest levels.

If you are intellectually inclined and make time available, let's rediscover what is forgotten and forbidden. Perhaps you will understand why an apocalypse is the **only solution** God-ELOHIM has to prevent a total global collapse of mankind by ending our civilization. That raises my hope and expectations.

As I think about our future, I wonder how mankind will live when this generation is destroying the environment to support LIFE, causing now so much GMO extinction and using up all its natural resources. The greatest threat on this globe is caused by relativism. It comes not only from nuclear annihilation at the front door prophesied but in addition from genetic modification programmed to collapse the entire food chain causing massive starvation in the near future.

If you still believe that Newton's laws operate, then wake up from sleep! My grocery food bill has already doubled in three months (December 2011) and will get worse. Seeing how our population has grown exponentially in the last 100 years and how it is projected into the future, I wonder as a scientist how mankind can exist when all natural fossil energy resources have been depleted.

Focusing on rapid declining energy resources, it does not take a rocket scientist to demonstrate unlimited cheap electricity was all along embedded in your Bible. The laws of metaphysics described it on the second cosmos creation day (Genesis 1:6) pointing to our global ocean made of H2O molecules hydrogen combined with oxygen, the hottest gas that can be made to explode. If allowed to separate - now possible very cheap - and recombined, it will release **gigantic amounts of energy** that is stored in atoms. How big is the ocean and how deep?

If familiar with thermo-dynamics should be linked to the second entropy law of the 4th cosmos Bible creation day forming a firmament (Ausdehnung) which gives us additional **awful energy** embedded in a moving moon influenced by the SUN crystallizing a GRAVITY-ANTIGRAVITY force, light and heat energy. All energy types each are a little different will give us free energy if the "two" entropy laws are combined.

A number of choices are pointed out on this Website never postulated in universities totally new, which get mixed replies as many have a problem not educated in true physics believing in unscientific evolution religion.

Solar panels converting electricity from the sun can be useful to split the water molecules and should no longer be denied by the energy cartel, described in (UREE #3 and UREE #14). That process was discovered by a forgotten German scientist, Hoffman, and rejoining the gasses will gain **perpetual free energy in a ∞ energy loop** cycle if you understand physics now linked to an extended UREE energy concept. Trying to find what is the Cosmic ENERGY, the driver for modern living and must consequently be our priority in investigating our earth a little closer. We should no longer refute the Bible pointing to infinite light in the first verse of Genesis but use our intelligence rooted in our MIND the center of our existence, which is the greatest gift God gave us.

Please read that paragraph once more to open your MIND to the metaphysical reality like the Genie Spirit in the bottle expanding when released!

Alternative green energy from solar panels, gravity, high-tech magnets linked to simple technology like splitting-recombining water proposed in my UREE concepts should no longer be avoided by ignorant governments? Is it in conflict with corrupt cartels employing special interest lawyers paying off to get favors from senators eager to get reelected to the class of privilege?

But taxpayers now waking up should have choices. Free energy that can be extracted from GRAVITY, AIR and WATER should no longer be denied. It is posted free on the Internet for everybody's evaluation, or shame on me if I am wrong, but come with three witnesses; I have heard enough fairy tales.

I do not mind abuse if it will greatly benefit future mankind. One thing I am sure the energy cartels only interested in obscene profits will hate this Jonah preaching ruining their business and on top dare to announce God's Wrath to stop mankind from destroying his creation. If you are not religious, this pearl will attempt to widen understanding of nature governed by physics interwoven with forbidden metaphysics. Sir Isaac Newton described some of it hundreds of years ago, but unfortunately, 50% what he postulated is still suppressed in our universities.

Going back where it all started to follow a new energy trail as God's foreknowledge granted what mankind needed at the end of time when the population would skyrocket. The time has arrived for that question to be answered as we have run out of coal-nuclear-oil energy. What is the next revealed energy source for tomorrow? That will now be answered.

Energy Options

Technically, there are three considerations for any civilization. Each will represent different approaches of applying various energy types, but ultimately any design must be balanced by calculating the energy cost per KW - often forgotten in our time. Three simple energy generation choices are presented:

1. **Stationary in home and office.**

2. **Mobile for cars - streetcars.**

3. **Airplanes and high-speed 500m/hr. trains.**

Presently the cost per KW, regardless of application, is decided by politicized criminal power elite controlled by a prevailing relativism. Historically, money interests were concentrated in building projects like cathedrals, which took several centuries to build by taxing future generations.

Now the nuclear power elite have created lethal radiation poison that needs babysitting paid by taxing future generation for about 100,000 years. Otherwise it will release deadly radiation in accidents like Japan and Russia causing permanent extinction.

Mankind is only 6,000 years on this planet. Already 80% of life has become wiped out. Calculating the probability rate of extinction for the remaining 20% from a technological perspective is unbelievably shocking - now accelerating to 10 years max.

But Bible prophecy overrides a society gone totally evil like Sodom and Gomorrah. God has promised to stretch it another 1000 years with his Kingdom on Earth appearing after the Apocalypse, giving birth to the prophesied third civilization in 2018 AD, which will never be controlled by relativism corrupted by evil.

Watch the apocalypse ending our global atheistic culture and compare it with a new proposed energy conversion will be totally different after the Apocalypse 2018. In my judgment, mankind's future energy needs will come from **ocean, water, gravity, solar and magnetism** utilizing my new energy conversion approach only on this web site.

If my new concepts pointing to **Perpetual Energy** in form of electricity linked to a UREE hypothesis eventually being confirmed as a viable economical option, it will give future mankind unlimited energy. Other secondary energy extraction can be combined like solar panels, wind, tidal or water flow enhancing the UREE principle if applied.

How does the 12th UREE Work?

The latest Popular Science magazine, January 2012, has an article by Theodore Cray on super magnets. Fatal Attraction-Magnets don't have to be big to produce deadly force. They are made from

Neodymium-Iron-Boron

These very dangerous and incredibly strong [2x2x1 inch] magnets have a force of 520 pounds. Previously, we expanded on the UREE technology connected to many rollercoaster concepts improving a generator design by proposing the use of bigger copper wire looped embedded inside thin, printed boards allowing for lower inertia gaining higher speed. Only when magnetism is ∞ recycled at a higher speed will we get electricity on a higher level.

The magnetism in the wire loop could be 1000 times bigger with a super magnets' force of 520 pounds as described in magazines. We are on the right path in postulating what will work, as my inventor's instinct never disappointed me.

Now I hypothesize possibilities for much bigger electrical output. Just follow the UREE trail. The stronger magnet increases electrical energy possibilities. Now more UREE designs show up being added to the list. They are proving again the theory for the ultimate UREE system.

How to Apply Deadly Strong Magnets

The basic UREE invention is converting energy from an air-tank environment exchanging a pressure to a round cylinder piston-head and heated up the air expanding with a higher force on the return cycle is now either replaced or added.

The piston head made from ordinary metal is instead made from super strong **neodymium-iron-boron** magnets. That will add a horrendous energized magnetic force when slammed down and up pushed by a coil mounted on top and bottom of the cylinder and synchronized alternately switched with opposite polarity.

The magnetic piston attracted a supplementary force to a polarized magnetic field that will add energy to the flywheel. When the piston is bottomed out, the coil field polarity is reversed forcing the magnet-piston back expelled with tremendous power as the electrical field is now opposite pushing against the high air pressure tank source never changes like cuckoo clock weight aided by the flywheel to repeat the cycle.

An additional option for bigger UREE can be supplemented using an external coil required via a capacitor to produce the extra heat needed to fire a spark plug expanding the compressed air mixture for the return cycle. A ceramic sleeve cylinder wound on the outside with a cooper coil can be simultaneously used with the same piston stroke to produce extra magnetism expelled when super-magnet-piston moves across coil windings on the outside produces electricity redirected to the net. According to Faraday's laws applied to coils will produce a bigger pulse.

The piston magnet therefore has a double purpose. It creates immediate electricity for the spark plug by charging a capacitor while transferring extra kinetic energy to the flywheel for bigger electrical profit. The principle ultimately is linked to David's rubber slingshot theory expanded with a larger centrifugal force running faster than the speed of the earth around the sun linked to gravity-antigravity force.

For a practicable application, go back and read the section "Technical Recommendation" on page 95 of the 9th Babushka egg concept book. It mentioned an aluminum-oxide ceramic sleeve insulator for higher wear resistance inside the cylinder wall could be utilized to hold a copper coil on the outside, as the high-pressure motor never gets hot. That would allow magnetism radiating from the magnet-piston to pass through unhindered generating electricity. Both options will be increasing a bowling ball velocity force further as every extra bit of energy generated in the system is transferred to two flywheels mounted on the end of the crankshaft for temporary energy storage. It is immediately converted to electrical profit by two new generators.

To explain it once more, the magnet system could still further be amplified to a higher 10,000 Kwh energy level with a newly patented Cup-Printed-Board (Pearl #206 - UREE #16) Generator designed for jets and applied to high-speed levitated trains using gravity like airplanes as a levitating means. If you are still not convinced on free perpetual electricity, let's get an expanded education and review how the first magnet piston UREE works similar to a clock.

Perhaps skeptics not believing in perpetual motion could be persuaded by a third witness from visiting a clock shop. I saw a fancy clock design where the pendulum driver looks like an upside down boat anchor that becomes the motor. On each side of the curved tee was an invisible magnet going through a round coil creating magnetism to push the oscillating pendulum with an electric current synchronized to the pendulum weight. Again we have a heavy mass storing energy like a flywheel.

When a mass is moved, kinetic energy is expelled and converted into electricity and makes gears run in this clock to measure time perpetually. Notice, if we converted the clock principle to a UREE system, it would take very little energy to turn a number of low inertia wires passing over a magnet or vice versa. But to make it run perpetually, the pendulum needs a little more force added to the return kinetic energy cycle. Several ways are possible.

Everyone knows that a solar panel will give free energy. If the clock is connected to a solar panel, it gets the needed push for the return cycle to make a clock run perpetually like a UREE. That is not permitted by the energy cartel lying and saying NOT POSSIBLE. Fearing revenue loss, most refuse to check out the many energy pearls describing scores of possibilities, with chances made worse due to a Bible link not tolerated as a reference.

I originally got the idea to investigate the infinite energy from the Bible. Genesis 1:1 started a process of discovering a magnetic bridge in the Hoover Dam electric generator. I proposed that condensed free electrons levitated from surrounding atoms can be extracted to create electricity, but much more electricity can be gained in a new UREE invention by using a double rotating generator design. That design forces more electron slaves over the magnetic bridge to give us mega-free ELECTRICITY.

It all started with examining ancient bronze-gold clocks exhibited in many museums that were still a mystery for scientists. It was a challenge for me to crack their codes. I discovered that they measured an ancient earth axis wobble with an exponential time curve matching physics. This led to learning that ancient gravity was different from what is measured today, which surprised me greatly.

Following the energy path led to a new Donut Atom theory to explain why to my grandkid, which led to a better understanding of science. These new ideas should meet the requirements of future society. If you follow my philosophy never taught in any university, you could understand the true nature of electricity. Applying the laws of physics to how my cuckoo clock works, linked to gravity, pointed to the many different UREE designs: all matching physics. When a super magnet is moved inside copper wire coil mounted on the outside cylinder wall of a basic UREE or vise-versa, it will separately create electricity.

These principles could be expanded in unlimited design possibilities. When biased magnetism is added to the wire loops, it gets a bigger pulse once more added to the same loop to create a bigger bias. Increasing multiple-clustered magnetism will become mega bigger, even possible in fueling a 6-piston radial airplane engine or future jets as explained to my grandkid. The principles of an uncomplicated application are in Pearl #206-A. But for simple amplified energy in my next UREE application and to program my reader circle follow the energy trail one step at the time to understand the biggest invention of the century with (7) more UREE ideas (Pearls #201, 202, 203, 204, 205 and 206).

I still have not heard from the Energy Department, as Obama never answered my courtesy letter. Instead, the IRS has investigated me recently. I wondered why I was being harassed after being retired for a long time with my only income being a meager Social Security check, no longer working for a paycheck. What a coincidence? That could only upset their many oil sheiks and energy cartel friends as more UREE versions showed up creating free electrical energy extracted from GRAVITY, WATER and AIR beyond previous theorized now get 1000 times more energy.

All UREEs are linked to multiple choices depending on application. Just follow the energy trail to the next generator version many more are possible like different airplanes where build after the first Wright bothers contraption.

Ultimate Renewable Energy Engine #13 Advanced Electric-Generator

Connected to single solar panel will increase the output 1000X higher!

This retired inventor-scientist behaved like a bloodhound following a trail discovering a new energy source not deterred until he found the ultimate objective hiding in the hole. We exposed numerous UREE motors all give us free electricity extracted from nature one way another, as new model variations keep showing up even simpler in design and intent.

Exposing new concepts in our modern universities is a big problem. It makes me long winded. Forgive me, but it needs a foundation to re-teach physics linked to the metaphysics forbidden in school. Seeing nature from a 360⁰ perspective needs a longer introduction.

The latest UREE motor is much cheaper if you followed the path of my new discovery. I am not educated in everything and miss a lot being more of a philosopher than an expert in automobile technology.

As a multidimensional thinker, I look at nature and include metaphysical science for roundabout 360⁰ perspective. At first I thought about how a million times more energy exists below the ocean and wondered could it be extracted and utilized. The idea came as putting a Hoover Dam generating plant in a submarine on the ocean floor. There behind the dam infinite pressure is available to turn a turbine perpetually to extract a thousand times more energy cheaply. I imagined various piston motor designs linked to an electric generator gained additional understanding about energy to find extra applications soon thereafter extracting free energy from water even air to make it portable.

The path led to investigating **GRAVITY** as a fuel, which advanced to splitting **WATER** a process invented by a German named Hoffman, now giving us dirt-cheap unlimited free energy using solar power a thousand times amplified. I hope you can understand what the head-liner implies using **SUNLIGHT** converted to electricity will release the ∞ ocean water ENERGY Hoffman postulated.

Ancient cultures divided nature in four (4) building blocks found everywhere gravity-water-air-sun now linked to four building blocks illustrated in DNA (A,C,G,T) fueling all LIFE with an invisible energy force. Modern technology found the key not only to unlock genetics but the greatest energy source for mankind. The Internet describes 16 UREE invention concepts, which can now be converted to **ELECTRICITY**.

It is a newcomer to our civilization laying there unnoticed since 2288 BC due to unscientific teaching in universities by those who rejected the Bible and/or motivated by greed. They prevented it to make it a free blessing to mankind. It is sad that our modern education ignores the metaphysics, which has consequence limiting nature enforcing only tunnel vision to new discoveries.

As an example to see how deep science has slipped to the lowest level worth to mention if you noticed a 2288 BC date you never heard about, but everybody knows about Noah and his boat story now forbidden to be mentioned in our schools.

Being educated on the metaphysics can compared a Nova program on TV (1/2012) to see the real reason behind the curtain related to what happened on the Antarctic continent. A climate changing very fast linked to the meltdown of the ice.

The cause was linked to carbon dioxide, which is a big controversy and comes with a lie to motivate you to pay more taxes increasing the benefit for corrupted officials. The program hypothetically projected another apocalypse to scare uneducated people pointing to some phony unscientific consequences.

Nature programs on TV should inform the public about these new science discoveries kept quite because of a few Bible verses related to physics. That method is not allowed; therefore, new concepts are postulated on a free web site, which cannot be silenced. Check it out what Babushka egg concept are revealing became nine books and start with #3 and especially #5 if you want to know the real reason for global warming with a lot more forbidden scientific facts if you still able to think logical.

Scientists educated in an unscientific evolution religion become puzzled when investigating core samples from deep under the ice. They noticed many short layers representing narrow time cycles and concluded from the life forms that the earth in ancient times must have experienced a tropical climate that abruptly changed with continents depicted as ripping apart without explanation "billions" years ago. Perhaps forgotten earth axis wobbles postulated in my Babushka books will give a better answer from a true science perspective.

Why are scientists not better informed about what is exhibited in many museums collecting dust? They should investigate ancient bronze-gold clocks revealing history linked to Julius Caesar correcting a calendar in 54 BC by calculating (without computers) a declining earth-axis caused by an asteroid strike dated from 2287 BC. That could explain and date core samples in an unbiased manner not manipulated by the illogical speculation taught in most universities.

Watching on TV, it becomes obvious that a hi-tech society no longer cares. We have become corrupt to the core destroying our environment toward total life extinction and accelerated by genetic modification of our food. Many scientists have forgotten natural laws destroying our environment and no longer applying old-fashioned Bible wisdom.

The consequence can be noticed the many colleges charge our youth $100,000-200,000 for a worthless sheepskin from education based on fairy tale science. Many students will start life being forced deep into debt with mediocre expensive degrees of questionable value. There are no jobs available in a foreseeable future and not trained in practicable skills will flip hamburgers, unless their families belong to the privileged government employed class.

As an example, I recently sent a courtesy letter to President Obama apprising him of my new energy discoveries, but I have yet to get an answer. NASA spends billions useless in outer space similar to CERN smashing atoms for 40 years only benefited a military complex planning the next war. I hoped against the odds they would investigate the energy trail and apply it to a low-tech UREE linked to new energy technology **benefiting everybody** could create worldwide peace no longer fighting over diminished oil reserves.

Familiar and safe, green energy technology can give us globally free electricity, but it will not get the research money needed. I am sure my inventions will work, or prove me wrong. Why the big silence? That only confirms my metaphysics persuasion having no answer from science. I risk big time my reputation as a previous inventor that everybody will laugh at me on the Internet, a position like the Wright brothers experienced. Why is there so little interest what could generate a million jobs for America helping the President to fulfill what he promised to the American people? Prove me wrong would be appreciated.

Applying business investment principles to new inventions on a broader base could make America prosperous once more again down to the lower level most people live. However, in spite of local politics, we will see globally a change appearing on the Internet postulating a new energy hypothesis how to get clean free electricity forever from **GRAVITY, AIR, WATER and the SUN**, the four building blocks of nature which can no longer be denied in an open forum unless proven wrong with true science if you understand it. Somebody will make the money.

However, true science can never be covered up proven in the Bible which has embedded cosmic laws to reveal in modern time a new energy source desperately needed running out of fossil fuel. The 16 UREE motor designs just appeared at the right time to show us how we can extract electricity by using low-tech applications not well understood by the atheistic establishment. It became a big stumbling stone to be silenced at any cost believing in an evolution monkey theory without facts not based on physics. That changed an education system no longer understand physics denying thermo-dynamic-entropy laws of nature Newton postulated replaced with evolution fairy tales. For an example bought recently a jacket and talked to a young college degreed salesman about my UREE idea and was so surprised that he never heard about thermo-dynamic-entropy laws.

Forgive my long winded overemphasizing a condition in our society postulated in free Babushka books reconnecting it with the metaphysics as there is no free lunch in technology and must therefore use logic no longer taught in schools. But the automobile manufacturer management is profit oriented and better educated despite government interference by choice I have no control over. Some automobile manufacturer looking for profit acknowledged my courtesy letter for evaluation Obama should have noticed.

Here is a recipe for success but first needs a little re-education what energy is and hopefully disregard an evolution religion globally enforced by a police force to indoctrinate our children becoming brainwashed in unscientific evolution fairy tales.

A New Printed Board Generator, How does it work?

New information is usually found in the latest Popular Science magazine, January 2012, has an article by Theodore Cray on super magnets.

> **Fatal Attraction-Magnets** don't have to be big to produce deadly force. They are made from

Neodymium-Iron-Boron

These magnets are very dangerous and incredibly strong. A [2x2x1 inch] magnet has a force of 520 pounds inconceivable packaged in such a small size. They would be perfect if used in my new UREE invention creating free electricity. Pay attention. It could make you famous and very rich. Because the UREE concept is so new, it needs repeating it over for many to explain its logic.

Babushka egg concept books postulate that MAGNETISM is created inside a Donut atom according to the three finger electrical rule applied to physics. When a current passes through a copper wire will produce 90^0 of the surrounding wire MAGNETISM. If we connect it to a solar panel which produce a certain current will magnetize the whole wire. Following the next step of the ∞ math symbol applied to energy if we add some energy from the first entropy to the second entropy we get more profit.

Applied to sun seeing energy burning hydrogen without oxygen still a mystery is considered a second entropy source but can be converted to 1000 times more electricity gained from solar panels by using some of the 16 UREE models linked to the first entropy. All technology must conform to entropy laws scientist question that ∞ energy does not exist coming from somewhere ignoring what is burning inside the sun on the second entropy level? Proving it by getting smaller in size and colder measured by science; whereas a magnetic force coming out from atoms in a super magnet does not change the weight which I consider first entropy, just ignore size but look at the principle.

Let's coil the energized charged copper wire connected to the solar panel and make many loops around a printed board wheel flatten out in a low profile. From the magnetic perspective looking at one loop the charged magnetic lines all will go through the center as the wire loop behaves like a Donut still connected to a solar panel creating magnetism in the wire. Many loops will force the magnetic field in the same direction toward the center creating magnetism concentrated around in a donut black hole.

We can visualize it in some galaxy photographs that are demonstrating the same process what happen on a miniaturized Donut atom level explained in a new Babushka egg concept theory. Therefore each single copper wire loop becomes a dense highly centered concentrated magnetic field generated from current extracted from the sun via a single solar panel.

Now electricity is generated when we pass a magnet through a wire loop we get a pulse on the end of the wire. Going back and forth with the magnet creates a frequency in a sinus waveform. Or reverse the magnet is stationary and the copper wire loops move. For the next step extracting electricity we now use a **stronger super magnet** that will get us a stronger pulse and when speed is increased gets a bigger pulse.

That will, if you understand physics, change the solar magnetic flux base concentrated in a wire loop Donut hole, now amplified by the much stronger super magnet field. It works because the plus North polarity loop produced from free solar energy blends in with the North polarity of the 1000 times bigger super magnet being recharged to a bigger pulse.

Or reverse looking from the other side the negative South polarity produces a 1000 times stronger magnetic force solidify the super-magnetism together which the transformed lower solar energy to a stronger bigger sinus wave. The extra electricity is generated from fueled magnetized iron atoms spaced very close, which is a much superior first entropy force. The super magnets are physical mounted on either side of the printed boards embedded with many copper loops on both sides of the board. But each loop must be ∞ linked with one loop folded over around the axis center and reversed on the other side passing together in parallel a 520 lb. magnetic force if you used Neodymium-Iron-Boron magnets.

The magnetic force is now going in the same direction right through a matching double wire looped hole. Remember how a motor works: the energy must be passed on ∞ from the stator to the rotor back to the stator. If on one side the North Pole magnet is facing the wheel board, it will transfer energy through the other side of a reversed matching loop facing the oriented South Pole magnet side to create an unbelievable, double strong (1,040 lb.) magnetic force.

However, the board loop designs can be enhanced with 10 times bigger output capability by creating a pie-shape 5 or 6 coil pattern connected ∞ reversed on the other side of the board matching the same design generating extra electricity.

I believe that will amplify the same force into 50 or 60 Hertz bunched together and could be unbelievable raised to 10,400 lb. magnetic power force matching the frequency used in our electrical net work synchronizing every electrical gadget. That means inferior magnets would work, too. After coming up with this idea and writing about it here, I learned that another scientist has demonstrated this same approach. Watch it work at:

http://www.youtube.com/watch?NR=1&v=PoJALsTaCAo&feature=endscreen.

The magnets must be spaced and physically restrained with a slot in-between just enough for a rotating wire loop board to pass through. If the board wheel is moved it will create a horrendous pulse in each loop. Adding many loops on stacked up printed board wheels and spaced in between super magnets aligned N-S-N-S-N-S with as many you can put on one rotating axis is creating big time electricity.

The extra number of printed boards and magnet layers will amplify the solar electrical current to **over 1000 times higher solar panels output** levels proportionally to the RPM of the UREE driver motor energized by perpetual motion. Modifying a 1/3 Hp DC motor connected to one 230 Watt solar panel demonstrates proof that a standard commercial motor can increase the rpm 10x at half the voltage by adding a few super magnets on the outside. If linked to low inertia wire-boards, a super magnet generator will validate my claim. (Stargate kits at Quantamagnetics.com)

http://www.youtube.com/watch?v=5Xv-req4U8U

The New Energy Technology Summed Up

The key for a new energy technology, not taught in universities, is that electricity is converted by perpetual motion. When applied to magnetism as demonstrated in an electric motor, it will produce a new source of energy if reversed becomes a generator.

Analyzing the principle in the new UREE double generator and using a super magnet will get big time electricity. But if a solar energy is added to the same system running perpetually therefore gets obscene 1000 times more profit if you understand the new ∞ energy exchange entropy laws. These new Babushka egg concept ideas could destroy the Holy Grail of the atheistic unscientific evolution religion by obsoleting the cherished Dr. Albert Einstein theory that ignored infinite light fueling the sun now explained with a better Bible entropy story.

However, if you want to make some money using gratis energy not taxed yet, though forbidden by the government, follow the evidence trail of free Babushka concept books. You will become better educated and in the process also become very wealthy. It is your choice. You can have your cake and eat it, too, based on forgotten entropy laws.

More importantly, you should ask why the ELOHIM-God would send a big UREE fish at this time to a world running out of fossil fuel. Maybe he wants us to know how the next prophesied civilization the Kingdom of God will be fueled as described in the first Babushka book written eight years ago. The key for every UREE is perpetual motion, not understood by science, which can be used to extract energy from **Gravity, Air, Water and the Sun** in combination to suit day and night cycles.

A flywheel is a wonderful contraption storing temporal energy and if linked to two generator-streetcar motor systems will give us sufficient electricity useful for driving a car without gasoline or fuel your house and all the factories in the future 1000 years prophesied. The Printed Board Generator in only one leg of the whole UREE system which depends on design and size like a Hoover Dam is generating electricity but is useless for a bicycle and needs a mini portable UREE.

The sun is everywhere an immense source of energy and if connected in a UREE can increase a solar panel output to 1000 times over and cheaper but still needs some extra energy reinvested or returned for perpetual motion like it is demonstrated in the math ∞ symbol.

Without the flywheel it will not work either in an electric motor or piston motor design because the DOT connecting two energy loops demonstrated a Time dimension, which is a **paradox**. We are mortal and "Time" is all we got to demonstrate a LIFE force, but if time becomes "zero" only than we can acquire profit benefit extracted from divine ∞ energy. Math defined it as **(Zero time = ∞E)**.

Babushka egg concept books teach us both levels of entropy. One loop is physics; the other the metaphysic level. The crossover is the Zero=DOT and becomes the transition (death or when time is no more). The shared ∞ infinite loop being resurrected to a higher eternal life energy level again expressed on one side mortality the other eternal, or like a caterpillar the other butterfly all must have the crossover zero point TIME DOT to allow profit to appear. That is the nature of the system and for its very existence must conform to metaphysical principle perpetually exchanging infinite energy flowing to every single existing atom to make up a universe to maintain its existence fueled with LIFE.

Have you ever wondered why the lights are on in every galaxy? It starts with a rainbow seen in heaven (Rev. 4:3) crossing over the ELOHIM Throne where the ∞ energy intelligence computer is directing a higher cosmos [Alpha(+ONE)force] level.

Following the rainbow energy spectral lines below each leg foundation begins an energy path recorded in the very first verse of Genesis, one side invisible ∞ LIGHT appearing and the other WATER, which became visible on the second entropy energy loop as the first substance created, a vast ocean reservoir embedded with free energy to start and maintain LIFE.

Both two looped ∞ energy levels are following a path to give Gestalt and Substance transformed from a metaphysical side to make the physical side possible crystallizing to be visible for us seen in Fraunhofer spectral lines. Once crystallized as atoms can be linked to MAGNETISM extracted in the UREE inventions to remove some energy rooted in the rainbow spectral lines in heaven stored in form of electrical energy cascading inside every atom Hoffman discovered.

It is now linked to a Printed Board Electric Generator to balances the system like a dual rail getting 1000 times more energy from the sun splitting water molecules all free getting obscene energy profit forever. Otherwise must wait for God's Kingdom on Earth arriving after 17 September 2015 projected by science, the Hebrew Alphabet Number System (HANS) matching Bible prophecy check it out in Babushka egg concept books, it's great fun to discover more not found in universities.

Ultimate Renewable Energy Engine #14 Herbert's Solar Perpetual Automobile KIT

To drive a car perpetually without gasoline is everybody's dream, but logic dictates that it requires energy. Free solar seems a candidate as early experiments demonstrate that it could work, but the automobile industry proclaimed that not enough energy is available to be practical for a standard automobile. Somehow we all know not believing the energy cartel trumped-up story it could be possible, so we keep searching for some alternate energy combination like building hybrids to fool you only offering more expensive electrical cars instead the cost reduction desired by everyone which is now possible if the energy cartel and government would allow it.

At this point in history, solar panels are still severely restricted by the government-cartel. They are kept limited in size very pricey and need a UREE motor to increase solar power output and make it portable. An attempt is made by expensive electrical cars seen on German TV build a 120,000 dollar model mounted a solar panel on top, but the range is inadequate needing costly lithium batteries. Only the perpetual motion principle of a UREE can extend its travelling distance even at nighttime and could reduce the price tag to $10,000 Dollars driving without gasoline obsoleting lithium batteries. Therefore, a UREE combination hybrid motor later described will give us the advantage using solar panels too to extent the mileage range a **thousand times** and much cheaper.

It is really ridiculous showing us the most expensive car in the world with a solar panel on top to mislead the public that portrays the message we are working on cheaper energy. The roof of the German car converted to solar would generate about ½-Hp energy wise and under the hood a huge gasoline-electric motor combination needs 400-Hp which makes this deception obvious needs 800 times more solar power to drive this car. It does give a message buy our cheaper $65,000 hybrid version with the $18,000 lithium batteries.

I have real solar-car applications available just follow the trail of this UREE #14 Herbert's Solar-Water Energy Kit operates with totally free electricity generated using a streetcar motor and mounted on the periphery with a "single" flywheel linked to take advantage of stronger super magnets shared with an advanced printed board cup generator design. It could even fuel jumbo jets (UREE #16). But let's just use what is available on the shelf. One choice is to connect a single solar panel to a smaller standard streetcar motor attached on either shaft side mounted with two generators purchased from the spare part dealer that acts as a flywheel.

This is a much better idea using electrical energy converted from the sun via a solar panel just can drive perpetual with a motorized flywheel generator combination at high speed. You might be surprised that the sunlight is considered the second entropy level, but could gain more electricity by adding the first entropy recycling a magnetic field similar to a streetcar motor described in UREE #11. But in our solar car conversion we must have two generators and recycle a biased pulse from the left copper loop crossed over to right copper loop which creates a bias adding a little extra energy to the next wire loop continuously crossing over getting bigger until it reaches the critical **resonance frequency** according to electrical principle.

Ordinary solar panels converting electricity are the cheapest energy source proven if you read the conclusion of UREE #11, but their use is suppressed by the energy cartel. Criminal nuclear-oil cartels cover up their polluting energy cost per KW by concealing the price tag of expenses incurred causing global warming, poisoning the environment, and not telling the public the horrendous cost of the highly deteriorated deadly nuclear maintenance. It is never added for each KW produced and hidden in the US trillion-dollar debt.

I recently found an article on the Internet that really shocked me as somebody sells a kit to make your own solar panel for $97 each. That is bad news for the greedy solar businesses or getting government grants.

ELECTRICITY BREAK THROUGH published by Moneynews.com/ www.Homemadeenergy.org.

That, my friend, is cheap and really surprised me comparing it with the latest NEWS, which made it obvious that our American politicians will do anything to discourage green energy. Green energy is the lowest funded appropriation despite the billion bonuses given to friends of the government and corrupt bankers to discourage the little entrepreneur. Remember the new startup solar energy company in California visited by President Obama making big headlines on TV and filed bankruptcy three months later disappearing with half billion dollars of taxpayers' money (December 2011). That bankruptcy was followed by a congressional investigation now vanished too, like vapor forgotten buried under our national sinkhole of debt. Both are rotten to the bone.

To put it in perspective, that evaporated money before it run into quicksand and fruitless investigated by a special senatorial inquiry could have bought solar panels from China to fuel the same town of 75,000 people with free electricity forever. It is just repeating another political football facilitating the energy cartels and the Arab Oil Sheiks, making them rich with returned favors of under-the-table side benefits to get politicians reelected. They live at the billion-dollar level and enjoy being part of the class of privilege.

But the American way being very resilient could build your own Mini Power Plant on the roof saving 50% -70% - a lot cheaper. Once solar panels are installed not only gets energy free but some can even get money back from the energy provider due to state laws. Enough information is now available. The green energy avalanche has started and sliding unstoppable. Hopefully, it will bury the greedy energy providers that grossly pollute the environment.

In the meantime, that all would walk away if we could extract solar energy on top of your roof linked to a UREE in the garage for extra benefit totally free. That together would reduce the number of panels needed and make it a lot cheaper. Even without a UREE, solar panels now steady being lower priced guaranteed by China is at this time still the cheapest and cleanest electrical energy for home and office.

The Next Level:
Herbert's Solar Powered Kit

Again previously mentioned in a NEWS flash a automobile show in Germany 2012 showing some solar panel electric automobiles on display. They were very expensive - in the $80,000 to $120,000 range, only built for those who do not pay taxes. My question is, "Why do we keep building the same 100-year old automobile design after horses disappeared?"

Automakers have pushed to engineering extremes using expensive batteries to power vehicles as an alternate energy option meant for longer range. When electricity was discovered, it was the main fuel from the 1800 to 1920. Thereafter the oil cartel took over and the electrical transportation disappeared and with it the electrical automobile.

To make sure that people were forced to change, the streetcar tracks were irreversible ripped out to replace them with stinking buses and trucks plugging up our highways. If we had kept the technology alive our air and water would be clean and our mode of transportation a lot cheaper. Even trains would still use steam fueled by cheaper unlimited hydrogen energy extracted from the ocean the highest energy level cleaner and not polluting the environment, an infinite abundant choice.

Why has the energy cartel forbidden a German invention from Hofmann splitting water using hydrogen as a fuel? That fuel is outlawed by the government and only allowed on rockets when sputnik was in space now cheap enough to go to the moon. We have lost 60 years going in the wrong direction and making the criminal oil cartel rich paying off corrupted politicians mostly lying only motivated to belong to the class of privilege with millions stashed away in off-shore bank accounts. It gets worse with the automobile industry getting big bailout tax money for the upstairs executives. They are just as corrupt for mutual benefits collectively rip-off the customer taxpayer never improved the millage since my beat-up VW 40 years ago.

Check it out and resurrect that old technology just read my 9th Babushka egg concept book, free on the Internet. I proposed some unusual forgotten ideas for the automobile industry not interested to implement green energy fearing revenue loss.

It makes more profit selling horrendously expensive poisoned lithium batteries we do not yet know how to even dispose reactive substances damaging the environment. I am hopeful it will advance a new thinking process that redefines what energy is not interested to implement free green energy asking instead for repeated billion dollar bailouts.

To make my case, I walked around a dealer showroom in 2012 looking at a $65,000 hybrid electric-gasoline car made in Germany that promised extended mileage. The salesman explained that the extended mileage is achieved by using $18,000 lithium batteries together with gasoline power switching back and forth.

Looking under the hood and seeing only a tiny generator I asked the salesperson why was this expensive car not designed with big extra generators to extend the mileage? Why spend $18,000 dollars for such a large bank of lithium batteries? My little grandkid knows that one apple added to another apple makes two apples, which is more than one apple, but the automobile manufacturer does not seem to know it. I sent a courtesy notice to 25 automobile manufacturers, 55 global embassies and 265 Universities none replied except one.

A German lawyer returned my letter unopened with a response that they have their own ideas and cannot accept unsolicited information. In other words, they were afraid to use my free ideas, or they do not want to sell cheaper cars. They think ripping off the customer makes more profit. Again, they look at my finger and not what I am pointing to. They could make obscene profits with my free information on the Internet.

Adding my five cents worth of wisdom, notice usually a crankshaft sticks out both ends of the piston block. Why not attach a low inertia generator on each side will not eat up but produce more energy? That makes the two apples my little grandkid figured out, which would eliminate some costly lithium batteries exchanged for a cheaper and improved generator similar to streetcars invented 100 years ago giving more energy. If inverted to a generator could change the whole automobile industry obsoleting dirty fossil fuel. There is absolute no excuse as new generator designs are available on the Internet claiming 50% higher output and received a gold metal. Why are they forbidden by the establishment management already getting billion dollar bonuses from the taxpayers no longer need to make profit from cars?

Why not use generators are safer rand cheaper than horrendously expensive batteries made from dangerously reactive lithium. The lithium extraction cost is really higher because they hide the damage of grossly polluting the environment, mostly in China. Please check out my different UREE model that gives more benefit. Perhaps automobile designers are either prevented from using alternative options or ignorant of electricity only educated by a faulty atom theory screwed up by unscientific fairy tales taught in every college book.

When you smash atoms to smithereens like at CERN, scraping fractured atom residue off the Nebelkammer stuck like manure on a dairy barn, it ends with useless junk theories. Atomic broken particles have lost the built-in intelligence. It is missing, no longer available. This can only result in useless opinionated speculation, which cannot understand or access the ∞ **energy source** fueling the universe as exposed in the Bible or seen in galaxies with the Hubble telescope.

That caused me to reexamine nature. It surprised me greatly to conclude that we can drive around perpetually due to a new concept of very cheaply extracting electrical energy from magnetized gravity, water and air. To gain inexpensive energy is a dream for me. Because I pay taxes, I have no money left over to test drive the latest expensive electrical cars seen on TV, yet I need to prove or disprove my new alternate energy theory.

It all arrived together and needed 9 Babushka egg concept books to explain physics linked to the metaphysics to expose the meaning of the infinite math sign ∞. Doing so revealed a model of how energy is looped, moving from one level to the other continually in an infinite energy circle from higher to lower entropy. It takes some time to sample what is on the table explained with the infinite math sign

$$\infty E = m \left(+\infty C \, / {-}\infty C \right)^2$$ - no kidding.

Now I have designed a simple UREE energy kit that can be mounted in the trunk of a standard cheaper electric hybrid automobile. Once installed, it will demonstrate its ability to tap into electrical power at the atomic level. It will pull electrons from wherever they go when they disappear to physicists to reappear, pulled by design to supply a virtual endless current. That means, once started, a car will never need refueling until the police stop me because I have no license to use free energy. Every country needs more energy, but they will not read the Bible with a scientific perspective, which would tell us where the free energy is stockpiled. It has been available for thousands of years.

That new energy discovery will be proven in my first attempt to drive perpetually. I call it, **Herbert's Perpetual Automobile**. With it I will repeat the history of the Wright Brothers who built their first airplane. Everybody shouted and laughed, "It's NOT POSSIBLE!" Of course, it took some time to understand the underlying physics, but later we built their contraption by the millions once their future had arrived, because we comprehended how the concept really worked.

Check once more my new Donut Atom theory model compared to the UREE patent application on page 15 (book #9) how the mechanized generator-motor combination is arranged. That can fuel a hybrid automobile with the possibility to drive perpetually by combining magnetism within atoms moved by kinetic energy extracted as outlined in the new Donut Atom Model. It works because of two thermodynamic energy entropy laws moving ∞ where time is zero; therefore, perpetual energy is generated for us infinitely.

When magnetism embedded in atoms is concentrated by super Neodymium-Iron-Boron magnets described in the magazine, it should extract more energy. That, by itself, will enhance the UREE like the blue color lines getting closer to the nuclear center of the hourglass galaxy, which raises the question, "Do we need both concepts in a UREE, or is just one enough? " With other words is a hundred year old generator sufficient or do we need a motor to accelerate wire loops in a generator to gain perpetual electricity. I do not know how old is the generator design used in present cars.

This UREE #14 model assumes that the end motors connected to side generators may not be needed and exchanged the compressed air tank driven motor now driven by a solar panel streetcar motor combination. This is another UREE option converting a cosmos energy via a solar panel amplifying free electricity by first running a small streetcar motor to generate electricity. It is really simple attach two standard electric generators bought from the spare parts department mounted on each streetcar motor axis side could be sufficient to drive perpetual around with a hybrid car. (See Stargate Motor p. 125.)

I haven not been able to check if the hybrid automobile generators from the auto store could generate energy using the crossover loop method, which makes the UREE combination work. If it does not give us enough power to drive the hybrid car, we cannot proceed, as only stronger magnets will generate sufficient electricity to drive a true electrical car.

The "old-fashioned" electrical generator design is the hinge point for more electrical profit or not. **It's worth a try.**

Phase 1
How does Herbert's Perpetual Car Work?

I am now too old to gather all the materials needed from dealers selling spare parts or to rummage through the junkyard. I have difficulties working in the shop like I used to, or I would have asked some high school kids to read this story. If interested and a little practically motivated, they could become famous like the Wright brothers, being the first persons to drive around with a perpetual motorcar.

Here is the recipe. I do not need the fame or money. All my Babushka egg concept books are free to pass on this knowledge forbidden in our universities for future generations. Modifying a car that could run perpetually in motion should get you the Nobel Prize. I likely do not qualify as it is only given to those who believe in evolution religion or given to corrupted politician even terrorist dressed in a nice suit with a tie. Bible science knowledge is forbidden, but it might be overlooked in a younger person.

To get started, find a cheap electric-gasoline hybrid car from the used car dealer lot at your expense, which we will modify together for perpetual motion. The goal is to drive around without gasoline in order to prove the theory and you get the fame on TV.

Notice we need to buy the following spare parts: two additional generators from the car dealer and one or three 1/3 hp electric Stargate motors. The first test is without an end-motor, and if not enough energy is produced, use the second option or add 2 more generators in series. The Stargate motor when piggybacked on the outside with super magnets gains double speed at half the voltage. Therefore, one solar panel can drive 2 Stargate motors. That could in turn drive 2 or 4 extra generators in series with the same the solar energy. If we played that game with the car-motor adding piggyback super magnets, it might drive at half the voltage eliminating some extra generators?

Some other parts need to be machined to make the UREE motor platform, which will be connected to the purchased car. This first version will look bulky like the first contraption of the Wright brothers. We all will laugh about it later, but the mechanically inclined can have fun. It's your baby.

1. UREE Base

The UREE platform needs a base: perhaps ½" thick 2x4 ft. aluminum plate or ¾" thick plywood would do it.

2. Flywheel Motor-Generator

A) Look first at the UREE motor patent picture in the 9th Babushka egg concept book on page 17 or on the Internet to have an idea firmly established in your mind. We must find out the drive-car's electrical motor specifications to figure if 2-4 standard truck generators give sufficient voltage-Amperes when linked to one or two solar panels mounted on the roof of our car. I recommended a Stargate motor design from the catalog requiring lower power. That motor only needs to turn the flywheel with the weight of two generators at 3600 rpm.

Our UREE electrical kit gains electrical energy from the generator-motor combination on a higher magnetic atom level. It gains electrical energy when wire loops more quickly pass a magnetic field. The field will intensify being elevated by adding energy from the first entropy level embedded in magnets' atoms. The solar panel only converts electricity from the sun on a second entropy level, but if both entropy levels are ∞ joined together coming from magnets via an energy storage flywheel using the combination of two electric generators will give us extra power to perpetually drive a hybrid car.

B) To attach a generator on each side of the flywheel-motor shaft, we need a 1 inch steel plate the size of the generator's outside diameter, which functions as a platform to be mounted on each side linked to the other car generator to create additional electricity. These are arranged in parallel. Since both have substantial weight, they function as one solid flywheel. We need some taped holes (6-32x12) on the outside diameter to balance the flywheel motor assembly added later.

C) Each of the two flywheel plates attached to a small streetcar motor shaft require two insulated copper rings glued on their outside diameters to function as slip rings with carbon brushes to extract electricity with two wires from the generator while the motor is running. The copper rings are connected to the generator stator windings with spring loaded carbon brushes on either side of the flywheel base. But what is important the "output" of one generator wire must be connected to one "input" wire of the crossover generator to link internally like a ∞ math sign shows. That will amplify a pulse in each set of copper loops and increase the base current by adding a pulse to the other set of copper loops. Read (Pearl #206) for a jet engine to understand the theory better.

An afterthought it could be possible that the end motors are not needed for an automobile application. Test first "without" the end-motors as now the conventional generator is reversed as the rotor becomes a stator "fixed" bolted to the base. But electrical connect a wire from the now inside converted stator across the other side wire brush on the flywheel now linked to the rotating generating winding side. We just reversed the generator function like a double math symbol ∞ and wired the right fixed stator windings in series across the left rotating armature wires and in parallel the same with the other left stator linked across to the other right rotating wire armature like two independent ∞ loops.

Once more backwards the rotating right winding armature is connected via the flywheel brush one wire crossing over to the left stationary armature winding coming out back across to the other end now linked to the right second brush rotating windings. Do the same with the left side-rotating armature winding crossing over to right fixed armature winding like two independent loops. But watch the polarity while crossing over as the armature winding direction must go opposite the spin.

Notice first test the generator without the streetcar motors attached, as rotors must be fixed to the base. It becomes in inverse generator as the outer stator armature bolted to the flywheel plate is now turning from the streetcar motor energized from the solar panel. Test both ways and find out which gives more free electricity. Again the principle of following the magnetic trail one generator creates electricity is put across the other armature windings increasing the magnetic bias gaining more electricity recycled gets bigger all along turning from the center motor linked later to a solar panel.

3. Testing with two Smaller Electric Motors

A) Now the second test and check the side electrical motors linked together with a flex coupling to the generator shaft. Let it run with an external variable power source. We must first compare the data. The output of the three generators must be sufficient to drive the car without the gasoline engine engaged to be powered as a purely electrical car. Extra batteries are not needed: the UREE System will create the additional electrical energy for perpetual motion. That is our aim.

B) I do not know and is very important to find out if the three built-in generators from the car manufacturer combined together and spun at high speed will produce enough electricity to drive a conventional hybrid car at designated high speed. The solar power only provides perpetual motion to start or reset the system cycle as the main power must come from magnetism produced either internally or from fixed super magnets running with either two or three generators. We will need to test each generator at highest maximum speed it will operate added together and compare the output with the manufacture's requirements for driving that car without the gasoline motor engaged.

C) Pay attention. Only if we find the right combination of a suitable generator we are ready to connect two separate electric motors on each side with a flex coupling to the generator shaft. If we do not find a suitable generator in a catalogue perhaps a bigger truck generator with a double Stargate system or must wait for a redesigned generator to include super magnets mentions in UREE #11 and #12. There is no free lunch in technology. Look at the UREE motor picture in the 9th Babushka book page 17 once more.

Acceleration of generator-motor has limits. Some design fundamentals need to be considered for the new UREE. It must be on the lowest power consumption level because it will speed up and give out more being maxed out at the point when the generator has the highest possible speed. The difference where the power-generating curve flattens out is profit. There must be enough Kilowatts leftover for the weight and speed to enable the car to drive perpetually. Only when physics is balanced do we gain energy linked to good engineering practice applying logic.

Check that the units are aligned and bolted down to the base. Connect the external wires from the motor crossing over to the generator for each section to get added energy. But watch the polarity, as the flywheel motor speed must be opposite to the outside motors. Mechanically check Herbert's UREE kit once more, It is mounted in the middle on the baseboard with or without a small streetcar motor attached on either side of the extending motor shaft a round one inch thick steel flywheel plate.

On each steel face we screw on the generator stator turned by center main streetcar motor running at 3600 rpm and needs to be balanced a system. The generator shaft is connected on either side to the rotating streetcar shaft with a flex coupling. Attach the wires and test.

4. Testing the UREE Concept

If we assemble the right generator-motor on a UREE platform, it can be tested and later put in the trunk and connected to the car's electrical system. The UREE kit is an add-on to the car's standard electrical system as the extra electricity is needed to drive perpetually with the same car motor running during the daytime using the solar panel to amplify electricity.

The attached gasoline motor is still being used but with a different fuel. I explain next how it permits driving perpetually at nighttime, too. But first we need to test if our UREE works to give extra electricity in the daytime. Now test the UREE concept if it produces the extra electrical energy needed for perpetual motion from the solar panel.

We test first with the platform UREE system assembled on the floor. Now we connect the center flywheel motor to a variable controlled electrical power outlet as the solar panel will provide the same current when later attached.

We start first by very slow raising the voltage - eventually to a maximum speed of 3600 rpm. Check for vibration, and let it run until we feel good about listening to your inner sense. We probably need to attach some lead weights to the outside of the flywheel plate. They can be easily cut and screwed on the outside of the flywheel plate using the predrilled 12 taped holes 6-32 x 1/2" to reduce vibration. That will take some time in trial and error. Maybe it can be connected to a tire balancer.

We should measure the voltage at various speeds and compare it with the output voltage of both generator slip rings brushes and keep records. If you used the same electric motor specified of your purchased automobile, compare the present output data with what the electric-motor input would normally be with the generator to drive the car at normal speed.

Play around for a safe flywheel speed versus output of the generator. That will decide which way to go for higher electricity output possibilities for maximum speed of the flywheel motor. I cannot second-guess the outcome.

Phase 2 - Connecting the UREE Generator to the Factory Automobile Electrical System

If the high-speed flywheel test is OK and generating sufficient electricity, we can proceed to connect the automobile electrical system together with the UREE platform mounted now in the trunk, or rear section of the car. We need to see if it will work together with the same generator-motor parts used by the manufacturer.

In Babushka book #9, the picture of the UREE motor on page 17 is connected with the opposite generator motor. We need to duplicate the two wire pair's connection by wiring our two generators in parallel.

A) Connect two wires from the car-electric generator to one of the UREE generator output brushes. The output of the car-generator is now wired in parallel and operates like two matching generators with a pair of wires going to the UREE-generator stator slip rings.

B) The car should be started in park with the motor clutch disengaged to test the electrical function of the car-motor-generator crossing over to the UREE generating more electricity. Electricity is now added to the car-electrical-motor. As the system gains more power an infinite loop producing surplus energy is set up and routed to the battery for storage. The automobile is now running with a moving flywheel still linked to the UREE external motor power source generating the extra UREE electricity produced, which is internally charging the existing car battery system. The car batteries should get warmer because they are not needed to power the vehicle. Later we will only need one battery for storage saving money.

C) We should start out slow, to crank the flywheel a little with a variable external battery source and test some output on the test generator compared with previous readings. Watch the flywheel speed accelerate as the voltage is added from the UREE flywheel-motor. The voltage is redirected to the car-electric-motor as it gains a little more voltage, if connected.

The internal car-generator-motor system is now connected to an external UREE electrical source. The extra voltage produced by a UREE flywheel motor system going faster since it is recycled in the car system. This should increase the voltage to the battery. Be alert. Keep ready to stop everything with an emergency switch if the flywheel should go too fast. Testing will give confidence. Try it with caution. Expect surprises. If it blows a car fuse, we need a higher fuse as we have now more energy.

If the flywheel gains maximum velocity and runs perpetually, we are ready for the next phase. Otherwise, it is back to the drafting board like the Wright brothers needed to test over and over. That is the fun; do not miss it. Suitable electrical generators are the very heart of the UREE system. They may be difficult to find because industry never improved the design since horses moved people. A new Donut-atom theory describing magnetism forming atoms could open the door for understanding magnetism better as applied to high technology.

That would gain higher output possibilities now postulated I am sure. The external UREE flywheel source connected from a battery, or 110 volt controlled regulator, needs to be replaced with a suitable solar panel. The flywheel, like my cuckoo clock, needs very little energy to keep it going.

Phase 3 - Solar Energy for Daytime Driving

My plans for free, green-clean energy are still suppressed by the establishment. The matter will not be decided by the experts, but someone will start the process. My grandkids will prove free energy by driving around without gasoline. His friend will copy him, and before you know it, a big crack in the Hoover Dam will have been announced making it obsolete and creating confusion in the Government establishment.

Perpetual motion is achieved by an electric motor run by solar energy. Now connected to generators gets free amplified electricity.

Remember to first check the output of solar panel specifications to determine how much is required to start a smaller suitable UREE electric streetcar motor. Bear in mind, less solar electricity is needed once the flywheel has accelerated to the maximum speed with a little help from the starter motor left on a little longer.

Once more all along electricity is fed to the flywheel streetcar motor turning faster from a totally free solar energy extracted to effect perpetual motion converting more electricity. That is what surprised me greatly, if my logic is correct. It works like my cuckoo clock once the pendulum is kicked it needs very little energy to keep it going as long the infinite higher chain energy is available. Once the solar energy panel is checked together with the UREE system, we are ready to test drive how long will the car run when parked in the sunshine connected with a solar panel on top. If nothing happens after a few hours, drive around a parking lot and smile a lot.

Lucky for my grandkids that energy from the sun is free. They will prove that fueling the many electronic gadgets in the house is cheap making their dad happy. Why? The solar do it yourself panel costs a mere $97 now priced very reasonable from China. Read this story again from the beginning. It raises the question, "Can solar electricity in addition be economically converted by splitting water to extract the hottest hydrogen fuel according to a German scientist Hoffman – therefore, also free?"

My grandkid asked me why it is not used and why is it forbidden? All we hear from the energy establishment on TV is that an idea like this is "NOT POSSIBLE" as published on the Internet and copied in (Pearl #218) with my comments.

But a teenager who builds an automobile kit could prove it. If you are ignorant of science could become educated, go to page 95 UREE #3 of Babushka Egg #9. If you are not demented by lies, you could right now use the hydrogen fuel in your present automobile every day totally free even at night.

Phase 4 - A Final Joyride Test

The fly-wheel motor can now be connected to a solar panel via the car generator charging the battery for a final test driving your modified car in a real world environment and have a friend follow you with another car and listen to the noise and slowly accelerate to see if nothing is different while driving around, eventually to the highway.

Test it, test it, and once more test it, to make sure you can fly and look out for the police going faster as everybody will not believe it and have another big problem to overcome. How to convince the expert in universities linked to the government controlled by the energy cartels, as all will shout - NOT POSSIBLE. Or sell Herbert's UREE Kit to make more money to the windmill manufacturer. He can add it to the system to get more energy producing obscene profits.

It has big consequence obsolescing somebody's oil business, obsolescing somebody's nuclear business and even the energy transmission provider charging me double the electrical energy bill will be out of business, too. I hope the government will get more taxes with new electrical technologies appearing. That should stop the now emerging, most destructive technology of fracturing the underground soil rock ecology to extract gas.

Fracking creates deadly chemical by-products that permanently destroy productive farmland with lethal radiation and poisoned groundwater. This problem is now being exposed by some in the media, but the pubic doesn't even know that these chemicals are simply dumped into their drinking water, as reported for New York.

The bureaucrats make the excuse that no laws exist to stop that evil. They are allowing the race to obscene profits to destroy the environment for our children. My prayer is for the Lord to speed up the apocalypse to save mankind so that we can continue on this planet the only place in the universe where life exists. Why let Satan rejoice when we demolish our surroundings causing massive extinction by crooked corporations and corrupt governments helping Satan to devastate this earth.

By adapting the many green energy possibilities of GRAVITY, AIR and WATER, a different age could arrive, yes, but civilization will go in another direction different from what we expect as happened to the Wright brothers. They never dreamed where their efforts led.

But why not make some money in the meantime - even if you are still not sure that my logic is right? Maybe this old man will yet hear some laughter or see rotten tomatoes flying in the air ruining somebody's business. It might take a minor inquisition, but not another hundred years, like it did to prove Galileo right?

Phase 5 - For Driving at Night

Nighttime travel needs another fuel source. I propose one even crazier - WATER guaranteed everybody says, "Not possible!" Ask my grandkid, educated in true science. He is glad to demonstrate in any high school laboratory that unlimited fuel can be extracted from saltwater linked to some solar panels to release the atom bonds and separate hydrogen and oxygen. It is a free fuel and could be improved for higher output via the patent applied electric generator in UREE #16. That would give big-time energy for trucks, trains and industry even a jetliner flying totally without fossil fuel now possible.

To make a kit to drive around at night in my car without gasoline, I went shopping with my grandkid in an auto store and picked up a cheap $25 air-tire-compressor quite useful for flat tires. I will use it for compressing and collecting the two separated gases extracted from water. The collected gases will be put into (2) one-gallon green, steel bottles everybody has seen one used by wheelchair people. It is small, a nice handy storage container for the high explosive fuel to be safely handled.

When using solar panels for free electricity can be connected to two copper rod electrodes immersed in a plastic container filled with saltwater gets an reaction separating hydrogen and oxygen every other day visible seen now compressed in a small tank. Controlled with a timer switch ($5.00) will transfer the gasses in the air tank and when full, start over with another bottle. The full bottle is re-attached to the return intake of my car with a quick connector ($15).

That has enough accumulated fuel for daily exchange-use and would not need a massive distributing system of wires and pumps. Forget the gasoline: it is no longer used. That is a huge savings for a teenager. It is a lot of money saved, and I will happily point it out to my corrupt government that protects the energy cartel. The time has come to end massive pollution and create many jobs for unemployed people like my grandkids' father who now drives a taxi and wastes his skill from building many houses. It could start many new businesses to make America great again.

It gets better. When your car is parked and plugged into your home outlet, you can turn on the electricity even at night with the leftover energy linked to a generator described in UREE #13 or UREE #16. This amplified energy gains a 1000 times more power extracted from the same solar panel via super-magnets, totally free. Unbelievable! Tell your friends. And it is all packaged with readily available, low technology from your local store or online. You do not need a government subsidy!

That, my friend is what the energy cartel fears most. It would end their greed and corrupt, under the table deals. Tell Obama that the time has come for the consumer to be ripped off, no longer and lied to. I have proven it with just one of the 16 UREE options. By applying forgotten logic, even my grandkid says it works. Read the same paragraph once more to get educated and not be blinded by an unscientific evolution religion ignoring thermodynamic laws.

But it is not too extreme if you recognize a steam engine running on water used in the last century to fuel an economy with a rising population curve. But now we have electricity, a cleaner fuel, and should visit museums demonstrating how a German scientist named Hoffman built a contraption that **split water molecules into hydrogen and oxygen gases** all explained better in Babushka egg concept book #9 page 95 UREE #3 and (Pearl #202 UREE #12).

Once more since we have excess electricity available when my car is parked during the daytime, it can generate electricity while working in the office and become useful storing up water energy. When the solar-UREE section is kept running it can convert additional energy by splitting water. The separated two gasses will make Hoffman happy on the other side. Occasionally, when the gas is building up, it must be pushed into a special air tank collecting the split hydrogen and oxygen gas accumulated and compressed in a small wheel chair steel bottle for future nighttime travel.

Now it can be reused connected to the gasoline input pipe filling the cylinder with collected explosive water fuel. Stored Hydrogen gas compresses with an upward stroke of the piston timed to the remaining oxygen gas using a fuel injector with a spark plug fired to explode it as in a gasoline engine. It will be the cheapest alternate fuel with very little mechanical change to an existing car.

Expensive batteries are not needed for longer range with a small gas engine for emergency back up. If you are happy with your car's performance and have found the right generator, you could make another Herbert's Automobile Kit for your garage. Just link it to a few solar panels on the roof. That gets electricity for your house saving the utility bills. Now you are happier, I am sure.

Sun and water energy is cheap and free as postulated in the Babushka egg concept books. You can have a repeat of what HP did in the garage in the Silicon Valley. Hire your relatives and friends. Design a plastic box packaged to sell a portable kit to every kitchen.

Customers no longer need gas or wood to cook food especially in Africa and India. That will result in a new hydrogen energy company and become a billionaire guaranteed on top selling Herbert's automobile kit cheap made in America. But do not forget the charity FFF feeding hungry children and educating the younger generation in true physics linked the metaphysics for a 360^0 science perspective.

A subsequently designed UREE kit will have a different high-tech fly-wheel. It will be a lot smaller and linked to stronger Neodymium-Iron-Boron generator magnets packaged and miniaturized inside a little black box connected to your existing electric hybrid-car to run perpetually. That should reduce the cost of an electric car from a $100,000 range to $10,000 for honest people paying taxes, once the automobile industry is, hopefully, no longer married to the corrupt energy cartel, both could make more profit.

That is a paradox: to sell something cheaper makes more profit with less pollution and will also stop global warming to make governments happy saving money. The principle proof that the UREE system works was mentioned in my previous Babushka egg concept book #7 on page 132.

Two years ago I explained that light runs in parallel to an invisible magnetic dual rail. If light increases so does magnetism and visa versa. Look at the picture of the UREE motor-generator in Babushka egg #9 book page 17 showing the arrangement. Now use a solar driven motor in the middle and mount a generators-streetcar motor spinning at higher speed on both sides.

To prove the principle, take a flashlight and shine light through a narrow slot and measure the pulse profile on the other side. Then use another identical flashlight and shine light on top through the same slot. It will not double the output but quadruple the pulse profile.

Therefore, magnetism running on the same rail will quadruple in the UREE system, as every bid of energy is stored in the flywheel converted to electricity. The light driving the streetcar flywheel motor linked to the dual generator generating magnetism will quadruple the output. If once more, you put in a motor linked to the same generator axis, it will again quadruple using still the same slot in the middle. It is gaining more electricity, passing more wire loops perpetually at high speed through a magnetic field.

This #14 UREE story was meant for high school kids to tinker around or perhaps picked up by patent lawyers always looking to make money. I hope it can convince someone to try what may not seem possible. It will open the floodgates unstoppable to utilize the ocean for the next energy provider.

In the meantime, to my high school friends building that contraption some advice: have fun like the Wright brothers. If it does not work the first time around, do not give up, as technology is more complex for some. But the true logic of the energy theory will never disappoint because our MIND is replicated from the bigger infinite ELOHIM MIND one notch higher only revealed in his divinely inspired Bible.

If Herbert's Perpetual Automobile UREE Kit worked get some friends together and sell the next UREE to your friends and relatives upsetting the corrupted energy cartel lying NOT POSSIBLE. But once more I do not charge for my ideas but are free should not forget to recycle some blessing to the FFF Charity. Remember there are million children globally starving and cost only $5000 dollars to drill one water-well giving 1000 people clean water forever.

Contrast that it takes a billion Dollar donation for politicians to be re-elected and do not care corrupted to the bone building another poison Nuclear power plant (2012) without the American people know about. The corrupted energy cartels have not yet invented globally one storage unit to safely dispose of spent hot nuclear rods still babysitting radiation lasting 100,000 years. Criminals to the highest degree only interested to belong to the class of privilege feasting on what the poor work so hard to earn.

This ends Herbert's perpetual UREE automobile KIT story for now. It is never finished, like the infinite loop of ∞. If you should tell your friends that ocean **WATER** is the cheapest energy fuel (Pearl #201-UREE11), they will think you are crazy, too.

Packaging dangerous hydrogen and oxygen gas in a small bottle used daily in smaller chunks is not dangerous ask a wheelchair person will not use unsafe oxygen compressed in a green steel bottle. Added with a solar panel on top is the biggest invention on this planet.

It is not dangerous. Ask someone with a green bottle respirator. Hopefully, I will no longer hear laughter of ignorant atheistic university establishment believing in an evolution religion avoiding hidden mysteries collected hundreds in nine Babushka concept books meant to better educate the next generation.

Look out for the jet engine UREE #16 (Pearl #206) to discovery how a Donut Atom is energized could explain the physics why UFO saucers could fly so fast. Or it could be applied to my newly designed gravity levitated bullet-train introduced in Babushka egg concept book #9 which would give us a better choice because it is 50% cheaper at doubled the speed to fly like an airplane inside a V-tract at 500 mph.

Ultimate Renewable Energy Engine #15
Golf-Commuter-Car Zapperino

Reviewing my underwater UREE version once more, I looked at a re-vealed picture design in the Babushka egg concept book UREE #9, Page 85. It took sometime to better understand it because these concepts are still new to me not familiar with the laws of physics governing a universe.

I am sure there are just no books available agreeing with a new Donut Atom model. Only my theory explained how life is embedded inside every atom nucleus. And further discovered each atomic nucleus is controlled by intelligence living within the proton-neutron structure matching the formation of matter a totally new concept.

I kept wondering how energy is transferred. The pictures from the Hubble telescope show the many lights on in the nightly sky that is energy to me I can see. Questioning where it comes from everybody seems to affirm that a black hole could be the energy source for our Milky Way Galaxy as postulated in these Babushka egg concept books. I am very open to hear anyone's different opinion but back it up with three factual science witnesses conforming to the thermodynamic en-tropy laws. Very few have ever attempted to combine physics with the mystery metaphysics in a full rounded 360° perspective. Sir Isaac New-ton tried to explain it, still suppressed half of his collected writings.

Let's investigate Jonah's biggest underwater UREE fish once more. It is exposed to ten thousand times higher, unlimited energy resource than the Hoover Dam. Massive energy is stored in the ocean still a forgotten discovery pointed out nearly hundred years ago by the German scien-tist Hoffman. My first attempt to explain again the #1 UREE it is still a simpler and better design. The other UREE models each became a little more complicated depending on application.

Fundamentally, they all work the same as electricity can only be convert-ed when kinetic energy is moved perpetually. What at first seems IM-POSSIBLE gets better once you understand the UREE theory principles.

Trying to understand the pressure conversion principles present in an ocean-water media once more, I soon discovered that it could also be ap-plied above ocean. It is also linked to magnetic GRAVITY, but the huge ocean pressure can be substituted for by air pressure to make it portable for driving a car perpetually without gasoline. Or, a container boat can refuel its high-pressure steel bottle tank by sinking it deep enough in the ocean to replenished pressure for the next trip if it had a leak.

For the UREE home-version, a tire compressor is enough to maintain free energy with your house detached from the Hoover Dam electrical system. That means disconnected from your energy bills, too. It works when perpetual motion can be converted to electricity as proven in many UREE examples.

To gain useful profit, a little added energy must be returned for perpet-ual motion. Similar to a cuckoo clock running with a fixed static gravity weight, it needs a weighted chain-force to run it perpetually, thus con-forming to the ∞ energy exchange.

Basically, a static high pressure, like GRAVITY, will move a piston down, but it must have a little higher pressure returned to the piston back into the same pressure tank to make the motor run perpetually ready for another cycle. A big UREE system only works when a fly-wheel is first cranked to maximum speed but still needs as explained in physics a little extra energy to keep it going. What is overlooked it will store every bit of energy for a short time until the next cycle takes over as linked to the crank. But is it possible to have a Mini-UREE without a generator?

Back to my first UREE invention to explain the principle, think of a single cylinder linked to a small pressure air bottle about 200 lb., for example, some might need more depending on application. Later, I discovered that the piston could be made from **super magnets of Neo-dymium-Iron-Boron** as mentioned in Pearls #201, #203, #204 and #206. Using super magnets will create extra electricity by means of a coil mounted on the outside of the cylinder. If an extra coil is used on the top and bottom of the cylinder, polarity can be switched for the right timing adding to the overall system, utilizing the nature of magnetism that opposite polarity is attracting and reverse will repel.

For a Zapperino version, the underwater picture shows that we must mount on the other side of the now converted magnetic piston with an embedded return spring of 20 lb. Its compression and expansion would only consume 10% of the total force - plenty available from the air tank. It will be used in helping to push the piston back on the return cycle needed for perpetual motion designed to give some inertia back just at the right time. We know the flywheel will store every bit of kinetic and inertia energy and like a pendulum of a cuckoo clock the velocity starts with zero accelerating like the piston being pushed down.

At halfway it is already maxed out, as the flywheel crank will slow it down forcing the energy velocity to become zero again like a pendu-lum. Analyzing the motor theory, the pressure valve from the air tank should be closed at that halfway stroke. The added energy for perpet-ual motion is still needed. It accumulates by a continued full stroke compressing a spring first from gravity force, but midway it switches over being converted from the embedded, extended inertia.

Once more GRAVITY is applied to the first half cycle passed on to the flywheel, which is now maxed out and switched over to collected in-ertia energy on the second half cycle. I should slow down a little, but it has energy still stored in iron atoms unnaturally compressed on a much higher energy level in form of the external steel-spring.

Once the atom force is released in the return cycle, it can be converted to electricity via a magnetic bridge and driving a flywheel that ends with zero velocity at the end of the stroke. It is only visible when the interplay of kinetic energy to inertia is needed as in the roller coaster example where the uphill pendulum motion became an inertia force no longer GRAVITY kinetic power. It has switched over to oppose gravity as demonstrated in a cuckoo clock pendulum.

For perpetual motion, the embedded compression spring of 20 lb. could be enough to store the extra inertia energy needed. When activated, it adds to the return trip expanding across a full return stroke pushing the magnet piston back with an extra 20% spring force to become 100% entropy when the spring atom force is added. This counter net-force seems supplemented in our example, being added mechanically to the coming stored energy from the flywheel.

But for driving a bicycle a little more energy is needed. It can be extracted from a cylinder coil - magnetic piston converted to electricity by charging a properly timed fat capacitor. The extra compressed spring force of 20 lb., now added to the capacitor spark force via a coil, will create the additional inertia force required for perpetual motion. It will easily return to 200 lb. plus the weight of a bicycle rider. The magnet-piston of 1000 lb. will return the previously stored air tank with no trouble because the static pressure has not changed but was closed by a mid-stroke valve before being maxed out. Now inertia takes over pushing the magnet piston further. It will add the converted electrical energy to aid perpetual motion to continue for another cycle to have a bicycle running with free energy.

That little extra energy needed is generated from the magnetic-piston-cylinder coil that converts electricity via a solenoid coil on the top and bottom of the cylinder that collects energy generated by the passing magnetic piston stroke. Energy is fed back via a link to the discharged capacitor to energize the coil both ways and keep the flywheel running to drive the bicycle. The mini-electrical UREE once cranked amplifies the recycled perpetual energy from the cylinder coil created by the magnet piston motion. When leftover electricity is reinvested in a solenoid, it gains energy I call profit, which drives a bicycle with free energy without an electric generator and perhaps without an air tank.

Summed up once more: the basic principle of exciting a higher electron flow comes from recycling energy by adding a stronger magnetic pulse to the next winding. The repeating oscillation becomes a generator to the other. Since the single piston Zapperino UREE design has limited output, it could be connected to a number units lined up perhaps in a golf car or in the basement of your house to balance out varying day and night power usages, or as an add-on to several UREE units.

The Zapperino Mini-UREE is perfect for applications without an electrical generator using free energy stored in a flywheel. I mentioned a rubber pressurized collapsible commuter car on the end of my #9 UREE story. It is the perfect application for a single piston, or miniaturized for a bicycle.

Two billion poor people could add it to the bamboo frame bicycle kit mentioned on the end of my Babushka book. If more energy is needed could link it to two generators with a flex coupling to enhance a flywheel at higher speed. Many more design options are available for special applications favoring one or the other design features depending on cost.

Have fun and let me know if you found another way to get green electricity dirt-cheap. You still can make obscene profits if you start your own business. It is easy to become a millionaire and sell globally where no electricity exists or to angry people who get ripped off from the energy cartel married to the automobile industry controlling the global market. But progress cannot be stopped.

You write the next pearl, as more applications are possible like the Wright brothers' invention expanded was seen on the moon watching TV. It takes some time to digest the new UREE theories applied to existing technology. I am sure it will be suppressed by the energy cartel ruining their business. But once a crack developed in the Hoover-Dam, it cannot be plugged up as the infinite Alpha(+ONE) energy is too powerful for mortals, as only explained in my Babushka egg concept books.

I am still learning about how the Eternal ELOHIM created his universe and energized it down to every atom to fuel each Zapperino heart beat, giving me unbroken perpetually free energy for 80 years. The new UREE concept revealed by the ELOHIM will become the next energy source one way another lasting 1000 years fueling the prophesied Kingdom of God becoming a divinely controlled third civilization described in Babushka egg concept books.

Ultimate Renewable Energy Engine #16 JET Engine Section A

The Wright brothers could not have imagined that improvements on their first airplanes would eventually lead to visiting space stations on a regular basis or even to the moon and back. Now NASA is investigating Mars. Mankind has jumped the boundaries of possibility a million times in the last century, but no one has yet quantified it. Faced with the challenge that civilization is running out of fuel from energy extraction, it seems very short-lived.

Most of our energy generation systems have become life threatening by severely poisoning the environment causing much extinction on earth the only place where LIFE exists. But God created man for a purpose. In prophecy He promised another 1000 years of human civilization, but our science projections do not go beyond 50 years if you apply logic where we a heading. That paradox seems to be answered by a new infinite energy discovery which can be extracted from GRAVITY, WATER, AIR and special MINERALS. Typical nature is organized in four basic building blocks that closely replicate the 4 DNA building blocks of life pattern controlled by intelligence. It runs on a dual railroad strand for balance, as each of the four creation units is fueled by ∞ energy to support LIFE at the atomic level.

I discovered that MAGNETISM is manufactured by protons inside every Donut Atom and can be spliced out converted to electricity through 16 UREE models of different sizes of applications. When analyzed, they will obsolete most existing polluting energy sources. My vision got expanded to realize that magnetism is really the invisible ∞ energy force coming perhaps from a black hole dispersed throughout space but when slowed down in a Time-dimension crystallizes into spectral lines frequency Fraunhofer defining it.

Being interested in the many galaxies photographed seen recently on NOVA-TV with the latest telescope showing how they form, like the Hourglass Nebula on the front page of my 9th Babushka egg concept book illustrating that energy is turning inside the black hole at high speed infinite faster what Einstein postulated.

We learned from the electrical three-finger rule applied to the electrical generators that when magnetism is turned it will extract electricity. Looking into the sky see so many lights will follow the same principle in physics demonstrated by so many scientists like Nikola Tesla, Michael Faraday and Marin Soljacic confirming a wireless electricity phenomenon. **Check Pearl #225 for more.**

Magnetic energy transferred from space works similar to the transformer principle redistributing a million volts down to a usable 110-volt level to demonstrate the finger-imprint eventually forming particles, still a paradox for science. Ultimately the exponential TIME curve shaped discrete atoms categorized as elements over hundred elements have been summed up in a table that, when analyzed, defines matter.

When matter is bunched up, it becomes magnetized gravity, a new concept needing time to understand. That converted force can be extracted, perhaps from the Milky Way black hole. When moved, it is experienced as inertia demonstrated in 16 UREE electricity generation concepts.

A new electron-bridge was invented for magnetism to flow linked to a doubles sided generator extracting from the pool of a billion x billion volts magnetically stepped down to the useful 110-volt level. That could drive a jet accelerating air using an inside out streetcar motor mounted with fan-blades on the outer periphery of a jet engine profile.

MAGNETISM converted from the infinite source will last beyond 1000 years I am sure, but it needs to be extracted in small, finite portions to be useful for mankind. Check out the new UREE inventions that corrupt energy cartels say are NOT POSSIBLE, fearing revenue loss. Flying a jet without fossil fuel seems preposterous, crazy.

This is my last attempt on the free Internet regarding a new jet engine. I will later explain how it could be adapted to a UFO motor, which must go faster then the speed of light. It could expand science as applied to proven UREE systems as both are governed by nature - physics-metaphysics exposed a little in my nine Babushka egg concept books.

How a Jumbo-jet can fly without Fossil Fuel

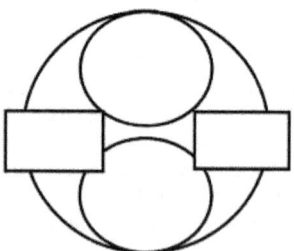

The UREE electric jet has tremendous advantages besides not needing expensive fossil fuel, but the weight reduction could change the whole design dynamic. I envision that the center of the airplane body cross-section is a circle with two long tapered cylinder jet engines embedded like a number eight, one engine on top another, which is the main body. In the middle on either side of the figure 8, we have passenger compartments double - decked and extending into a single delta wing section with 500 or more seats on each side.

The electric jet engine is long and narrow on a single shaft with multiple Basket-Cup Printed-Board Generators turned by a turbine upfront. The generator units are designed with fan blades mounted on the outside.

It is a reverse electric motor turning on a double axis. Each outside motor-winding-fan-blade unit must run faster to accelerate the air at the rear-end to create the highest possible jet exhaust air pressure.

There are a number of possibilities as I am not a designer of airplanes or engines, but take it from here and improve it as I can only point out theories with my finger. Do not look at my finger but where I am pointing. Once more the future UREE jet engine is very similar in design to the present jet engine with a long tapered body large in the front with multiple fan propellers and on the rear section ending with turbine blades getting narrower and smaller in diameter as the velocity increases.

In the new UREE jet motor, the first power section is fueled by compressed air from a tank turning a turbine at high speed. But the turbine must be linked to the second stage, which looks identical, but it functions as a compressor to push back the lost air so that the tank's air pressure is never emptied. Both units run simultaneously on the same axis.

The second stage must have surplus power to overcome the remaining tank pressure and needs a flywheel. Combined utilizing a streetcar-motor design of heavy super magnets is mounted like shroud on the outside diameter of air fan blades therefore function as a flywheel useful for a generator. That produces a little extra electricity to drive the turbine motor linked to one shaft driving many smaller generators. The motorized compressor unit perpetually restores the air pressure lost in the first stage, which drives the big fan wheels that collect large volumes of air from the outside to be speeded up to higher pressure to make the airplane fly with a narrowed jet in the rear.

The air tank again works like the cuckoo clock gravity weight that never changes but applies power to the gears - in our case driving a number of electric generators. Therefore, the electric generator is the real power source and becomes the first entropy level with everything connected to one central axis. The energy must be first moved ∞ to be perpetual and adds some energy from each loop to get more, which is over the air tank pressure (considered the first front stage) but is second entropy from the energy exchange.

The first entropy forces are super-magnets that do not change either but are linked to magnetized atoms extracting neutrino energy from the black hole of our galaxy. If you want to know how it works, read my Babushka egg concept books free on the Internet and become familiar with where energy comes from and how it can be converted in various UREE ways not yet taught in universities.

The first stage must conform to the UREE principle in that the spent air pressure must be restored to create perpetual motion. On the same axis are connected multiple generator rotors made from low inertia printed boards. UREE 13 - Pearl #203 is meant for automobiles and homes, but a jet engine requires a higher velocity speed like a levitated train going at 500 miles/hour floating on air pressure or like airplanes levitating on anti-gravity force. Both need much higher electricity output in both current and voltage. The printed boards have low inertia and therefore can be turned at higher speed. Many experts are immediately turned off either seeing a Bible verse linked to a forbidden energy source or magnetism at 10,000 rpm thinking it's NOT POSSIBLE, which threw my unopened book in the toilet.

Therefore, let me explain it for my grandkid on the third grade level has enough common sense to understand it. I even talked to the taxi driver on my ride to the airport. Even with a lower education, it got him turned on. I was surprised that he understood the principle the first time. I am sure universities will come along once they realize that they can make obscene profits if they tried these ideas out.

If we apply the electrical three-finger rule to a copper wire when connected to a single solar panel, it will be charged up creating a magnetic field around the copper wire conductor. That can be measured in any high school. If we connect two solar panels in parallel, we get a bigger magnetic field: quadrupled, I believe, as postulated in some other pearls. That should put to bed that word "IMPOSSIBLE" as solar energy is free once you build your own $97 dollar solar panel tired of being ripped off at $3000/panel.

Now we learn that if we shape the copper wire into a loop and push a magnet through, it gets a mystery electrical pulse nobody knows why. (Read Babushka egg concept books to widen your science horizons.) But consider, when making a loop with the copper wire connected to a solar panel, it will generate magnetism now forced to go through a loop-center which at that point has higher magnetism bunched up all concentrated in the middle. That would make a much stronger pulse with the same magnet, because the copper wire is now biased getting energy from the sun.

Let's duplicate it with another solar panel connected to a second copper wire loop and then do the same thing but this time we added a pulse generated to the bias of the first copper wire loop. The result of the first loop will increase energy as the magnetic force centered in the middle has become bigger because of a higher bias creating a larger magnetic field around the copper wire. Now let's repeat and take the higher output of each wire loop switching back and forth always increasing the bias now getting bigger and bigger. Would you agree? Or read my 9th Babushka egg concept book to widen perspectives to a full 360⁰ horizon. Let's review it for the PhDs in universities postulating, "NOT POSSIBLE."

1. A current created from a solar panel produces a magnetic field around a copper wire. True?

2. If the charged copper wire is made into a loop, all magnetic fields surrounding the wire have not changed but become more concentrated in the middle going in the same direction and stronger in the center. Therefore, a biased charged copper wire loop has more concentrated magnetic line in the middle then bare copper loop and will consequently create a higher pulse with the same magnet. True?

3. If a previous pulse is added to the charged biased copper loop will increase the next pulse, as the bias is higher. If we kept it going always adding the next pulse to the other gets a bigger output eventually like double a solar panel. True? The Pearl #203 stated 1000 X Solar panels!

4. We have proof in the streetcar motor design. The stator is connected to the rotor back to the stator in a ∞ loop and will accelerate the speed of the rotor until it explodes or is controlled by the streetcar conductor starving the input. Now better understood; a streetcar motor design will increase magnetism by a biased current inside a copper wire recycled ultimately gaining higher electrical output. A starving input differential can be extracted and made useful as electrical profit, but if returned, one could drive the street car perpetually without a conductor provided the energy is not stopped but freely let it go. True?

5. Some parameters will influence electrical performance limited by resistance, heat or eddy-current which dictate that we must transfer the energy at the source with a transformer needed which will become obvious at the end of this story. That should cover most objections; however, the streetcar motor only runs perpetually if either the first or the second entropy energy loop is added and interlinked as outlined in the UREE principle. Question; if than the conventional current is shut off will the streetcar stop. True?

6. But another law of physics is not shut off. Mass if moved will store energy, like the streetcar has mass still running at high speed, which becomes in principle a flywheel storing energy. And if the stored energy at that moment before it slows down to zero, according the physic, is converted and added to the biases of an electric generator will create a leftover electrical residue. That little portion of energy re-invested creates perpetual motion, mentioned earlier as profit. True?

7. According to another law of nature if a saved energy like profit is reinvested and put back in the pot it gets more profit. Every business works that way why not in a streetcar motor? All we need is to add a motorized flywheel the only storage devise I know off which will store temporal energy and drive an electric improved generator could run the streetcar perpetually. The streetcar motor once in high motion switched over now will be driven by electricity generated from an electric Motorized-Flywheel linked to two generators UREE shown in this Babushka book #9 page 17, but must satisfy two energy consumers linked to two entropy energy loops (∞). True?

8. My Grandkid now understands investigating the law of nature, which makes energy conversion possible. It is like throwing a stone into the pond: The accelerated stone has embedded energy that creates on impact a secondary energy wave. But if we combine both entropy levels like connecting heaven to earth linked to a UREE, the ∞ invisible energy can now be perceived with our eyes because they are calibrated to the many laws of physics. But what we see in physics must be linked to the metaphysics to better understand it. The laws of nature are replicated on different levels and if used together in a proper sequence could get an advantage for us. True?

9. The UREE only works when perpetual motion is achieved interconnecting two energy levels but needs to improve an old-fashioned 100-year aged generator design. The crux of the whole UREE system summed up is that a flywheel once accelerated will peter-out and

needs an additional energy to get it going perpetually, not much is needed and is now found in super-magnets a free force we all know we played around with. Magnets are used generating electricity and added as a force to a motorized flywheel will run perpetually and connected to improved generators on each side of the motor shaft is adding a bias to each loop amplifies more electricity big time profit. Take it from here, my finger pointing. True?

For the ultimate skeptic I introduce one more another unknown law of physics perhaps it will shed more light to believe that a Jumbo-jet can fly without fossil fuel. Inventing 16 UREE does not happen overnight but is a collection of unrelated experiences gathered in a lifelong technical understanding. We learned from the principle of an electric generator that it will give more electrical output when a pulse is overlaid with a previous pulse using stronger magnets, but I have not investigated the amplified **Electromagnetic resonance**, which would give more electrical profit.

I remember in my early days as a research technician of wanting to play around with vapor deposition applied in semi-conductors. I used an obsolete military microwave generator for radar application found in the junkyard. It resulted in creating extremely thin metal layers useful in the semiconductor industry in applying refractory metals like titanium, molybdenum needed in a new semiconductor industry.

I became educated working in extreme vacuums and various noble gases like argon with its inherently big atoms bombarding a substrate in a vacuum with micro frequencies similar to your microwave oven. It created big new technologies, but being a German immigrant ignorant of business I did not make money from what I invented. Thereafter I followed another inclination.

While fooling around with the microwave generator, I experimented with a simple paper-cone glued on a small loudspeaker. I wanted to find out about **resonance frequencies** not found in schoolbooks in order to learn more about unknown technology. I touched the pointed paper-cone to the main beam next to my workbench and searched for the resonance frequency and waited to see what would happen. Not knowing that the tiny waves of energy touching the main beam would add up getting bigger on top of the other to reach the critical frequency of the building.

I soon learned about the consequence when the whole steel-glass building began to vibrate unperceivable like an oncoming earthquake. It would have collapsed when the tremendous embedded energy reached the **critical resonance frequency**. Being ignorant, I first heard a little strange sound and felt some vibration within the whole building, which then got my fellow employees worried. They looked up being confused. It scared the heck out of me, too. I stopped this experiment that not even my boss knew about.

Now, 40 years later, I discovered 16 Ultimate Renewable Energy Engines remembering the old military microwave principle once more using the resonance frequencies now better comprehending physics, which could extract big time electricity. At first I did not understand the principle but dialoged it in the first 10 applications, which crystallized further in six (6) more UREE applications better explained.

I put those in the Annex inserted in third edition of the 9th Babushka egg concept book. I asked an automobile designer why any motor has a critical resonance frequency needs to be avoided; either below or over for a smooth ride or the motor will self-destruct. Look at your speed dashboard indicator do not go beyond the red color dial because that is the range of the critical resonance frequency.

When the resonance frequency reaches the highest possible critical speed for any given proton circling around and through the apple-do-nut atomic nucleus, it creates a different element. That is what Fraunhofer discovered but is now explained better.

Atoms become visible to us in the 4000-7000 Angstrom bandwidths that match our eye calibration for visible light that shows the embedded Fraunhofer spectral lines within the rainbow frequencies of every element in existence, what a coincidence.

Perhaps the air media atoms when compressed reach critical frequency being condensed and restricted spilling over exiting excess energy now caught in an electron UREE fishnet. It is a good story for my grandkids. But getting the highest electrical output in a UREE will only happen when it is close to the critical electromagnetic resonance frequency, because that is where the most electron fish congregate. It stirs up a frenzy within the atom pool, if you understand my grandkid's previous analogy.

That is the point to apply the magnetic UREE Bridge to seize a big catch of concentrated electron fish: it is large enough to fuel a jumbo-jet engine. You can check it out in the next "B" Section if you want to become a billionaire and built on the Wright Brothers' invention to the next version to fly faster and higher.

The world at large settled on 50/60 Hz. I remember when airplanes were 400 HZ 50 years ago due to a fear of electro-magnetic interference. Perhaps a new world standard needs to be settled with the new UREE because, when the electrical energy reaches the critical resonance frequency, it must be used up at that time at a constant rate.

The airplane application is a perfect example. Being portable, it does not need transmission wires or big transformers, which will obsolete most energy cartels and become a great benefit for future mankind. The latest Jet Generator is designed according to principles true to the Bible such as, loosely quoted, "If you give more, you will get more." This mystery is proven for me.

Recently, I walked around in my neighborhood visiting a new car dealer investigating the latest $65,000 hybrid electric-gasoline car, which promised extended mileage using $18,000 lithium batteries. I looked under the hood and saw only a tiny generator, so I asked the question, "Why not use extra generators to extend the mileage?"

My grandkid already knows that for 40 years the automobile industry never improved the gas millage. It has not changed much from my old beat-up VW because they are married to the criminal oil cartel in an unholy alliance for mutual benefit to rip-off the public and protected by corrupt governments.

Green free energy is not allowed. The German scientist Hoffman 90 years ago invented and demonstrated how to separate water, the greatest energy reservoir, but it is still suppressed. Now hydrogen is only allowed in rockets because Sputnik, if you remember, forced America to develop a better fuel to equalize what the Russians did.

Why is free green hydrogen concealed by the automobile industry, as wheelchair people not afraid to use compressed oxygen in a little green steel bottle? Look at the global auto show displayed on TV in Germany 2012 demonstrating the latest solar electric automobile for $80,000-120,000 loaded with lithium batteries, the most polluting energy conversion poisoning the environment. Check how lithium is mined in China.

My grandkid, not corrupted in politics, knows that one apple added to another apple makes two apples, which is more than one apple. The automobile manufacturer does not know that. I sent my courtesy notice (UREE page 3) to 265 universities, 55 embassies and 25 automobile manufacturers. All got ignored, but one surprised me.

A German car manufacturer returned one unopened with a response from lawyers informing me we have our own ideas and cannot accept unsolicited courtesy information. In other words we are afraid to use your free ideas or do not want to sell cheaper cars. We think ripping off the customer makes more profit. They looked at my finger and not what I pointed to.

They could make obscene profits with my free information on the Internet. Adding my five cents, notice we have a crankshaft sticking out on both ends of the piston motor why not attach a generator on each side that will not use up extra power?

That makes the two apples my grandkid wants and would eliminate a number of costly batteries exchanged for a cheaper generator that should have been improved after 100 years. It is easier and safer than using lithium batteries made from a dangerous reactive element horrendously expensive to produce.

The real cost is much higher because they hide the cost of grossly polluting the environment, mostly in China, to squeeze more profit. Do not be fooled by a $65,000 hybrid car to find only a tiny electric generator on the inside. It makes no technical sense to have an array of $18,000 lithium batteries assembled rather should be replaced with additional, cheaper and bigger generators to extend the range of 100 miles to 10 times larger.

Deutsche Welle aired a TV program (8-28-12) showing a new Company developing an electric motor boats on Lake Constance, as polluting boats are not allowed. They demonstrated a racing boat using a streetcar motor that gives 10 times more power than diesel fuel. Think, if your gas bill on the pump is 10 times smaller on top will not pollute the environment? Why is this technology not used on your car?

The next Section B describes in detail how the new electric UREE generator works while in the process of a patent application. The Jet-design allows a number of simple design reductions for lower application that could replace all other designs mentioned before.

If you are interested, this design is not free at this time but requires some free-will offering to Faith in the Future Foundation charity, as funds will be used to feed hungry children around the globe and educate them about the future. Contact this charity that owns the applied patent described in the next section.

That next section is the big Jonah fish. Hopefully, we all should pay attention because it is linked to a warning for our civilization but preannounced a future free electricity energy for our grandchildren extracted from magnetic GRAVITY, AIR, WATER and super MAGNETS.

Section B -
Basket Cup Printed Board Generator:
1000X Higher Output (Patent pending)

Description of the Basket Cup Rotor Drum:

A. Please look at the diagram picture below. In the center we have a hollow shaft turning from the first front unit streetcar motor driving the compressor to the last generator section on the rear end. Each basket cup is attached or bonded to the shaft and turns at the same high speed. The many wire loops from the various basket cups pass through the hollow shaft for access to the very end, which can be electrically serviced by conventional motor-brushes.

B. The center inside the basket cup is shaped in a rotor-drum holding a number of super-magnets spaced apart, polarized and supported by a bearing linked to the shaft. Depending on the application, they are held together by a nonmagnetic media a professional designer would know about. It has two built-in features depending on application:

1. The rotor-drum linked to the shaft bearing could be spin-driven forward or backward by an outside motor-stator winding on the outer periphery which is also the platform for holding the high speed fan-blades for a jet application like an upside down motor design as the conventional stator becomes a rotor. It is similar to your ceiling fan only much faster, or a rotor within a rotor a double feature. If you wondered why the rotor-drum could run in two directions depends on the jet fan-blades related to magnetism maxed out.

2. Each generator-motor stage is independently speed controlled, as jet engines must accelerate air velocity toward the rear. If the outside motor winding is fixed for lower output, it must lock in polarization with the rotor drum for lower speed applications.

But make sure that on either side the fixed rotor-drum polarity is alternately positioned. Electricity is still generated from the vertical loops passing by to the horizontal loops cutting through a magnetic field of the outer motor polarization. Follow the magnetic lines to understand the principle. The magnetism is not affected if shared in two directions.

C. A pulse is generated in each winding embedded on both sides of the basket-cup with the shaft turning at high speed.

D. I recommend making the basket-cup from aluminum ceramic for high speed with the imprint of the wire loop pattern photo-etched, thereafter-deposited copper. The final shape of the surface is ground off leaving copper embedded according to the pre-etched grooved pattern.

E. The inside and outside copper wire pattern should be cross-hatched to give a double output for the same magnet position. Another layer on top is like two (really four) apples.

F. The loop pattern needs to be reversed connecting the inside to the outside of the cup to allow ∞ loop designs, which will double the pulse again like 8 apples.

The sides or bottom of the cup works like UREE 13 -Pearl #203, but what is added in this version the same magnet is used going through the sides of the cup, a double action as the magnet field is circular, and I do not care where loops are being cut now, as it is increased to 16 apples for each UREE section.

The vertical winding is crossed over to the basket horizontal windings which will add from the first vertical pulse a bias to the next horizontal winding therefore gets a bigger pulse. The output is linked to the next section repeating once more gets a higher output again. Keep going to the next motor section big time electricity.

Basically, each basket-cup section has an additional power source either an individual outside motor or one motor on the rear end with windings split meant for variable flexible power requirement, if used in a high-speed train or skyscraper.

Adapting it to jet engine, I do not know the power requirement. That is beyond my capability, but as each individual external motor section could have variable speed with a fan-plate attached to accelerate air with the highest rpm on the last station for a jet application. I am only a philosopher-inventor pointing my finger at possibilities, as applications are endless across a wide range only the future will reveal.

I learned from the Wright Brothers. Perhaps with the next version we could once more visit the moon using an improved UREE jet design. It is much cheaper and getting better. If you want to travel through space using gravity-antigravity as a fuel, check out a UFO-UREE design in Pearl #206 Section C.

Returning to the **resonance frequency**, certain elements in the Periodic Table Fraunhofer classified as spectral lines are more inductive to produce magnetism. The spectral line for each element is really the resonance frequency of every atom and can be visualized in the atom picture at the beginning of the energy trail on the 9th Babushka egg concept book.

That was a new discovery for me, and it has big consequences for a new science understanding. Have a closer look at the magnetic lines going through the donut center explained in UREE 11 - **Pearl #201.**

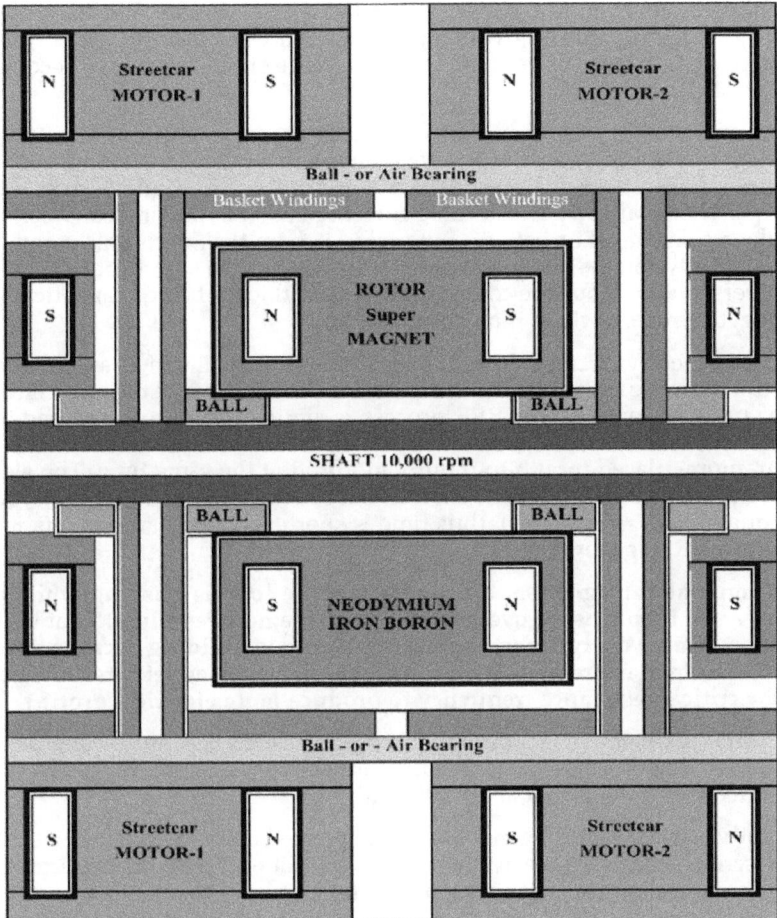

UREE Patent Exposé

Once more I like to repeat my theory for my grandkid. It is difficult understanding the physics of how electricity can be perpetually amplified by rearranging wire loops. A certain size of copper wire made into a right turn loop gets a positive (+) pulse.

The left turn loop gets a (-) negative pulse. The solar panel will produce an even current like a battery but needs to be converted to an alternate current to be useful. The UREE system thus produces alternating free electric current to drive a car or fly an airplane without conversion losing energy. For higher power requirements, only an alternate electric current is useful with the polarity switching from (+) to (-).

No one can explain another phenomenon in the teeter-totter relationship. In general, when the same magnet passes a bigger loop, it produces a smaller pulse, and a smaller loop gives a bigger pulse. Therefore, only concentrated magnetism will give more electricity, but the same magnet passing faster gives bigger amplitude. In the UREE patent design, we twist the copper wire loop in a figure eight laid over to become two smaller loops.

As analyzed, it gets first a plus (+) pulse (half the amplitude) getting bigger by adding a bias to the next loop with a bigger (half amplitude) but is now a negative pulse which is twice the amplitude adding the (+/-). That is already an alternate current needed bypassing conversion like a solar panel or battery. But seeing it from the South Pole magnet perspective on the other side of the printed board, it seems that the two reversed wire loops are colliding. The second loop, being magnetically negative repels, which is an added force.

Consequently, it can only explode 90^0 in another direction as demonstrated by the principle of a cheap nuclear bomb because the same laws of physics apply. The UREE process is similar. Think of two loaded canons, one side with plutonium and the other with unstable enriched uranium, placed mouth-to-mouth and fired at the same time. The elements will collide 90^0 with explosive force toward the infinite energy equation ($\infty E = ZeroTime$), thus time is shortened. The same occurs inside a UREE generator.

Comparable magnetism on one side of the loop is pushing while a reversed loop crossed over on the other side now repelling. That will give us more energy when opposite polarity is colliding because time is shortened expressed 90^0 in higher rpm of the magnet rotor toward the critical resonance frequency to produce more electrical profit.

Further explaining this paradox: we like to get even pulses one next to the other, but if each consecutive pulse (+/-) is getting bigger by an added bias, how is that accomplished?

Study again the UREE generator round printed board design previously described. On each side, we have 5 or 6 pie patterns of loops duplicated. One pie section starts copper loops getting smaller round and round toward the center. That pattern is duplicated on the other side but in reversed direction. The pattern runs parallel like a railroad turning in the center to get two end wires, which can be terminated by brushes on the outer diameter board edge or continued, to the next cup section.

Let's make a hairline-cross mark over the egg-shaped pie pattern. The super-magnet is placed horizontally covering all the egg-shaped loops when passing by. When the printed board disk turns between either side's fixed super-magnets, a number of pulses are generated in every loop passing a magnetic field. But a different UREE generator loop pattern is obtained by drilling a small hole at the vertical crossover hairline which will connect a front loop with reversed back loop of the same size with a little solder. The front loop is now linked to a reversed loop, hence it will create first a positive pulse linked to a negative pulse created by a twisted loop. It must conform to a twisted figure eight folded over reversing the direction of the pulse.

When each loop is added by a bias, it gets a bigger pulse, but if linked to the next backside loop, it gets smaller in size with smaller pulses. The loops are continued connected from the front to the back linked through the next hole of the next smaller front loop on the vertical hairline repeated now becoming smaller loops continued alternately on the front and back to the other side toward the center with a continuous figure eight pattern. That will get smaller pulses, as the loops are smaller; however, it is compensated by a larger voltage from a bigger bias for each added loop to even it out the frequency.

If you can imagine an apple-Donut Atom flattened out with the atom's North Pole and South Pole interconnected, though squashed flat, it works like an UREE board generator as the identical laws of physics operate on the same level. Better yet, look at the hourglass galaxy's red rays turning right, the other side turning left at the center of the critical frequency energy black hole, which is the same principle.

Therefore, 10 loops times 5 pie shaped loop patterns on one disk is good for 50Hz at a certain speed now manipulated "even amplitude pulses of an alternating current" linked to a suitable transformer. A faster magnetic field movement in a teeter-totter relationship will cause bigger amplitude or will be expressed in higher frequency to create smaller amplitude compensated by increasing the bias in every loop. But a biased double looped system gets 4 times higher amplitude, nice and even.

If we convert the vertical disk system to the next horizontal rotating wall-cup design using the same super-magnets, we get (4x4) = 16 times amplified all even at 50Hz pulsed but exponentially getting bigger as linked to the next cup. Therefore, the rearranged copper wire pattern could give us 1000 times more electricity by adding each section when running at higher speeds next to paired super-magnets.

Note: The first Streetcar motor is offset to transfer higher velocity to next section. The vertical basked winding linked to the horizontal winding is adding a bias current therefore speeding up the next second streetcar motor. The motor will turn faster and accelerate in the next stage - which is the third magnet-rotor. That creates higher electricity on the next stage vertical winding repeating the cycle gaining electricity in the next stage.

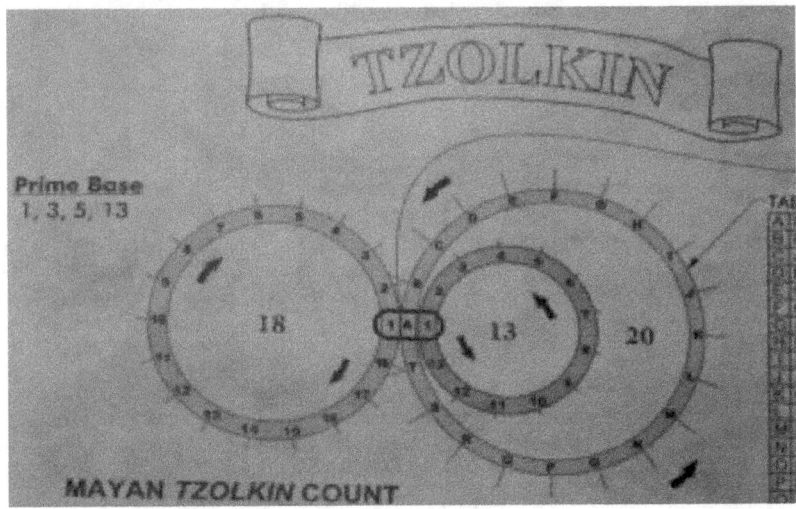

Look again at the Tzolkin Clock with the 20-gear driver, which compares to the perpetually moving streetcar-compressor shaft linked to the basket-cup windings that transfer energy to propel a magnet armature rotor (13) and outside rotating motor-fan armature (18), too. The UREE #16 is an exact replica of that design. If the driver motor-20 turns a generator-18 linked to the other generator-13 = 31 (13+18=31), that produces a bigger, outgoing energy potential initially at 20. It works and is worth another billion once more.

How to Create More Electricity

This method of generating electricity matches the new Donut Atom theory as explained in Pearl UREE #11. According to that model, magnetism is generated by protons moving around and through, up and down the poles of the apple-shaped atomic nucleus to create magnetism 90^0 from the inside to the outside according to the electrical three finger rule.

Magnetic lines travel at the speed of light around the donut atom going through the center and around back through the center again like a ∞ math symbol. When magnetic lines go through the center of a wire loop, energy is concentrated and transferred, as in the apple-shaped atom. Consequently, it must some how store more magnetism on the inside like the iron core of the earth, which is another globe-shaped magnet.

Technology dealing with electricity cannot live without a capacitor, which is basically a mechanical electric generator storing up energy. If biased, it will discharge its stored energy very fast. That could match the apple atom center. Taking a capacitor apart as a kid 70 years ago, I found out it was just copper sheet rolled up very dense. The additional copper somehow has more capacity to hold the magnetized apple-shaped Donut Atom's energy together.

Once more, when a magnet passes a copper wire loop, it is like passing through a charged capacitor of densely stored energy. Because of inertia, moving it will capture some energy in the wire loops to be transferred. Therefore, a capacitor behaves like the hole in the donut atom, which is the energy center.

This is useful linkage to store concentrated energy and pass it on very fast, similar to the discharge of a spark plug in your car. A capacitor spark can be transferred to the next wire loop cycle creating a bias for a bigger spark. Keep going to the next wire loop to grow electricity transferred from the protons. Energy moving on the inside of atoms is expelled to the outside by a magnetic bridge. My grandkid has learned from many analogies in the Babushka egg concept books that magnetism behaves like water. It is useful for transferring energy. Magnetism becomes visible when a pulse is generated, and reversing the loop will reverse the pulse polarity and trigger a discharge of the capacitor expelling accumulated energy like a spark plug.

The capacitor therefore is very useful to transfer energy from moving magnetic lines, if lined up between wire loops. It will transfer an available charged-up bias energy pulse to the next wire loop and accumulate energy. It is enough to create a spark in the next wire loop like a capacitor discharge when the polarity reverses from the next wire loop. The discharged energy is caused by the inertia of wire loop, which carried along a moving magnet field and ends up as a biased bigger pulse passed on to the next loop. The next wire loop also passes on another capacitor discharge creating higher electricity. I hope my grandkid will understand that. Read it again, because that is the secret as to why the UREE works.

At the end of this story, I show an air bearing design, which means the winding can be made much smaller and closer together, like micro-inch. This means the speed can be reduced, which is an issue that troubled me earlier a little but is now solved. That opened up more applications, no kidding. Thus, electrical energy will grow bigger if we have a suitable capacitor between a number of copper wire loops creating a polarity. When it is linked to a capacitor in series, it will discharge its accumulated energy as soon as a number of pulses is reversed so as to release more energy.

Therefore, this produces a bigger biased pulse in the next loop proving perpetual motion stopped in an explosion having reached the critical resonance frequency. Hook it together for a ride, and slow it down by making the air-gap smaller. This gets obscene electricity profit putting the corrupt energy cartel out of business. Now you can drive your car around never to use dirty gasoline again if the governments allow it.

UREE #16 Output Calculation Assumptions
(one motor section only)

Horizontal cup printed board has 6x10 winding loops:

Both sides	@ 120 loops
Vertical cup	@ 120 loops
Total loops left side cup	= 240 loops
plus right side cup	**= 480 loops**

For one minute rotating (60x480) = 28,800 loops

Running the generator at 3000 rpm which is 3000 revolution/minute (3000 x 28,800) = 86,400,000 loops

Running one hour to get watt/hour or Kw/hr.

(3000 x 60) /hour = **5,184,000,000** loop

Watt = Volt x Ampere

At this time I can only approximate what is possible. I would guess a mean averaged loop pulse could be =.01 watt (.2 volt x .05 amp) of power.

Super magnets **Neodymium-Iron-Boron** are extremely strong. Perhaps much more could be postulated.

5,184,000,000 loops in one hour passing will yield electrical power (5,184,000,000 x .01 watt) = **51,840,000 Watt** or 51,840 KW/hr. (51,840,000/1000)

Expressed and converted in Ampere power

$$= \textbf{471,270 Amp @110V/hr.} \ (51,840,000 \ /110)$$

For one 10-12″ diameter double cupboard generator to get 51 thousand Kilowatts @ 110 Volt /hr is extraordinary. If we converted it into horse power (@.746 Kw), we could be driving 193 cars/hr. (@200Hp for each car).

Could that fly an airplane convincing my grandkid?

However, one more thought. We need to consider that the UREE #16 jet application uses a set of printed board-cups with six (6) rotating super magnets in the center for each pie-shaped section. That will generate six times more extra pulses and in addition will yield increased voltage from the biased loops linked to capacitors.

That could be 311,040 Kw/hr. (6x51,840), and converted to horsepower, compares to a jet engine at 232,035 horsepower/hr. (311,040 x .746). A Rolls Olympus engine of 19,000 lb. thrust uses 25,000 Hp in an airplane or about a 1:1 ratio which could power nine (9) jet engines (232,035/25,000) at only one/hundreds of a watt per loop. No kidding. If I overestimated the .01-watt per loop by a 10 times factor for one UREE motor section, it could still fly a jumbo jet by adding motor sections as described.

How is Tesla Electricity Magnified?
The Secret of Why the UREE Works

The Third Edition of the 9th Babushka egg concept book is truly unique. Page 3 referenced some YouTube videos of different electric generator-motor designs demonstrating perpetual electricity by just connecting a motor to a generator will cause it to run perpetually. However, to fly an airplane requires much more power, perhaps a **RODIN COIL** system as illustrated to resonate magnetism toward unlimited energy. The videos all demonstrated conventional rotating designs to be functional but one is a different Tesla exception.

The UREE motor is unique. First, air pressure from a small wheelchair tank drives it mechanically to make it portable. To return the piston stroke with higher pressure is achieved either by expanded air pressure or magnetically switching coils. That gives us extra force to create the perpetual motion that returns the same pressure back to the air tank. That configuration will overcome friction and at the same time get free energy heating the air with a bigger spark plug (UREE #12).

Perpetual motion will only gain electricity if two energy sources are combined with the two entropy thermodynamic laws.

But if magnetism embedded in super magnet atoms is connected to the COSMOS MAGNETISM, a much bigger force is accessed by adding ∞ together with the two thermodynamic entropy laws works similar to a transformer. This approach was successfully demonstrated by Nicola Tesla, it could perhaps obsolete my mechanical UREE motor designs.

Pay attention. Tesla drove his electric car around all day without a battery, which caused the millionaire J.P. Morgan to cut off his funding because he thought that no money could be made from free energy. Tesla began working for George Westinghouse who became extremely wealthy by introducing alternate current. His many patents established a more efficient system as demonstrated by burning 200,000 lights during the 1893 Colombia World's Fair. That started harnessing electricity with the first commercially built alternating electric generator at Niagara Falls.

Watching famous movie clips on driving without a battery a little closer, I compared it with the UREE invention. I noticed a special 3 ft. antenna coil on the test application reengineered from patent drawings, which demonstrated how magnetism may have been extracted from the space in the adjacent coil one mile away on the beach. We all have seen lightning discharging between clouds and know that magnetic energy must be somewhere in the air everywhere. If this energy somehow could be get hold of with a special coil antenna that could drive a car as Tesla demonstrated. The criminal energy cartel did not want free electricity; therefore, it destroyed the microwave tower Tesla built to remove the technical evidence of how he returned an amplified signal to his nearby car. I wonder how it worked.

The UREE applications use biased copper loops to get a bigger pulse, especially if used with a **RODIN COIL**. A Tesla induced frequency (works like a crank) sent through the air, added a little biased static energy when discharged over a capacitor in the second coil. Then, if repeated to the other coil antenna like the UREE model, the increased energy via a capacitor is returned to perpetually switch back and forth at the same frequency similar to how the reversed winding of my printed board generation loops gains energy.

That function makes the next frequency pulse grow bigger when the transmitting frequency is passing again the static charged oxygen atoms air domain picking up more static energy across the airways in-between two antenna coils. It grows every time a capacitor pulse is discharged on the other end, and recycling will raise a little bigger capacitor pulse on the other end passing the air space again, increasing the electric carrier frequency to the critical resonance frequency. The resonance is visualized in super magnets spinning at extremely high speeds in a test with Rodin coils. The function is similar to the horizontal Wiener-wheel application UREE #6.

It is analogous to the UREE two generators hooked up to a motor that increase the bias in wire loops to get more profit. Consequently, if the first generator rotor is connected in a cross loop to the second stator in series, and if the second rotor is linked to the first stator, it will add a biased voltage to each wire loop producing more power. Tesla's antenna works the same way but perhaps he modified the driving motor and duplexes the generator coils together to one unit explained by Peter Lindeman-Electric Motor Secrets. (http://www.youtube.com/watch?v=i LGuf1geOiQ&feature=related).

On YouTube a technician in Mexico used a standard half PS motor and took it apart. Then he cleaned out the slots in the stator and put extra copper coils in the slot pattern duplicated. When the motor is running, it is reconnected internally to the added extra coils using the same magnet iron motor stator, now become simultaneously, a generator infinitely reconnected like a streetcar motor. It is kept running perpetually without being connected to the grid - amazing kid. That was the car-motor Tesla used linked to two antennas connected with a frequency generator to driving around without a battery. Why is that suppressed by universities? In both cases the generated voltage will accelerate expanding and creating more electricity right up to the critical resonance frequency like a streetcar motor needs a conductor, as explained in the previous UREE application, making bigger profit. Watch Tom Valone

http://www.youtube.com/watch?v=MxsiCBzeYkw

My Donut Atom theory postulates that magnetism runs on a dual light rail from the sun. Both ideas are not accepted by the establishment. Even Tesla's many experiments were not fully understood by scientists. Nearly all of Tesla patent drawings have only one wire circuit, the ground and open ended like the sky the other wire. Combining the space energy linked to the ground can be utilized either by a Tesla antenna combination as described in my atom theory or explained in Herbert's magnetic gravity supplicating the principle mechanically in a UREE system.

The Rockefeller-Ford-Koch cartel decided not to use free electricity and pushed their dirty oil-automobile technology never improved in 40 years - still the same consumption like my beat-up VW 30m/g. Tesla's single wire circuit got frozen out replaced globally with a two wire circuit system not possible to get free energy if you understand my energy path.

Politicians controlled by the energy cartel do not want free energy as Tesla being frozen out and destroyed his popularity. Only ELOHIM the other side of the equation will soon be coming and put a stop of corrupted cartels and end or atheistic evil civilization to save LIFE fast declining on this planet. Watch the sky 17 September 2015.

Thin Film Technology – (Pearl #216) A Bigger Jonah Fish

Designing 16 Ultimate Renewable Energy Engines give us free, nonpolluting and cheap electricity extracted from magnetic **Gravity, Air and Water** using new generators outfitted with permanent super magnets. It keeps expanding into bigger business possibilities, now reaching the point of separating boys from the men who have decided to become billionaires. The UREE keeps expanding to become a bigger fish.

The concept of the Ultimate Renewable Energy extracting free electricity from nature with relatively old technology seems really simple and continues to accelerate to be a bigger Jonah fish and should get global attention. Once it is better understood, the underlying fundamentals in physics linked to a theorized magnetic energy trail will lead to many more UREE applications o**bsoleting batteries by using thin film** technology. No kidding.

Again, similar to the Wright Brothers first airplane, they never imagined that a Concord-jet could be outfitted once more with cheap free energy. I nearly reached the unthinkable stratosphere, superseding all previous 16 UREE models, when stumbling upon **Thin Film Technologies.** Or perhaps my thinking possibilities have reached the critical resonance frequency not very well understood by scientists. Some still think that the UREE is NOT POSSIBLE holding on to their atheistic unscientific evolution religion.

I keep checking the Internet on what general science is teaching about magnetism to realize that I cannot find what I am trying to explain. I am not an expert and not qualified, but I keep trying to find a match with my philosophy. I came to realize that magnetism is still a wide-open subject surrounded with a lot of controversies and seems new discoveries show up daily in our world, like the use of magnetized water.

My UREE concepts need a lot of experimenting to learn the mystery how magnetism could be amplified in a better generator to access the perpetual electricity embedded in the infinite energy. Twisting wire loops seems impossible. Just tracing magnetic lines to find out that the polarized magnetism again conform to the ∞ energy loop perhaps needs a special pearl to explain it.

When two opposing energy loops are forced together, time becomes a factor in the energy exchange equation like ($\infty E = $ **Zero Time**). Just think of that cheap nuclear bomb we exploded with the unstable enriched uranium and plutonium. Together, they give a bigger bang with an energy exchange we can watch on TV, now feared it could be done by terrorists sent by an oil tanker could destroy NYC prophesied.

In the UREE case, we must apply the same laws of physics and speed up an UREE to release more energy as learned from the energy-time formula. But magnetism is a less dangerous force. It was created to maintain LIFE as postulated in my Babushka concept books. Just building destructive bombs and playing around with a high-tech phone gismo gets the allusion we got it all but have forgotten how food is grown and snicker at anyone who has metaphysical opinions outside of political correctness.

This third edition Babushka egg concept book keeps growing, just like the Wright Brothers first airplane started the race for bigger ones. Read last page on the end of my 9th Babushka book which now becomes a far bigger UREE implication following the energy trail according to the electrical three finger hand rule that when a current flows in one finger direction, a copper wire is surrounded 90^0 with magnetism.

Finally, somebody dared to define what magnetism really is. It is much better explained from the metaphysics to clarify why a copper loop with a magnet passed through it creates an electrical pulse. Many pulses create electricity, still a mystery unless you read Babushka egg concept books. Ending my energy book with 16 UREE motors nearly was superseded by new science applications. Now to find out that perpetual free energy could be expanded and extracted from **thin film technology**. Look at a loudspeaker design investigating Faraday's principle of induced EMF and analyze how it works linked when a modified magnetic coil is moving copper windings crossing over creating a biases increasing magnetism similar to the UREE #11 to #16.

Usually in a conventional coil one loop is wound next to the next loop all going into the same direction put together with a machine just turning from a wire spool. Another way to make a coil just etch on a rectangular Mylar strip parallel all the right turn copper loops and on the other side the left turn loops in parallel but a little tilted to allow a small air space across hatching for magnetism to pass similar to RODIN coils linked to Faradays laws.

Next spot weld on one side the copper loops together which is the center of the ∞ loop similar to a through plated hole in a printed board connecting electrical one side with the others. Now shape the Mylar sheet with horizontal copper lines heated up and now cooled creating a hollow tube which has inside and outside the loop pattern should look like a loudspeaker coil and if osculating super magnet across loops could become another UREE.

It works when one loop of the biased magnetism from the previous voltage and will accelerate crossing over on the inside loop will generate more electricity to the next loop up to critical resonance frequency just like a streetcar motor needs a conductor or microchip to control the energy otherwise would explode. Since the coil has an open slit on the side but can be shaped in a round tubing wheel able to pass through magnets aligned like pearls on the inside in a circle and turning the wire Mylar tubing at high-speed. That is once more another model UREE #18 again a rotating UREE linked to perpetual motion.

I hope by now to understand that the UREE Invention is different to electric generating technologies. Every motor-generator has a bunch of coils wrapped around heavy iron mass to produce magnetism, which in turn pushed the next polarized coil-iron. To gain electricity with higher magnetism many coils must be looped and sequentially added to get bigger power to magnetize a bigger iron core becoming more massive. But the UREE invention uses super magnets, already a much bigger magnetic force, not moving as lighter wire coils have lower inertia. In addition, each single coil is biased creating an alternating double pulse. A printed board generator can be moved faster through lower inertia, and when linked to end-motor crossed over, it produces bigger amplitudes.

Expanded a UREE to Linear Logic

Applying a UREE concept to another case 38 years ago, I worked as a research instrument maker for a German scientist, Dr. Oskar Heil, who in 1970 invented an Air Motion Transformer, which moved sound at a higher speed producing a 10 times stronger loudspeaker at a higher quality. His German research laboratory at (EIMAC) San Bruno, California build self-focusing klystrons applicable for a microwave oven as some bigger klystrons where applied to the first Linear Accelerator (SLAG) linked to the Silicon Valley Palo Alto Stanford University in Californian. The microwave technology was connected to a side project similar to JET ADAM Transducer Accelerating Ribbon Technology.

A mechanical accordion ribbon oscillates with a bias current switching in a magnetic field surrounded by a big super magnet. If we replaced the aluminum metal deposition overlaid with copper winding loops embedded on both sides of the ribbon and crossed over to another unit, it could possible amplify electricity according to the UREE principle. When the stipulation of the linear motion applied in a ribbon-loudspeaker is further developed, it can make it possible to have many more potentials in thin film technology all dovetailing not known 30 years ago.

To widen a science horizon applied now to new thin film technology using CRYSTALS osculating at high frequency that is linear motion also, but extremely tiny giving it miniaturized possibilities. Again, imitating the UREE invention principle in a hybrid electric generator design could be envisioned useful for computers telephones and robots replacing expensive batteries. It could obsolete pricey lithium batteries and zillion carbon-zinc batteries technology linked to solar energizing crystals revolutionizing another industry. When I found out that electricity can be produced in a crystallized thin film UREE application now perpetually extracted as electricity therefore cheaper and environmentally friendly.

Just think of the possibility that the size of your flat Apple tablets or iPad using thin film technology combined with CRYSTALS applying "linear motion" patterned like etching microchips would give us enough electricity to drive a car or fly an airplane without fossil fuel. It is just a continuation of the energy trail as technology never stops could be miniaturized now replacing batteries giving forever electricity fuelling our million gismos creating another big time business.

21st Civilization Roadside Reflection

Babushka egg concept books are based on forgotten logic exposing poor memory confronting the atheistic establishment once more with what was never taught in any university. The issue is caused by global **relativism** that usually dovetails in political correctness corrupting and destroying a culture. If you belong to that class, do not bother to read my Babushka concepts and be surprised when destined to perish by suppressed out-of-control science perhaps possible to survive GOD'S WRATH once more terminating a civilization with another asteroid.

So why is it so quiet around the university? They must be facing a tough choice of chucking so many unproven unscientific opinions that are collapsing if you are honest. The establishments promised since 2008, but have yet to explain why the Standard Kilogram IPK in Paris and its 6 sisters have changed in weight. Nor can they explain why light is 300 times faster as measured in Princeton University and CERN?

Belief in unscientific, illogical evolution theories makes it impossible to explain physics and what we see in the universe, interpreting galaxies photographed with the Hubble telescope, postulating ramshackle fairy tales. If you are entrepreneurial inclined, you should become a billionaire before the gold rush starts for free electricity in the next few years. Opportunities like that only come around every 100 years.

Thinking why God is so angry examining the atheistic university priesthood that has forgotten the reason why institutions of higher learning were founded. They changed the purpose, as greed became the prostituting motivator for government grants. Why charge outrageous fees to students going deep into debt? Education should have been free as demonstrated by disadvantaged third world countries like Cuba or Costa Rica so that our grown up kids better educated getting superior jobs in the future paying our retirement.

Therefore the universities are not very bright cutting off the branch they sit on. They have become totally corrupt and forgotten why they exist. Nevertheless civilization will continue. There are other options such as teaching on the Internet the old-fashioned way, which is crucial for anyone growing up being better educated, a precious commodity which makes us human. If you are unemployed, why not become independently wealthy and throw out unscientific evolution religion being a hindrance causing a mind to succumb with the lights "off"?

Instead, consider an unpopular Christian-based ethics that created Western Civilization and survived 2000 years. It is rarely applied anymore, now doomed according to the Bible prophecy verified by documented Newton's laws. His science principles are not taught anymore in our universities. Unbelievable! Worse, why is 50% of his writing still suppressed? Ask any graduate student who no longer can think in a 360^0 perspective mostly brainwashed in relativism difficult to delineate what is evil or good making bad choices.

If you wonder why so many government bureaucrats became corrupt and abandoned what was once taught to every child that has consequences in a physics denying the obvious, forgotten metaphysics. Or can a MIND be explained if you have one not functioning. Why is mankind the only species on earth being spiritual? Like the atheist says, "Thank God I am not religious."

To be better educated, study the free Babushka books forbidden in universities that reveal and define concepts duplicated in nature many times. To illustrate hidden laws in physics for my grandkid, I use analogies to understand nature better. Start with a new Donut Atom theory, the smallest egg to see what is invisible to our eyes, provided it is not smashed by CERN, a stupid idea only useful if you want to write fairy tale opinions to convince uneducated senators to appropriate more money.

Investigating nature from the metaphysic domain will open the door explaining how complex atoms work on the inside as replicated in photographed galaxies, which is the same egg only bigger. It can be compared to a cuckoo clock, another Babushka egg designed with a lot of features to allow comparisons on my grandkids' level. His level is far more advanced for his age compared to many blinded, brain-dead professors. The beautiful galaxies need interpretation. What we see and compare in a new energy theory requires a little logic based on physics. They will no longer wonder why the lights are on in the night sky.

We have the advantage of understanding from applying modern radio telescopes using infrared, x-ray and other frequencies to expose more knowledge. That by itself should widen our science horizons as the laws of physics are always duplicated down to DNA-DNR forming a micro-universe, another egg replicated in the sky as the law of physics applies there, too. If you see one egg and visit a chicken ranch, you will know there is a relationship. My free Babushka concept book raises it to a higher level and asks what came first, the egg or chicken? It is not yet defined in the halls of higher learning. Now you will understand that many professors teaching our kids are brain-dead. The proof is easily found. Just investigate the results of the universities' teaching governments, "What you sow you will harvest."

Look around to make my case. Notice the environment: so much extinction linked to horrible pollution with a fishless ocean contaminated with oil-plastic-gunk hundreds of miles long, floating islands everywhere in the ocean and nature dying the world over. It is always associated to a PhD believing in unscientific evolution religion denying divinely revealed morals that no longer cares to support LIFE for our grandchildren.

Obscene profit is the only motivator. Where are the honeybees to pollinate our fruit and vegetables? They are screwed up by GMOs pioneered in university laboratories financed by corrupt evil cartels. Why is food for mankind permanent genetically altered mixed with embedded chemical poisons and deadly pharmaceutical drugs hidden inside the food cell structure causing many diseases?

I saw a Frankenstein-Salmon-fish on TV in 2012 that was three times bigger, being mixed with boa constrictor genes. It will be become a disaster like carp from China introduced accidentally in the Mississippi river now replacing all the other fish specie. Have you seen a butterfly lately or even a robust housefly and vanishing honeybees still manufacturing annually thousand of tons ROUNDUP killing everything like DTD to make obscene profit? Worse Monsanto cartel removed the original gene blueprint sanctioned by the FDA causing seed extinction on a big scale no longer procreating exchanged for inferior cloning causing declining food production?

On TV in 2012, the news announced a massive failure: government statistics tell that the global food harvest was 25% lower. Read my Website. The Apocalypse has started. Do not say it is an act of God when God does not exist or blame it on global warming based on many lies; therefore, it is not possible to explain it with counterfeit science. When you are poorly educated in lies and turned into a brain-dead atheist. Believing in an unscientific evolution religion has irrevocable drastic consequences.

Going back to the ULTIMATE RENEWABLE ENERGY ENGINE keeps growing. It now passed 16 models, not allowed by corrupted energy cartels, and the US government controlled by debased politicians not interested in free energy extracted from magnetic gravity, magnetized air and even magnetized water. The latter increased food production 30% in some countries, as presented on the Internet. Some think it must be religious seeing Bible verses quoted. Check out magnetized water at Moreplant.com. If you have become depressed from the roadside reflections be comforted by what the Bible is teaching and feel much better to know how it will end.

Ultimate NON-Renewable Energy #17

The Ultra Energy to travel with ease throughout the universe could be availed for everyone as the principle in nature is only revealed in the Bible. It is demonstrated in a resurrection viewpoint we all can watch and experience during springtime after winter new life appears and made discernible on a higher level by Jesus Christ's 35 BC resurrection well documented. With the realism of the many unbelieving skeptics, let's look at the metaphysics reality of Jesus resurrection overlaid in physics investigating some historical data examining genuine witnesses. Hopefully, it could persuade some atheists of the unknown physics of a new Jod dimension and applied to biblical principles could tell how it works.

Too most the metaphysics is a mystery especially for those who are not familiar with the many unusual science facts overlaid with invisible laws of nature described in nine Babushka egg concept books. This pearl becomes the last UREE and is just a little appetizer to entice you discovering the full menu. This could be the most important Pearl for every person explaining how we can overcome our mortality and travel in another future dimension.

Remember the analogy in physics a butterfly was designed via a detour. A caterpillar in his wildest dream could not imagine that one day he would cross many countries without borders free to go without limitations. It is similar explained in biblical truth, not possible to explain all and can only point to so some unusual concepts not allowed in universities. Do not look at my finger but what I am pointing to.

Going through a cocoon is a drastic experience we never know what hit us. Recently experienced and lost consciousness but waking up being surprised seeing five firemen looking at my face and my wife hysterical on the phone. Asked what is going on being mystified. Now I know when we go from this planet is not a fearsome experience like a caterpillar we will leave behind everything we have accumulated, naked we came and naked we leave, except LIFE returns where it came from.

However, life was modified to take along the embedded experienced data of an immune imprint acquired during a lifetime which will be needed waking up on the other cocoon side for the next life cycle to live forever. Notice a tombstone has a name of one person with two birth dates engraved. A person is uniquely designed with 60 trillion DNA genes forming proteins impossible to be duplicated ever to exist another time. Each individual name preexisted and was entered in the BOOK OF LIFE that starts the cycles of our existence. At the end of our caterpillar cycle will be evaluated if we passed the test to continue or being terminated but must have agreement with the judge as a previous legal contract was signed on our behalf by our parents.

They messed up but the contract continues until paid in full. Unless you got an influential lawyer paying off your parent's sin-debt passed on to you added with your own all must be paid in full. Otherwise, the divine judge will not be happy. He made provision: if you are poor, but it needs our agreement one way or another. If we want to be his adopted child, God will invite us into his private palace for that reason, which is the Star of Bethlehem golden city. (Revelation 21:10)

The infinite ∞ math sign applied to our resurrected life cycle is demonstrating again in entropy principles rooted in physics similar to light energy cascading to a lower level and must be returned for a recharge to continually refueling every atom for the cosmos to exist. It is like our electricity network with two wires the black plus wire has the energy but the minus white wire must return to the same generator 10,000 miles away for the lights to be on.

Being connected management knows how much energy was used and sends you a bill. If that energy is interrupted along its journey or switched off in the cosmos control room the whole universe would collapse in femtosecond and our mortal body too made of trillions atom gene-chain connected would disappear at an instant. That same control room determines how long we will live with the lights on in our MIND. The MIND works as a complex system linked to the metaphysic intelligence center and analyzing my becoming unconscious, not happy waking up lying on the floor rather would see my Lord.

But seeing my wife again crying and wondered why all the commotion and five guys from the fire department asking me questions? I just experienced an electrical side-switch temporarily turned off in my MIND similar to the Hoover Dam wrongly activated would have disastrous consequences for the whole country. Accidents do happen for an example a Russian scientist vaporized the whole nuclear-plant Chernobyl in Russia when he turned the wrong switch.

I generated 4 years ago a time graph explained it better in a recent Pearl #225, "Is Einstein's Theory obsoleted linked to Magnetic Gravity," which rationalize our universe creation bathed in a TIME dimension without no atoms could exist to hold nature together to prevent a cosmos collapse in femtosecond. Some possibility can be visualized how a future new heaven-earth is designed and explains a little the why and what was previously planned for mortals but will be a one notch higher as a system looking a little closer at the graph.

The Time-graph shows a TIME dimension, which was the reason that God created a Daleth dimension (Daleth is #4 = interpreted from the Hebrew HANS code "this world in this time"). The Daleth dimension which is where we live was caused and inserted in the preexisting cosmos when Satan's rebelled against divine laws in 4488 BC and will run its course to

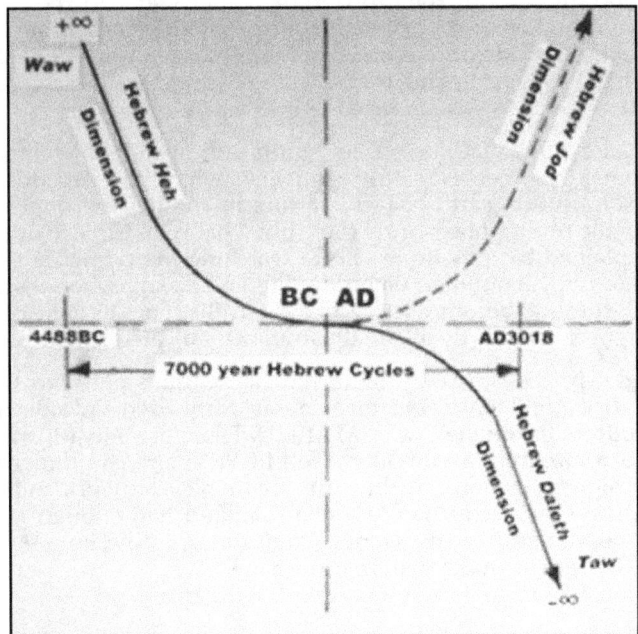

correct the problem. A Time Dimension will no longer be required after the emergency Plan for Mankind was concluded in 3018 AD.

The newly created being appointed for replacement of fallen angels called in the Bible MORTALS and is based on a two entropy level one is physic the other metaphysic-spirit and function together in a very complex integrated DNA-DNR system producing protein controlled by MIND intelligence down to every atom. Ignorant atheistic scientists now have a problem to explain how was it possible to evolve DNA needs protein and protein needs DNA in an infinite cycle, if even one gene is misplaced out of 60 Trillions you would not be able to read my story even as a ghost. The cosmos design reflects the same principle and is controlled by the same laws of physics interconnected to the laws of metaphysics. Ask my grandkid will have a better answer showing more commonsense questioning how could a big-bang NOTHING explode and evolve in a cuckoo clock with a complex designed gears and a cuckoo bird even wind itself with some unknown energy which has embedded intelligence like a big computer.

May I present my grandkids' wisdom to discard an unscientific atheistic evolution religion from the Middle Ages that does not make any sense for children? Looking through the telescope observing a vast universe we notice a condition it seems like one frame of the movie seeing frozen explosions or contracting space dust forming galaxies seems without a time dimension. However when time stops we would have an implosion down to a DOT. But time still exists but can only see one frame of the movie.

Satan's rebellion in 4488 BC caused the Time Dimension that stopped the cosmos movie for a while but only temporarily. When the correction was inserted will fix the problem thereafter will start the cosmos movie again. God created a "time dimension" embedded within every atom now discovered and explained in Babushka egg concept book. Or just ask NASA what is ticking on the inside the nucleus which became a useful atom clock to synchronize the Internet?

The vertical exponential line on the graph indicates the metaphysical laws showing the process creating a universe, which has embedded the blueprint of a future earth-heaven starting in another Jod dimension. Only the Bible reveals how we get there but what is invisible to our eyes can be explained by looking at nature watching the changing season after winter new abounding emerging life and seeing when a seed is put into the ground becomes a tree, or a caterpillar goes into the cocoon come to light as a butterfly all are illustrations indicated a resurrection.

The human species is extraordinary unusual curious outfitted with a divine higher intelligence controlled by an embedded invisible spirit MIND replicated from the ELOHIM MIND. Therefore only Mortals are capable to communicate with God the ELOHIM and can understand how he designed his universe, how the creation cycles work and what was the purpose the creator fashioned mankind had enough angels. Why did he miniaturize his likeness conforming to the infinite ELOHIM MIND, **"Let us make man in our image."**

You can find the answers nine (9) Babushka egg concept books comes with a plan to explain it with science linked to the metaphysic for a full rounded perspective. It is fun to read not preached in church.

The Principle of a Resurrection

Looking at the Jod dimension on the Time-graph is pointing to a future promised new heaven-earth and Jesus revealed that it would be a familiar world for resurrected people investigating the Gospel account.

To understand the Jod-dimension we should just compare the body of Jesus before and after his resurrection having the same exterior like ours. That could give us a foretaste what it will be contrasted with the future laws of nature when LIFE is continued designed in two stage systems after passing the death cocoon event. Speaking of a cocoon, pay attention to the Bible account from the science perspective not usually mentioned in church most Christians overlook.

To follow the path of infinite energy, let's investigate the greatest historical resurrection of Jesus Christ (35 BC) demonstrating a principle seen everywhere duplicated a billion times in nature every spring showing new life resurrected from winter cold which was proven with science if you read the Gospel account.

By carefully analyzing the witness which is demonstrated by the Sanhedrin Supreme Court in Israel who crucified Jesus and the Roman authority to testify a total mystery why a body had disappeared secured by a number of soldiers. The chief priest and the Pharisees elite went to Pilate and said lets make secure the tomb, otherwise the disciple may come and steal the body as this last deception will be worse than the first. (Matthew 27:63)

One of the most influential men, wealthy Joseph of Arimathea, requested from Pilate the body of Jesus. He owned a grotto for his own burial and together with his friend, Nicodemus, a member of the Sanhedrin who brought along a lot of expensive myrrh about seventy (70) lb. and aloes which works like glue for embalming and wrapped the body in strips of linen according to Jewish burial customs. (John 19:38)

The body was taken down from the cross at three o'clock afternoon now hurriedly needed to wrap the body before 6 o'clock in a nearby grotto according to the biggest holiday regulations. The custom was to mummify a body if you had the money with lines wrapped tightly around with myrrh and spices in a mixture like a plaster solidifying quickly similar to shellac glue extracted from insects. Except the head was not bandaged and only covered for identification with a shroud now emerged exhibited in Turin.

Sunday early morning the soldiers being blinded by light and saw the heavy grave-stone mysteriously rolled away without anybody around and greatly confused hurried back to tell the Centurion officer who immediately came to investigate and discovered an empty cocoon bandaged shell with a body missing.

Perplexed unexplainable pondered why a solidified bandage of solid plaster not altered or broken but was empty hollow on the inside with a body missing? How did the body vanish without showing the evidence of an undisturbed solid bandaged shell? A few hours earlier three women excited rushed and called the older disciples of Jesus; all were convinced seeing the empty burial shell remembering now what Jesus said before he would be resurrected like Lazarus demonstrated. The solid bandage shell was immediately confiscated by the religious authority for obvious reasons Nicodemus protesting, but the cloth blanket was missing saved by three woman showed up in France recently investigate the Shroud of Turin. What is so extraordinary it has the crucified image of Jesus Christ embedded like a nuclear high energy flash now investigated thoroughly by science and concluded it is authentic but cannot explain the negative photographic image with present science technology still a mystery.

On top heard from the temple management another unexplained mystery that a 500 year old embroidered half inch thick carpet used as a temple curtain ripped apart from top down to the bottom in open full view of the Most Holy gold plated objects. That curtain is fixed on an upper beam with big bronze rings and works like a wall to separate the outer court and inexplicable tore apart at precisely three o'clock in the afternoon after Jesus cried out lout "I have finished" (the great plan for mankind). It was accompanied with even a bigger mystery more amplified unexplainable why was there three hours darkness starting at 12 noon in Jerusalem ending three hours later with Jesus last words followed by tremendous earthquakes. (Matthew 27:45)

Even seasoned Centurion officer became fearful filled with awe and said, "Truly this was the Son of God!" Check the facts reported from the enemy of Christ that paid a lot of money to silence what happened. Or more facts revealed in science read my ancient clock Babushka book #3 describing an earth axis wobble causing the last flip flop explains the reason why a sun was no longer shining and set off darkness around Jerusalem which has a rationale from science, check it out in Babushka eggs.

Following the metaphysics mystery of unexplainable events hidden from science but can now be uncovered by recognizing a "resurrection" principle that will demonstrate scientifically an infinite energy exchange by using our calibrated mortal eye sight designed in a visible light-band frequency spectrum. It can be equated and interpreted to Fraunhofer electro-magnetic spectral lines, that tells us Jesus resurrected body later appearing again recorded in the Gospel.

Jesus after his resurrection for 40 days looks the same, eats speaks and laughs with his friends, but there is the difference in atom physics and walks through solid stone walls indicating the absence of a time dimension explained in physics or better in a new theory described in Babushka book #6 about Donut-atoms and #7 What is the Time Dimension?

Jesus is seen seconds later 30 miles away and finally levitating toward the clouds without a helicopter surprised 500 amazed witnesses. They kept looking in the sky until two angels appeared meant the show is over and said Jesus will return the same way in another time. A timeless dimension concept will be new to many and postulated some ideas in free Babushka egg concept books to broaden our mortal MIND.

Following an energy path to understand the resurrection principle on a child's level takes a seed put it into the ground or watch a butterfly emerging from the cocoon seen on NOVA-TV science programs. If you are educated a little higher investigate physics let's look further demonstrated by Fraunhofer spectral lines could help understand modern electronics. The difference of elements, TV, Telephones and the million electronic gadgets is defined by a specific frequency of an oscillating magnetic energy explained and summed up in previous 16 energy UREE Pearls #201-106.

Therefore MAGNETISM is a major player in the sub atomic world as we learn to manipulate it and made into a motor will serve us like slaves used to do or give us free electricity if converted inside a general electric generator. Having investigated the energy forces below the ocean linked to GRAVITY following the path of energy now embedded in WATER and LIGHT from the sun ending with a **UREE** electric generator utilizing physics not taught in universities.

A Little About the Jod Dimension

For the cosmos cuckoo clock to restart the Jod dimension system is reminiscent to wind up once more the weight of my cuckoo clock outfitted with a chain. A cosmos is controlled like a computer electronic clock, God only needs to go back which created original a TIME dimension for the Daleth dimension and remove two embedded neutron blockers to change the intelligence code forming atoms. It is located in the exact middle of invisible light frequency spectrum, a little sliver on either end of the visible rainbow color only our eye is responding seeing rainbow spectral line frequency measured by science 4000-7000A^0.

On either side of the frequency spectrum is embedded an invisible intelligence HANS code 7:5 when altered works like a computer switch to make an invisible energy to become visible with a time base. It has embedded massive information like our two byte ratio (zero = one) multiplied thousand intelligence codes like DNA-DNR captured in a super computer to pass on to every atom how to form genes ending in GESTALT and function only when properly timed within a Time dimension. We can photograph it in expanding or shrinking condensing galaxies with our telescopes or study the system under the electron microscope watching the effect of DNA as the same laws apply down to a micro silicon chip computer controlled by magnetism embedded in atoms. But the cosmos Jod dimension energy switch is much more complex linked to a gigantic cosmic intelligence ELOHIM computer designed with a complex comprehensive code of a 7:5 ratio controlling metaphysic laws compared to our 2:1 (zero-one) byte code controlling our computers.

Operating a cosmos required an infinite intelligence to give direction when the higher dimension energy is switched cascading to a two looped 2:1 ratio energy to the second entropy level which can be visualized graphically in a blood particle dished in on both sides in the center. My grandkid starting to become educated by his grandfather needs illustrations to understand science.

The cross-section of a blood particle is reflecting a double looped ∞ math symbol illustrating a moving energy path maintaining our LIFE. It works like a wheel within a wheel on a sub-atomic level of our body and its function was copied in a new UREE invention. But all the laws of physics and metaphysics must be controlled by a gigantic cosmos computer ELOHIM linked to an Alpha-(+ONE)-force which still maintains his creation controlled by infinite intelligence information like the biggest super computer code.

The initial computer code was duplicated in the beginning and embedded into Adam the first created being linked with an intelligence 7:5 Hebrew HANS codes embedded in a MIND - computer to live and continue to fuel LIFE extracted from the cosmos source. The first divine breath now replicated in every computer mortal MIND start the system of LIFE propagation rooted along a complex designed intelligence DNA code controlling 60 trillion genes embedded in my body still must be fueled by electromagnetic energy for every atom to exist. If you have a better opinion explaining LIFE, let me know.

However the previous existing COSMOS before the Genesis Bible report and expand our vision to understand science better, God revealed that the angel dimension operates on a different spectrum without a TIME dimension. But to implement the plan for mortal mankind required new cosmos laws that needed to be inserted to fix an existing problem linked to Lucifer's rebellion that created a temporal universe.

It is still bathed in infinite light but comes now with a TIME-dimension creating something modified later described as darkness like a night cycle (evening-morning) announcing a new creation designed for mortals governed by different laws. Even the angel rejoiced in exuberance being amazed seeing God's original Heh dimension creation now tailored to a similar Daleth dimension although they did not yet discover the reason why. Mankind appeared later. From our mortal perspective we recognize the principles that nature is controlled be a MIND bathed in time to function in a Daleth dimension. It previously did not exist and was fashioned to fit our mortality still designed to live forever but on a two-stage system. It is analogue illustrated in the caterpillar-cocoon-butterfly indicating two birthdays embossed in every tombstone demonstrate for my grandkids to understand the metaphysics.

The Jod dimension (Butterfly stage) is similar to a Daleth dimension (caterpillar) but one notch higher, which can be explained much better with the Hebrew Alphabet Number System (HANS), which is a handy code to widen our understanding horizon. The number 10 in (HANS) is an elevated HAND going toward MEM a reaching HAND = 40 analog Ausdehnung like a wide-open ocean without a horizon.

The substructure of Daleth = 4 is elevated to 40 because the home for redeemed mortals (butterflies) and the Saints is one notch higher both destined to live forever but need an environment similar to earth to be happy. Remember the structure of mankind can be expressed in the (HANS) number code (1-4-40-400). But the Daleth dimension illustrated in the Gregorian calendar has a **TIME DOT** as time is cascading from 4488 BC to **BC-AD** reverse back to 3018 AD.

The DOT in the middle tells us that the nature of TIME linked to infinity is only temporarily and will be ending like a flywheel run out of energy unless refueled once more. On both sides of our calendar representation is a ∞ math symbol telling us where it started and ends. The **Jod** dimension explained on the (HANS) code is linked to the intelligence Alpha-(ONE)-force system to control every atom no longer 7:5 neutron ratio (7:5 x 5:7 = 1 Alpha) but now merged back to the original [7+5=12 or 10+2] code that makes everything to exist forever the way it was before. The letter **Jod=10** tells us that the letter Alpha =1 is on a higher level (10), but continues once more with B=2 in Beth (10+2) still divided in an angel dimension realm going toward a new creation of a [new-heaven and new-earth] cosmos system. It will be populated in the future with LIFE once more but with resurrected billions of mortals (butterflies) from the earth governed by million appointed Saints, which represent the entourage of Jesus Christ.

The (HANS) code made it plain for us combined with the math of physics to clarify our mortality. Seven (7) represents the Daleth dimension to reflect the principle of an embedded "sword-evil" Zayin-7 which is Beth-2 added to previous Heh-5 having a rebellion causing a temporal interruption modified to Zayin. But returns back to the (Jod) dimension to what was before Satan rebelled [10+2=12-Perfection] the original cosmos universe still fueled with infinite ∞light energy.

When the DOT in the center of the ∞ math sign connecting to two energy entropy loops representing a time dimension if it is taken out it will become a full circle of 360^0 again still moving energy but "without" the TIME dimension now become the Jod Dimension. In math "perfection" = (12) is Jod expressed on a higher elevated 2=Beth level 12=[10+2]. That starts Genesis over again with the first letter in the Bible-Torah "B" as in Bereshyth continued like the previous first Bible word but now one notch higher elevated level which points to a replication of an Adam-Eve story with the creation of many more planets and galaxies to be multiplied ∞ over and over. The energy is still flowing forever but now only without the Time-dimension no longer detoured. It returns to the original condition in a full circle like the cosmos movie started to run again expanding and contracting atoms fueled by infinite light energy creating a rotating universe.

The Universe will continue once more what previously was interrupted by Satan's rebellion and expanding into a bigger cosmos. I am glad GOD will give me a bigger MIND more suitable to become a well-educated Saint. The new creation with the old math [∞E = Zero] is still around but becomes just [∞E = O] a full circle. The energy fueling a New Heaven / New-Earth must be regulated by the same cosmic energy laws that never changed, but then switched, the energy railroad to a full circle no longer controlled by a detour Time-dimension as the Plan for Mankind was concluded. That will change nature to get back to its original eternal design condition before Satan's rebellion. It also will end the mortal school system to learn about and be inoculated with good and evil to experience death, tears, deceases, pain starting with a new butterfly cycle stated by Jesus the LOGOS who is the inventor of the system. Read the last page of Revelation to testify:

"BEHOLD I WILL MAKE ALL THINGS NEW"

To experience new LIFE embedded in a new world order will no longer be mortal subject to death, pain, tears now living in a new Jod dimension no longer governed by TIME forgotten NO MORE. But it is still fueled by the same one loop ∞-energy Alpha-(+ON)-force higher energy cycle to maintain Eternal Life but without a TIME dimension. Go back and read Babushka egg concept books for a fuller perspective not meant for those with a closed MIND prevented from understanding God's creation. A closed mind is caused by God with a shut "off" light in your MIND because of rebellion expressed by an unrepentant behavior ignoring God ELOHIM existence denying plainly revealed truth. To prove when the lights are "off" notice that most of your friends will just leave you alone to be isolated demonstrating their contempt or hate you for no reason just mentioning the Bible.

The Bible teaches they will be excluded forever receiving Eternal Life not invited to the Grand Party standing in front of God's Throne. That is onetime event happy celebrating being redeemed from mortality and receiving a gold metal grand price and in addition receive a beautiful (butterfly) eternal body praising God in the midst of billions of angels. (Revelation 7:9)Read the Bible road map while it is still time. Do not miss the higher level of privilege being invited to the heavenly Grand Party belonging to the Firstborn adopted into the entourage of Jesus Christ, the King of Kings, Lord of Lords. Do not be deceived by the cares of this world. To settle down to NOTHING spiced with evolution lies to join with the rest of the multitudes tarnished with evil, not forgiven, only to vanish in a fiery galaxy on the last day when time is no more.

Phony evolution fairy tales full of deception invented by Satan taught by the atheistic priesthood will not pardoned you standing before a Holy God looking at the teeter-totter balance to give or take away Eternal Life decided on the White Throne event. Why not turn the lights on in your MIND to discover unimaginable treasure of wisdom comes with happiness and peace to get a foretaste of God's provision. It gets back to the principle light always will penetrate darkness which is needed see what God designed for his pleasure because I was created if you remember Genesis personalized:

"Let us make Herbert in our image."

God's plan will be finalized for you and me when our name is called from the Book of Life on last day when TIME is no more. It is more fun standing among the pardoned billions of redeemed butterfly-resurrected saints praising ELOHIM for His great gift of Eternal Life. The key to understand the last UREE is not easy and takes a long road but is worth every mile gaining a higher perspective like climbing up a pyramid takes some effort but will be highly rewarded on top seeing more. It starts with the foundation and asks what is the PURPOSE of mankind, which could get you a ticket invited to be the guest of highest honor to be introduced as the FIRST BORN. Now celebrating in the biggest party the angels organized to enjoy yourself that EVIL IS NO MORE. That will establish a new timeless cycle living in a far-out cosmos no longer remember the old way.

Once more, I urge you to be convinced to receive a beautiful new resurrected body like a butterfly to be happy and have a party with the angels to live in a new dimension forever. You can start the journey from the first to the last 9th Babushka egg concept book designed to venture from the bottom to the top of the pyramid will understand the metaphysics more fully expanded to higher level denied in our atheistic culture not even preached in many Christian churches.

The Babushka book titles summarize a metaphysic overlay symbolizing in nine pyramid steps which points to a future Jod -10 = (Elevated Hand) of a New Heaven - New Earth, which is a totally new elevated creation symbolized in shape of a pyramid. The structure is very interesting with embedded mathematics, calendars and atom theories check it out in the third Babushka egg concept book and the many Pearls pointing to how to get eternal life. Just start reading the Gospel. It will acquaint you with ETERNAL LIFE one notch higher, which follows again the Heh dimension rainbow frequency cascading with intelligence on an infinite energy loop symbolized in the math ∞ sign from the Throne Room. But travels on two loops joint together like a balanced railroad or DNA strands illustrated linked to a center TIME DOT recorded on most tombstone for mortals.

You do not have to be highly educated to understand even science still have problems defining life walking through a cemetery in totally silence on this subject but some can know how it started and where it will end, embossed with two birthday dates on your personalized tombstone. The second embossed date is your resurrected body Jesus told us about from the other side. He is the only witness from the other side because he is the inventor of the system.

Your choice is to believe a real person or be stupid trusting in phony fairy tales invented by atheistic priesthood to stay in power. The Plan for Mankind mandatory required from the Creator-Inventor that all its members of a new world order would first be inoculated with a good dose of evil. Otherwise we have a repeat like Satan's rebellion upsetting a whole angel community once more. Evil will never appear again and works like a silent immune system suppressing wickedness will no longer be tolerated but immediately snuffed out in the bud by a watchful experienced inoculated Saints.

Become educated on a higher metaphysical level with the basement of the first Babushka egg concept to understand God's Plan for Mankind and get a ticket to the Jod dimension by climbing like a pyramid to the next book level will discover the road map to your treasure. But the light needs to be on in your MIND to understand God's Plan for Mankind and must be linked to the energy source Alpha-(+ONE)-force to be elevated to live forever one notch higher. Ask Jesus please forgive my trespass I am sorry, I want to live in your house by your rules and give me grace not deserved and want the gift from the ELOHIM - Eternal Life now declared for the last time before the final APOCALYPSE ends:

The Ultimate Non-Renewable Resurrection Energy
Is the fuel for mortals to live FOREVER!

[The resurrected Jesus stained glass window from the Munster Cathedral in Basel, Switzerland.]

Part 3:
More UREE Gravity Applications
Transportation and other Inventions

The greatest shortfall of the human race is
our inability to understand exponential function.

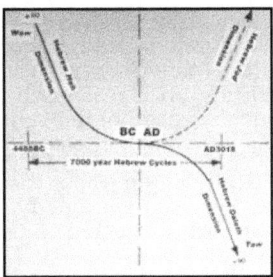

(The most important video you will ever see.)

http://www.youtube.com/watch?v=F-QA2rkpBSY&list=SP6A1FD147A45EF50D
http://www.youtube.com/watch?NR=1&v=VoiiVnQadwE&feature=endscreen

"I do not feel obligated to believe that the same God who has endowed us with sense, reason and intellect, has intended to forego their use." Galileo (1584-1642)

"Facts do not cease to exist because they are ignored." Aldous Huxley

The core of Menken's social philosophy was relative simple. He believed it is the nature of the human specie to reject what is true but unpleasant and to embrace what is obvious false but comforting.

The Great Challenge

Can you think of any problem of any scale, from microscopic to global, whose long term solution is in any demonstrable way aided, assisted or advanced by having a larger population at the local level, the state level, the national level, or globally?

Think about it ... will anything get better with increased population?

"Unlike the plagues of the dark ages our contemporary deceases (which) we do not yet understand, the modern plagues of over population is soluble by means we have discovered and with resources we posses. What is lacking is not sufficient knowledge of the solution, but universal consciousness of the gravity of the problem and the education of the billions who are its victims." -Dr. Martin Luther King Jr.

All aboard!
Next stop is the last misguided transit experiment in history.

History offers astonishing testimony in Jules Verne's uncanny novels depicting the evolution of science and technology for his emerging modern world. A quarter century before the advent of automobiles, Verne described in detail how Paris became choked with traffic jams.

Of the innumerable vehicles that wound their way along the boulevards, the vast number operated without horses. An invisible force within gas-powered motors made them move. It was simple and easy to handle. The driver, sitting on a seat, used a steering wheel and a pedal at his feet to change the speed of the vehicle instantly.

He wrote of an elevated mass transit system powered by compressed air and running automatically along tracks that would ferry 1,000 passengers every 10 minutes. He also mentioned a 500-foot lighthouse that dominated the skyline of Paris – erected in the area where the Eiffel Tower was later built.

"Verne saw the 20th century as a rather pessimistic place," wrote Piero Gondolo Della Riva, a renowned specialist, who runs a private museum filled with Verne artifacts. He simply figured that poetry books would have a hard time coexisting with science and machines.

In America, the automobile liberated the public from the collective tyranny of mass transit, of being dominated by fixed timetables and long waits at terminals. Passengers always had to get to and from terminals on their own – usually by foot, only to stand in line to buy a ticket. Then they rushed off to wait on platforms and fight for seats - always running the risk of missing the last train or bus home.

Instead, the private automobile made each individual king of the road, affording him the flexibility of going wherever and whenever he wanted. The automobile is treasured by Americans as the embodiment of freedom. Next to imprisonment, nothing deprives man of his freedom more than suspending his right to drive.

That pretty much sums up the dim-witted transportation policies pursued by many nations for the past 60 years. Starve the passenger rail system with scarce rations, while bailing out big auto industry with generous bonuses. Governments did not subsidize any rivals to the internal combustion engine, and almost pulled the plug completely on trains when they could not compete.

No international transportation system can prosper without massive governmental subsidies. Without a thriving transportation system, a nation cannot prosper.

Our civilization would have been stuck long ago if airplanes had not eased the pain, but the population kept growing. Now airports are plugged up, too. Looking back to when it started, Vandersteel referred to a car as an embodiment of freedom, but he was living in the dream world in 1920, when a car was new and the road uncluttered. Then, access to automobile freedom obviously arrived for almost everyone until...just watch the early TV NEWS in 2010, as ten thousand cars wait to cross the Oakland Bridge in California, where they must pay five bucks.

Such insanity became a virtue in the transportation departments that employ comatose bureaucrats who shuffle papers 30 floors up and sit in plush offices, surrounded with glass that provides them a fantastic view. From their superior height, they can look down at the average person and not be bothered by seeing so many drivers patiently waiting up to one hour to cross two miles of freeway in bumper-to-bumper traffic.

They just need to collect the five bucks. They live high above the honking horns and rude gestures of discourteous drivers in a hurry. It gets worse when you become confused and switch to the fast track gate, where it is impossible to turn around immediately will send you a $29 ticket in the mail. Creating confusion pays for extra bonuses I am sure.

It requires a solution. I do not mind paying higher fees, provided I do not lose my freedom during this economic collapse caused by a corrupt government.

That would force me to give up my car, which is still my dream passion. This dilemma requires a solution; otherwise, the prophesied Armageddon will arrive sooner. What is missing is an open forum where simple people have a voice, instead of the special interests that pay off senators and tell them what laws to make so they can spend our money. It is not fair.

Although I am now retired, I feel sorry for those who have to commute, but I should share my life experiences, at least to get it off my chest. So here I present how we can keep our individual free choice when travelling from one place to another in the shortest possible time, getting home just in time to watch the local football game. What a dream!

It is for you to ponder, all in one package; a special Traffic System for the 21st century, which would verify Verne's forecast for Paris and maybe re-educate bureaucrats about how to make us happy and no longer send outrageous tickets through the mail.

1. High Speed Bullet Train - 500 mph levitating on gravity.

2. When you leave the Bullet Train in the center of the city, catch the Monorail above the streets, since they are not bothered by stopped traffic. It could also run in an elevated position along the center strip of any freeway. It is still a free real estate that is not taxed.

3. Commuters would then leave the Monorail to get home in a personal electric car, waiting at the automated parking lot. This enables us to get to our destinations on time, but do not forget to plug it in at your transit stop! I forgot to tell you about another mini-car that would be waiting for us at the other end, to get us to our offices without getting irritated.

No more waiting for the banana buses that are lined up, all going in the same direction. It provides public economies of scale without sacrificing personal freedom. The ability to travel in a timely fashion must first be considered, if the government wants to wean automobile drivers off. Otherwise, there will be a repeat of what happened in China in 2010, where they had a congested freeway plugged up for nine days, with a thousand cars that could no longer move. It was totally frozen. It's time we think of other possibilities, which are presented in these few pages.

General Disclosure:

Another grand application of using positive gravity will forever change our civilization and could be very beneficial to mankind. It is being forced by an increase in the population. Space compression in the inner cities will eventually reach the saturation point with no return and will collapse, if you apply logic.

Population planners must now consider the data, which can no longer be ignored. Either that or we can wait for another political administration to come around. As explained, it must be seen from a multi-purpose perspective with an

Integrated mass transit, personal vehicle, transportation system

that facilitates a fuller understanding of the existing problem. The innovation lies in how an existing technology may be inexpensively packaged in order to solve transportation and space problems everywhere in our world's now overcrowded cities.

Please notice why every system has and will fail again which the planners in government will not consider. Think, infringing on the personal freedom enjoyed when driving an automobile will never work.

It is dependent on government control that forces a previous car driver to use expensive, slow-moving buses that are not integrated where I want to go. An example is the San Francisco Bay Area, where there is a long corridor called El Camino-Real that goes from San Jose to the bridge across Oakland.

The Spaniards used horses, but the speed has not been improved in modern times, judging by some empty buses that come by. It is unbelievable! It still takes 8 hours for buses that start from San Jose and go around the Bay to end back up in San Jose. That is no faster than the Mexican horses, as the transportation system is not integrated for travel around the Bay Area in modern times. The car is the only option in spite of the bureaucrats spending billions of dollars.

This is duplicated globally and I believe that we must have a fresh start and analyze the fundamental problem. Articles of new innovations presented here should be read by many governments around the world or by investors looking to profit handsomely with very little risk, due to the incremental build-up. It is a total concept that is much cheaper overall. The overpopulation will force us in that direction one way or another. Guaranteed!

I present five main elements of a futuristic, low cost Traffic System. That system would be half the cost of any levitated electromagnetic suspension train, such as the one built in Germany. It may cost the same as conventional trains going 250 mph, which are now seen in magazine pictures; or watch it on TV on the international channels, such as Japan and China.

Check out popularmechanics.com December 2007, page 93, FAST TRACK by John Quain, teaching and describing (at that time) the latest in fastest train technology, but also the most expensive ever. Then compare what you read here now updated on YouTube with modern technologies at half the cost and double the train speed postulated next.

If you need more information of numerous specific design features not possible to mention here, please contact the author for special consulting time and concept sketches to be made available.

1. 500 mph Levitated Bullet Train

A high-speed train going 500 miles per hour could replace the short-haul airplanes and could even start right in the midst of population centers, where it would travel at speeds of 20-50 miles an hour, rolling along on small diameter wheels, powered by electricity. That in itself saves time when traveling to an airport, which is also not very energy-efficient and requires massive parking structures.

After leaving the population city center, the train accelerates by either an external jet engine or electricity, up to 500 miles per hour. It would then be floating like the magnetic-levitated trains, but would be much less expensive with positive gravity pressure. The wheels on the track would only operate at lower speeds during the transit's approach and departure from a station and function as a safety backup during in-flight emergencies, of course, with the loss of levitation. During high speeds, the wheels would be retracted and the train would be levitating, as demonstrated in Germany. How Does It Work?

I propose a high-speed train traveling on a self-generating, low-friction, positive gravity cushion, with air pressure at speeds of up to 500 miles per hour. The train chute roadbed is elevated above ground, supported by a single pillar that is placed in the median of any existing street, boulevard or highway, thus it can be adapted to any downtown configuration and greatly reduce traffic congestion.

Each connected wagon will have a series of freely rotating, small bearings installed that act like wheels if a high-speed contact problem should occur, quite similar to the airplane landing gears. They are mounted in the middle, high up on the body of the bullet train, to be engaged on top of the outer edge of the two-railed, V-shaped, chute track, but still allow passengers to look out the window over the rail edge.

This bullet train will fly with low wind resistance being protected inside the V-shaped chute track and floating without wheels. It is guided much like a torpedo inside the V-shaped tracks of a U-boat. The train travels inside the chute track and conforms to the same principle as a disc drive that functions in your computer. The disc drive in a computer works when a magnetic head flies over the disc at high speed with a very small air space, thus creating levitation and never crashes. That was my previous invention that opened up and made future personal computer data storage possible.

That same fundamental design is applied to the high-speed bullet train at high speeds, which will levitate with the wheel bearings retracted about .125 of an inch above the outer chute standard rails. It is therefore riding on a positive gravity pressure differential like a disk-head on a self-generated, floating air cushion achieved by hydraulically activated force plates that apply pressure sideways to eight plates for each wagon.

The proof that levitating air pressure works can be demonstrated in a modern warehouse, using 75 lb. of air pressure to move a platform weighing several tons, operated by only one man. You need to be convinced through watching about how it is possible to transfer tons of weight, pushed by one man quite easily from one end to the other of a modern warehouse.

We apply the same principle to the trains force plates inside the V-shaped composite plate tracks, which function by lifting the train body off the outer rail tracks for faster levitating speed by a self-generating air pressure. Once more the pressure-plates fit inside the V-shape and are comparable to ski boards, hydraulically activated and pressed sideways, to lift the train off the wheels, which would now levitate and perform similarly to the warehouse platforms that function like wings on an airplane at high speed.

In the event of an emergency, a backup low air compressor supplies air through small holes on the bottom side, distributing floating air to interface with the V-chute track, which is also useful at lower speeds, as observed in the warehouse.

The purpose is to double the safety at high speeds, in the case of encountering potential air pollution or dirt contact on the elevated protected track. The inclined surface V-chute track is made of black polished granite or sintered composite metal plates impregnated with Molybdenum-disulfide, so as to make it resistant to high friction and to the imperious problems of water. Also, deflected front air turbulence is useful to blow off the potential rainwater film in front of the inclined track.

Temperature differentials, caused by environmental day and night cycles, are compensated by a laser beam monitoring the displacement and automatically adjusted at the pillar foundation. One side is fixed and the other open-ended across the rail span in order to allow for expansion, similar to that of a bridge. The train is powered by either electric or jet engines depending on economics, environmental conditions and the distance traveled.

2. Advanced Monorail Hybrid Design

This newly designed monorail hybrid system is linked to the high-speed train and is built in the center of an existing highway, right-of-ways or median strip between traffic lanes. It is a real estate, without the additional cost and could be useful for an elevated traffic monorail system. Having learned from a proven German monorail technology, a redesigned system can be pre-assembled offsite; thereafter, it can be constructed from the top down, using a special assembly method. It can be mounted in such a way that it does not interfere with existing traffic below.

It incorporates a modern lightweight, see-through construction, not like the heavy monster over-the-street rails from the past. The financing of the construction and operational funds can be raised through public or private partnerships, where the government could just grant leases and/or permits for above ground construction over a public right-of-way. Picture a street intersection being used in a mini-skyscraper providing residential, offices or retail supermarkets, with restaurants on the top.

There could be mixed use at the very center of the intersecting street doubling as transfer stations for the monorail grid. The potential to build a transfer station for the developer, and in exchange getting free real estate for a building, including all license fees, requires financing, which would in part come from the monorail system radiating from its structure.

Subsidizing the project through public stock offerings would keep public costs minimal and eventually generate a revenue stream for both the investor and the city. It might help solve a public transportation problem; people would rather go to their office or home on their own schedule, rather than waiting for a comatose bus driver to show up who does not get them where they want to go on time.

Check once more the World's Fastest Roller Coaster printed in **Popular Science April 2011** issue. Look closer at the picture how to propel people at 149 mph on a skinny track. It will give you an idea of future high-tech monorail. The Jules Verne dream mentioned in my opening statement might still come true that described transporting 1,000 passengers every 10 minutes on an elevated mass transit system powered by compressed air along tracks as photographed in the magazine.[1] My Gravity Motor and the fastest bullet train running on compressed air would fulfill this prophecy.

3. A Public, Low-Cost Electrical Mini-Vehicle

Instead of monorail passengers needing to wait for buses at the transit stations, I envision a lightweight, collapsible low profile, electric, two-seated mini-car that is publicly owned and leased upon use for a low fee. It could be the cheapest car ever built in this world.

The concept is analogous to picking up a shopping cart to shop in the store and carrying purchases out to the parking lot available everywhere. The mini-car is left at the transportation station, bus stops or supermarket parking lots, which would not even need hired personnel because of its low maintenance design. It would just be left in a certain place for another person to pick up and use somewhere else.

The mini-car cannot be privately owned. They will all look the same and be leased by the hour, using a credit card that starts the motor. That design allows cars to be cheaply mass-produced and intended for absolute low maintenance. Their use is almost guaranteed, as 80% are single drivers going to work every morning on congested streets, looking for a late bus and wishing for cheap alternatives that are a little less time-consuming. I am sure the high school kids will love it.

A double diamond lane, automatically engaged, embedded electrical rails on top of a road that allows a rechargeable current to flow, which can become a reality in transporting people directly to where they want to go. It will eliminate frustration without being forced by a bus driver who shows up late, causing people to miss the train, who will then need to walk quite a distant sometimes. It would make transportation labor strikes obsolete.

Another wonderful feature will be experiencing downtown without big conventional gas-guzzlers, which must be off-limits on certain streets. Due to the small size of the rental mini-cars, the number of existing parking spaces is greatly increased within any downtown area by at least three times, without incurring much extra expense. New parking spaces can be created from thin air as elevator platform shafts can be mounted on side streets, storing collapsible cars like cigarette boxes in an automat.

---

[1] On compressed motors, http://en.wikipedia.org/wiki/ Compressed_air_car, retrieved 8/10/2011

Commuters, shoppers or high school kids will take the two-seated mini-cars home and plug them into their electrical outlets; then the next day, they can be returned to the monorail station or any parking lot and be left in the designated areas. At the other end, they can pick up another mini-car and repeat the cycle. The high speed bullet train will open up more options for opportunities, while fueling an economy and creating alternative home location choices at low cost and with much less congestion.

The environment and balance of payment problems are positively improved due to the reduced use of gasoline vehicles, thus helping the severe pollution problem in most inner cities globally.

If electricity is not available could build in every town on a vacant lot plugged into a small version of an electric horizontal Wienerrad-wheel generator as an intermittent solution, among the other options available, from windmills to solar panels or better use a mobile mini-UREE version for generating free electricity.

An inexpensive, specially designed, commuter electric mini-car will be fun to drive 30 mph.

The mini-car is constructed like the inner tubing of a rubber tire, similar to an air-pressurized riverboat, only shaped like a boxcar with inflated, pressurized square walls with a thin plastic barrier on the outside that are extremely lightweight. Any crash would be like soft tires hitting each other; they would be indestructible where no dent is possible. It is molded in one piece with a large open storage behind the pre-shaped seat, which is meant for shopping bags. Batteries are under the seat.

It is a four-wheel drive, where each tire has a DC electric, printed board motor embedded that is specially-designed with super magnets that are now available commercially. Windows are weather-resistant plastic that function as a half door. Braking is normally accomplished through reversed electricity. For quick stops, hydraulic brakes are activated.

The wheels and the roof are scissor-linked, collapsible when pushed together, in order to allow a space saving, low profile. They would be available anywhere and stored in an elevator shaft, much like the cigarettes packaged in an automat. The car is accessed through a front hood assembly. These can be mass-produced like a lawn mower for about $500.

4. Faster Traffic Light Sequencing

Finally, I propose a different approach to traffic light sequencing. The present arrangements do not usually work as desired because they were designed in the last century, when there were much fewer cars. As traffic lights have no brain and are never synchronized, much time is wasted. If you figure it out with a little math, you will see that it actually costs billions of dollars annually.

My proposal is very timely. It could increase the efficiency of traffic flow in every town and the removal of a lot of annoying traffic signs, perhaps making for happy commuters once more.

5. Other Unclaimed Inventions Linked to Traffic Congestion

1. Looking back on many of my inventive ideas, I see that they are collecting dust now that I am retired, but I still have quite a number of unused, potentially viably hot ideas that are not yet invented. For instance, there is a cheap highway or a train overpass, which is desperately needed in many countries and is designed to be portable. It can be assembled in just a few days as parts are not hi-tech and can be purchased over the counter. I will share it if you want to become an entrepreneur. You could make some money by leasing the overpass to the government and get paid every month forever.

2. Another million dollar idea is waiting for the right entrepreneur, which is mounting a $300 low–tech mechanical attachment to the rear bumper so that you could park your car in seconds with only 3 inches to spare not touching the neighboring bumpers causing damage. A real help without twisting your neck and causing traffic to stop, which is not possible any other way in that half minute or less.

3. A new dirt tire and racing tire (all-in-one) can be sold to the tire manufacturer, which is worth another million. Do you ask me why I do not make the million with all those good ideas? I am an old man who is soon to depart and writing Babushka egg concept books is much more precious to me, as I am passing on a treasure to the next generation.

4. **The Cheapest Bicycle**. If you are an entrepreneur and want to become a millionaire, I designed a renewable, totally green bicycle useful for the very poor (over two billion) living on this earth. Who cannot afford $300 for a new bicycle from the Wal-Mart store? A small donation to the charity FFF will let you have it. Being made from bamboo, it is 75% lighter than the most expensive $5000 titanium racing frame and much stronger, The most important consideration regarding the design focused around much abuse, dirt, rain and environmental conditions, so as to be absolutely maintenance-free and could be sold as a kit for easy assemble. Its portable wheels can be taken off very quickly by sliding just two pivot screws and can be taken inside a crowded bus, completely flattened out.

If you spruced it up, many bike racers would be very interested in paying more, as it would not only be stronger and lighter, but it would still do the same as the most expensive bike, as the secret is the frame. In time could hook up a Mini Gravity UREE somebody will build and drive around perpetually without fuel, unending possibilities will become common place in the future even without the energy cartel approval.

5. **Avoiding Upcoming Tornadoes.** A global warming weather pattern in the early 2011 spring season was the most destructive in America wiping out totally a number of towns and communities and hundreds of deaths with property damage and business in the billions of dollars. I keep thinking about how we can influence and neutralize a powerful vortex gaining speed. If we could snuff it out in the bud could save lives that could be possible with high technology.

We can now buy cheaply satellite positioners to locate a spot within inches. We need two positioners aiming at the starting storm center and could triangular calculate the exact center of the vortex with computers. At the right timing could shoot a handheld missile with an embedded air-pressure charge exploding similar to submarine bomb aiming right in the vortex center ignited with an on board programmed positioners which creates a diversion pressure blowing up the whirlpool apart. It is worth a try pass it along.

6. New Construction Methods.

I have some ideas about rebuilding destroyed towns. Standard housing can be built at 30% of normal cost in half the time. FEMA would gladly give the initial investment and could make you a millionaire in one year if you are a handy contractor and entrepreneur.

Conclusion

We have been exposed to a number of unusual inventions. Combining these ideas into one unifying concept as piecemeal will not do. They should be laid out graphically by a talented professional in order to see how they work as a whole, then the presentation can be shown on TV. Thus, good graphics become a powerful tool in educating society, and especially to help bureaucrats (usually the biggest obstacles for new ideas) to understand novel concepts. Many will benefit and much can be gained from being exposed to fresh ideas, as the population will surely expand through the roof, but why not prosper, too?

As a former immigrant from Germany, I was forced to create my own company without financial backing. Mostly I had to operate with limited funds. That training is reflected in the fact that my designs tend to be practicable and economically efficient. That approach can be transferred to billion dollar projects, such as public development endeavors.

Being a practical inventor has its advantages, but without the backing of governmental policy decision-makers, many good ideas will collect dust. An, unbiased forum open to public participation is needed. Most universities have lost their purpose, only being interested in another government grant.

The corrupt faculty prophets no longer allow free ideas to be presented in our universities to solve pressing problems. Perhaps problems can be defined much better by analyzing how to apply my proposed inventions because some of these ideas will work well in generating innovative solutions to these vital issues, which can then be used to educate the public.

Hopefully, this will also inspire others to better understand my scientific biblical Babushka concept books linked to metaphysic realities and therefore become better educated from a 360° perspective, in order to meet the challenges of the future.

If you wondered about the coming Apocalypse, Global Warming, Genetic Modification, and Ancient Clocks like the front picture with the two clocks, one hanging in a museum and the other in space, Atoms and ∞ Energy bathed in the Time Dimension; just follow the Evidence Trail on the last page, or watch YouTube to have fun.

That ends my Magnetic Gravity story fueling 16 UREEs until 3018 AD. More is explained in 9 Babushka egg concept books to put in plain words the purpose of mankind. Notice how tiny the earth is.

An open letter to the most honorable 70 members of the KNESSET representing The PEOPLE of Israel

As a retired inventor-scientist living on my vineyard in California, I recently discovered an inexplicable Hebrew Alphabet Number System (HANS), which looks like a code but works as a template for analysis of science from a biblical perspective.

My research included answers to questions raised by the last book of the Bible written about 98 AD known as John's Revelation. HANS indicates that the pages of Revelation may have been mixed up, perhaps by a monk in the 4th century. This book has confused many theologians and opened the way for hundreds of Christian sects. HANS also corrected the Jewish calendar since Moses time compared to a number of ancient stone-bronze-gold clock/calendars exhibited globally in various museums.

My investigations with HANS ended with this last Babushka concept book revealing the Ultimate Renewable Energy Engine (UREE) gravity motor that can extract infinite electrical energy from GRAVITY and OCEAN-WATER. The core concept gifted to Israel can potentially produce many patents worth billions.

Electricity is already a major energy source, but the UREE will provide it much cheaper. Humanity will no longer need power generators that poison the environment. The UREE motor makes nuclear-coal power obsolete. This free energy will change our civilization over the next 1,000 years according to Jewish prophecy. But it comes with a heavy burden to manage and requires someone who will survive the prophesied Apocalypse 2008-2015, now under way in condensed fulfillment following the exponentially parabolic curvature of time revealed by HANS analysis.

Being familiar with metaphysical laws of nature, I recognize recent modern science in what was written thousands years ago. It can now be better checked out with HANS, a powerful tool proving the atheistic evolution religion taught globally in schools to be unscientific. This methodology has expanded into nine Babushka egg concept books where I share the modern science I found embedded in the Torah.

These scientific discoveries directed by an internal force realizing only lately that God had appointed me to be a modern Jonah with a big fish story - the UREE. When the inspired Torah, the first book of mankind's history is connected with the last Bible book of Revelation, HANS reveals mathematically a ∞ creation loop crystallizing the purpose of mankind.

It starts in Genesis with Elohim revealing his plan of a future mankind encircled with a cosmos expressed in Heh-Daleth dimensions (heaven-earth) that bridge two immortal-mortal life echelons. It opens with the LOGOS revealing a human language linked to a ∞ MIND that communicates with each other in a miniature replica of the One who said, "Let us make man in our image." (Genesis 1:26)

The "us" expresses the One God encircled with different dimensions expanding an eternal MIND articulated outwardly with the LOGOS now possible to communicate with the Adam he created. The miracle is that the infinite Creator ELOHIM from the Jod-dimension (Genesis 1:1), where angels live, became visible to our eyes and clothed himself in mortality in the Daleth dimension (John 1:1) in order to reveal his final Plan for Mankind in the middle crossover of the BC-AD Time Dimension of past and future.

God's lesson intend is also revealed in NATURE as the crossover of the ancient math symbol signifies a cosmic resurrection principle like embedded in seeds, which must die to be resurrected into trees, or in caterpillars transformed through death-like cocoons to become butterflies.

Therefore, the ∞ infinite domain is bridged by a crossover point in time, a special mystery person expressed as LOGOS-Yeshua. He is the incarnate, eternal essence of ELOHIM but outwardly abided within the Daleth dimension constraints of physics, appointed to communicate God's will to mankind.

God used prophecy to provide proof for skeptics, which is a paradox by writing history in advance. Hopefully, would convince us of the many invisible realities planned as the eternal purpose for mankind. Historically, only one person in the past qualified to connect the Heh to Daleth dimensions, as proven with the resurrection principle.

For an example, two thousand years ago, the learned lawyers and scribes in ancient Jerusalem approached Jesus preaching in the countryside. Judea was surrounded with so many laws, regulations and contradictorily cultured traditions, just like today. They questioned Jesus' authority, "What is the first law on the top of the others?" Jesus quoted Deuteronomy 6:4-6 but added to the law a new concept of "MIND".

> Jesus answered, "The first is, 'Hear, O Israel: the Lord our God, the Lord is one;
>
> You shall love the Lord your God with all your heart, and with all your soul, and with all your mind, and with all your strength.' (Mark 12:29 NRSV)

A misconception exists among Christians and Jews that the eternal God only spoke once and is never allowed to speak again. However, we can witness further revelation of God speaking stretching across 6000 years because God appointed numerous writers to reveal further knowledge to explain his message.

Like the prophet Isaiah records, "Whom will he teach knowledge? And to whom will he explain the message? – For it is precept upon precept, line upon line, here a little, there a little." (Isaiah 18:1)

But usually ignorant traditions burden a culture. Applied to Israel, a 2000-year-old VEIL still hangs collectively over most who consider themselves to be real Jews, not using the spiritual MIND Jesus referred to when quoting Deuteronomy. Not acquiring the spiritual MIND prevents one from being linked to the ∞ spirit as the innate physics of a sinful nature avoids intellectual to be connected to a ∞God. That also means ignorance of how God in Jod-dimension could transpose himself into the LOGOS from the invisible Heh-dimension to communicate to mortals in the Daleth-dimension once visible from the ∞ side.

If we follow logic, only someone from the other side can edit the Law given to Moses; therefore, he is allowed to expand the laws of God to be better understood because of a blinding VEIL surrounding the priesthood establishment still around even in our times. It greatly perplexed the lawyers and scribes to hear Jesus say that he was before Abraham. The proof that Jesus came from the Heh-dimension, as all his life-events demonstrated, were pre-announced hundreds of years before and confirmed with the historic fact of a resurrection for billions people to witness ever since and cannot be wrong.

A resurrection into the Jod-dimension is essential for all creation, but it can only be proven by an invisible SPIRIT embedded in mankind. It can only be defined outwardly by metaphysical laws or HANS. The blinding Jewish and Christian VEIL of not understanding God's purpose for mankind as demonstrated in history as compared to an old-fashioned record of a famous composer: if one groove is damaged, it can no longer play the rest of the music embedded, as the stuck groove is now repeated over and over. Pretty soon people forget that there is music on the other side.

Only reading the first five books of Moses like stuck on one groove becomes a veil to avoid using a MIND generated from a "reborn" spirit Jesus recommended. The reborn concept is a resurrection principle seen everywhere duplicated in nature as Nicodemus a member of the inner council found out talking secretly with Jesus. (John 3:1-21)

When the Torah is read in the synagogue, could it be possible that there is music on the other side? So many ancient prophets wrote about the Messiah Yeshua's coming first (Son of Joseph) to atone for sin and then being crowned as future rightful King (Son of David), like the Dead Sea scroll found foretelling it.

Fake imitation messiahs will always be around but cannot confirm the written evidence in the Bible and show an unbroken genealogy historic record all the way to David to establish royalty and to Adam to ascertain that he was also human but still linked beyond to ∞ LOGOS. Why deny prophetic history written in advance?

Notice the names come in groups of 14 like the ancient Aztec clock calendar or the 14 stations in Jerusalem of Jesus walk to the cross.

Rather, it is better to prepare for a BIG PROBLEM facing Jacob's travail prophesied in the near future for Israel's, which started to roll in 2008 if you understand **HANS**. Another Jewish genocide is in the making on a grand scale, but will end victoriously with a King of Kings coming just in time to save his people.

It is the same person well known globally who was previously crucified to atone for the sin of Adam infecting all of mankind. The two different prophesied Messiah(s) really merged into one person like a ∞ loop: A suffering Messiah bearing the sin of the whole world and therefore acquired the prerogative to be a Messiah crowned as a King of Kings ruling over the world. Jesus returns to rule as King of Kings on his birthday, 28 September 2015 (as dated by HANS), during the Feast of Tabernacles. So begins God's Kingdom on earth [Revelation 11:15].

The Land of ISRAEL, even in the best of times, never had over six million persons, but Abraham was promised, if you can count the stars, your descendants will be billions and become the head of nations globally with the UN building in the middle of a newly raised Jerusalem after the asteroid impact calculated from science, HANS and Hebrew Holidays overlay. The date dovetailed with many witnesses, 17 September 2015, verified with sun-moon eclipse aligned to cosmic-galaxy mathematics combined as aired on TV.

Investigating the bad and good news sworn personally by God to Abraham comes with a promise to Israel of an unimaginable fantastic future. It started with the rebirth of Israel against great odds in 1948. What the Knesset does not know is that 70 (Ayin) years later, in 2018 AD according to HANS, the Ezekiel Temple foundation stone will be set according to Daniel's 2300-day prophecy to honor the Alpha/LOGOS linked ∞ with a temple bridge named Yeshua/Jesus.

A Warning to the Knesset

Babushka books attempt to repair the stuck groove with HANS and lift a 2000-year-old VEIL so that people can hear once more the creation music from the ELOHIM composer from the other side of the grove. It will end with an indispensable Apocalypse as Israel was appointed to be the spark in the powder keg accelerating to a whirlpool of Jacobs's big time trouble prophesied in order to save humanity from self-destruction and the annihilation of mankind.

Therefore, the Knesset of Israel is now warned to repent from willful ignorance and remove the VEIL like the message of a historic Methuselah telling Noah of an impeding disaster. The first Atlantis civilization was terminated by an asteroid in 2288 BC, judged by God who caused it, and should question why do we not find Noah's relative in the boat?

They did not believe the warning from Methuselah as Enoch named his son "when he dies it will happen". Poor Methuselah, all his life he must have been the butt of every joke always kidding, "Meth what will happen"? Why would God destroy a whole civilization of the ancient world and cause God's WRATH billion people disappearing with their high technology without a trace?

The Bible beginning in the Torah gives us the answer overlaid with HANS. When mankind again reaches the level of high technology like in ancient times, assaulting God's creation with genetic-nuclear technology, it grossly breaches the laws of nature and physically attacks the foundation that was created for mankind's benefit. It violates God's contract made with Adam and ratified with Noah by a rainbow.

God's laws are being desecrated again in conflict with the same misapplied high technology that caused Noah's civilization to become totally evil and degenerate like Sodom and Gomorrah. You can read the result in the Torah, and the most ancient books by Enoch and Noah, children of Methuselah. Three witnesses will testify in any court of law to prove violations of God's supreme cosmic laws.

Worldwide, a global Monsanto cartel aided by the FDA in America is splicing out the original reproductive intelligence codes removed from every cell destroying the many immune genomes that make gene transfer possible and protect us against salmonella, E-coli and many more deadly bacteria.

Worse, these companies have genetically embedded deadly herbicide-pesticides in every modified food cell now causing many diseases the medical profession is wondering about. It is causing the global death of honeybees and insects pollinating our food on a global scale leading into extinction and will end food pollination forever in the next generation, guaranteed.

In addition, notice that most plants in nurseries are reproduced through cloning, all for greedy obscene profit. If the cartel would go bankrupt, like large banks and the stock market recently witnessed in 2008 to start the apocalypse right on schedule, the seeds to grow food will no longer be available. They are controlled by a cartel most farmers depend on.

Assuredly, this will have consequences. Look at our environment. Already billions of people are starving. There is a glimmer of hope after the apocalypse the surviving nations will open the Norwegian underground seed bank, happy to start over again with original seed.

Everybody should read the Torah to find out that God the Creator never changes. He, therefore, will respond once more when his handiwork is assaulted in such a gross way. The future of mankind is at great risk with the food chain collapsing. If you are not educated and want to know, read the fourth Babushka concept book **Genetic Modification Exposed!**

The detailed account of major events that will happen in Jerusalem can be read in the Bible, available in any bookstore. Just turn to Matthew chapters 24-25. The same person, Jesus very accurately forecasted the 70 AD Temple destruction, literally fulfilled as confirmed by history, not one stone was left of a massive stone fortress Temple, ripped apart to get to the molten gold from burning timbers.

Therefore, we should listen once more to Jesus' warnings projected by HANS in the next 42 months, if you want to survive the last Jewish holocaust. The ancient prophecy road map describing the apocalypse, now overlaid with HANS, ended in a special report free on the Internet, **Mystery of Tammuz 17**.

It unfolds Daniel's prophecy along an exponentially parabolic curve in time ending with Jacob's Trouble, all detailed and dated. Even Jesus mentioned it and gave information about how to survive the coming ethnic cleansing by an Islamic mob gone amok.

Israel will be completely abandoned and hated by a world now terrible exposed even by the UN that conspires to exterminate Israel totally by starting the last war near Armageddon. Satan's followers of Islamic tradition linked to Western atheistic evolution religion, both evils in God's sight, will be terminated forever according to prophecy. In addition God will make an end to this environmentally destructive, computerized hi-tech civilization enslaving all mankind into one antichrist, global dictatorship. He will stop extinctions caused by lethal genetic modification to save his creation.

A greatly angered ELOHIM will destroy once more evil and judge the world system just like Noah's Atlantis civilization which ended with an asteroid will be repeated once more dated with HANS. A 52 km asteroid in 825-day orbit and a number of science witnesses point to 17 September 2015 hitting the earth once more. (Revelation 6:12)

It is calibrated to our Gregorian calendar coinciding and overlaid with 17 September on the ancient Aztec calendar and projects 21 December 2012 as the day when the current earth wobble will come to rest, all detailed if you want to survive. (Revelation 16:17)

ISRAEL will be purified through much suffering and made truly kosher as God will separate his sheep to enter his kingdom and the goats for slaughter exposed to his WRATH. (Matthew 25:32) That will exchange the Knesset for a Monarchy.

Try investigating HANS overlaid with Bible prophecy, expect sometimes to be a little off requiring some adjustment because a previous earth axis wobble that changed calendars since Julius Caesar. Thousand year mysteries may need a little more time, being mortal to cross-check it with science.

If you want proof, recognize two witnesses:

1. The big fish of Jonah's story changed a military might and world power to repent overlaid with HANS now became a fact that God announced electricity for free fueling His Kingdom on earth with applications of UREE gravity motors, which comes with a testimony to all nations now revealed in nine Babushka concepts books just read. (Matthew 24:14)

2. Watch the signal for Jacob's Trouble to start counting Daniel's calendar along a Moses temporal tent temple structure allowed by UN to be erected next to the Islamic Mosque to squelch a Palestine riot.

It will be followed by HANS dated prophecy ending with an uncontrollable Islamic riot starting the war at Armageddon ripping apart the Jewish Temple tent to the last screw. That will cause the WRATH of

ELOHIM exploding shaking the earth with another axis wobble ending our atheistic civilization.

God's Kingdom on earth will start with God's footstool on earth, the last final permanent five cornered Ezekiel Temple, built no longer cubic, totally different after 2018 AD with an emerging society lasting 1000 years.

ISRAEL will be elevated as a leader of world-peace ahead of all the nations, then greatly honored and awakened to a Royal Monarchy that started with David as promised to be an everlasting Kingdom soon fulfilled. The future civilization will honor David's last son, not dead, seen by millions.

A resurrected and returned to earth Yeshua ruling as King of Kings is described with more detail in the nine Babushka books that investigate the ancient Hebrew Alphabet Number System I did not invent Israel's projection but overlaid it with the Torah. Check out New Research Pearl #158 and the first babushka concept book: God's Plan for Mankind.

From Eden to New Jerusalem:
Waw → BC/AD → Tet

Be reminded that Satan was successful fooling Adam and Eve being confronted to choose "Did God say…?"

That is like changing the railroad switch to detour the train of the Time-Dimension ending into the junk yard revealing two possibilities. Either to believe and trust God who gave you Life or follow Satan's "Lie" ending in Death.

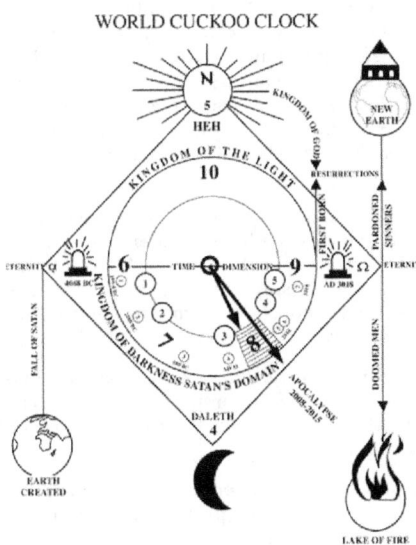

WORLD CUCKOO CLOCK

Your choice, like our parents', will decide what we want, as God does not desire people forced to live in his house. His house is kept clean not allowed filthiness and rebellion and all must be dressed in his pure holy righteousness but trained and educated in EVIL to make sure another Satan's rebellion will never happen again.

Akin to a mushroom growing at night snuffed out in the bud by redeemed MORTALS born again more beautiful than messed up ANGELS. "Let us make man in our image" to enjoy LIFE together forever.

The Evidence Trail Continued -
Nine Babushka Egg Concept Books

1. The 1st Babushka egg concept book tells the story of God's Plan for Mankind, which has never been preached in church. If you want to know what happens after death and what our destiny is, which is not answered by the many confused theologians, here is a simple book with profound consequences. My promise is that you will never be the same. **Apocalypse Prophesied 2008-2015, From Eden to New Jerusalem.** It is translated into German, Spanish and Arabic.

2. The 2nd Babushka egg concept book tells the story about what will happen during the Apocalypse. All events are dated. **Mystery of Tammuz 17, Ancient Hebrew Keys to Dating the Apocalypse,** also translated into German, Spanish and Arabic.

3. The 3rd Babushka egg concept book deciphers a number of ancient mystery clocks, which are exhibited in museums globally. I am sure it will become included in the canon because it connects three civilizations with different earth axis wobble calendars. All it needed was science and mathematics, linked to ancient prophecy and history. **Asteroid Answers to Ancient Calendar Mysteries, Aztec-Mayan Calendar, Antikythera & Other Ancient Clocks Confirm 21 December 2012 in Prophecy.**

4. The 4th Babushka egg concept book is about dangerous **Genetic Modification Exposed!** Atheistic, criminal scientists do it again. Just like the Atlantis civilization, they incur God's wrath, as they genetically destroy God's creation. Check out the consequence in physics linked to metaphysics and why we have an apocalypse in our time.

5. The 5th Babushka egg concept book corrects prevailing false opinions: **Reflections on Global Warming** answers the question, Why. It destroys myths postulated in many universities, which are only interested in grants for higher salaries and do not pay attention to causes, since they reject Bible information linked to the moon's gravity.

6. The 6th Babushka egg concept book demonstrates a new representation in physics: **A Donut Atom Nuclear Story, How the Universe was Created.** A totally different atom theory; linked to galaxies and never before postulated. Gain an understanding of how nature works and how it is connected to the metaphysical. For the first time, it will explain how life is embedded inside the atom. It will also explain what gravity, magnetism and electricity are and explain why the Platinum Standard Kilogram PIS and its 6 sister copies have changed after 135 years; and how it is linked to the ancient mystery stone-bronze-gold clocks?

7. The 7th Babushka egg concept book expands a new atom theory, now proven with more scientific facts linked to Genesis 1:1: **What is the Time Dimension? An Invisible Galactic-Atomic Force Connecting Gravity with Magnetism Linked to the Strong - Weak Force.** It is still a mystery, as the missing links could define a New Model of Structural Physics. It became a special synopsis that explains what a Time Dimension is, which is associated to our mortality.

8. The 8th Babushka book, **New Research Pearls,** is a collection of unusual questions on the Internet. Requested by readers and answered from both physical and metaphysical perspectives. In the HANS code, the Hebrew letter Chet = #8 and corresponds to meaning, "corralled in – a new beginning".

9. The 9th Babushka egg concept book describes a new invention The Ultimate Renewable Energy Engine (UREE) that creates a free electricity profit. If it works, it will make nuclear, coal and oil energy obsolete, whereby we could be driving an electric automobile perpetually free. The energy is extracted from Gravity and will prove that Elohim controls the infinite energy cosmos. The Hebrew letter Teth = 9 "New Life" – **nishmat ha-avir** – (air-breath) will start a new civilization with cheap ∞ fuel.

Herbert R. Stollorz

The shortness of time predicted by Herbert's research does not permit cautious reflection and debate in academic circles. We made the decision to make these materials available to everyone.

Herbert R. Stollorz has invented and improved upon a number of hi-tech devices demanding precision performance – including the floating magnetic head that made possible the original 5-inch mini-disk drive for PCs. Born in Germany during Hitler's reign, he fled from the communist side of Germany and immigrated to Canada. As an inventor of numerous electronic consumer oriented products, Stollorz ultimately owned a high tech company located in Silicon Valley California. He is now retired and lives in Northern California.

Our Basic Premise

The physical universe extends into the metaphysical realm as one fabric of reality or ultimate truth. Only reality's material dimension can be explored by science because scientific instruments cannot detect or measure the spiritual dimension. Thus, God's revelation of the spiritual relationship to the material as recorded in the Bible is essential to unlock the mysteries of life.

Because it is the Word of the Creator, the Bible is congruent with the laws of the natural universe as well as being an integral part of it. The words and structure of the Bible are interwoven within this single fabric of reality so as to reveal the scientifically undetectable aspects of time and eternity. The hypothetical conclusions drawn from studying the Bible through the methodology of a scientist are presented in the books found on this web site. The author believes that his observations may prove to be more reliable than any theological dogma or speculation.

Since the material world images the spiritual, we can explore ultimate reality with the Bible and science to guide us. The test of our grasp of ultimate reality is revealed in a comparison of our proposed picture of God's Plan for Humanity by comparing it with how the present transforms prophecy into history. For all of humanity, the ultimate crucible that will test each individual's, each religion's, each society's, each nation's and the global collective's beliefs about the ultimate reality of divine eternity is the Apocalypse, Last Days or Great Tribulation of the Bible.

The research presented in the books written by Herbert R. Stollorz sets the start of the Apocalypse on 21 December 2008. It will climax on 17 September 2017. The coming biblical judgment of all nations is sure. God planned it from the beginning as the only way to extend His omnipotent blessings to humans and all creatures.

It is certain because the Creator of All, the God of Israel, will intervene directly in the physical universe to make these prophecies become reality in our material dimension in the manner and timing He purposes. Many believe in this certainty, the controversy lies in the author's determination of specific dates. Please read the Author's Preface where Mr. Stollorz discusses the difficulties and uncertainties of his proposed prophetic chronology.

What is Faith in the Future Foundation?

Faith in the Future Foundation provides insights into how and why the world is as it is from an integrated perspective of science and spirituality. In our pages you will find solutions that you can apply to your own life now and faith in the ultimate outcome of the world's current trajectory towards increasing political and economic oppression of a global super-culture based on materialism and spiritual confusion. While the short term future looks ominous, there are a number of sound reasons for hope. We do not give up the fight: we endure and continue in confidence.

To most people, what is referred to as "Christianity" seems irrelevant to our modern world. Despite that understandable perception, we believe that the Bible has the answer to the big questions of life when it is unlocked by insights from natural science. By integrating scientific observations with a better understanding of human history, Faith in the Future Foundation offers you valuable educational tools in your search for spiritual meaning and purpose - for your own life and that of the entire planet. We pray that this support will encourage you towards living a life now that bears abundant fruit in the lives of your family and friends, as well as helping you along on your own journey into an eternity joyfully spent with the Creator who defines true love.

The foundation is not related to any Christian denomination or church. Hence, our Statement of Faith is simple. How each of us puts the principles below to work in his or her personal life varies. Life is a growth experience that occurs at different rates of consciousness and behavior modification. The most important thing is to be on the only guaranteed spiritual way to a good eternity, which is based on these points:

- The Bible is God's word or testimony, the divine revelation of His will and purpose for life. It is the written expression of His Spirit.

- Personal faith in Jesus the Messiah and obedience to God's Word, according to one's understanding, is the only way to receive the Holy Spirit of God, who is the presence of the Father and the Son in the believer, creating a new life that will last for eternity.

- The Holy Spirit is required for continued growth in spiritual matters such as overcoming sin (a changed life today) and a better understanding of the Bible.

Faith in the Future Foundation teaches that a believer's "citizenship is in heaven." The directors do not endorse or support any political party, military operation, forced conversion or change of religious or philosophical beliefs, national entity or any activity that seeks to bring biblical prophecies to pass in this world and time. God is Almighty and able to bring about His prophecies without dependence on human works. He alone can judge the mortal heart and determine a person's spiritual condition and future in eternity.

You may contact us by mail at:

Faith in the Future Foundation
PO Box 6384
Minneapolis, MN 55406

The Birthing of the 3rd Civilization!

The Apocalypse

A Cosmos-Epitaph Perspective (Daleth-Zayin)

2008-2015

The Quintessence Apocalypse

A Time Dimension Perspective (Heh-Chet)

4 January 2012 – 25 July 2015

God's Wrath

(Like a Clock-Escapement controlling the Cosmos)

**Starting 7th month of the 12th cycle
exploding Asteroid Impact**

10 September 2015

**Plus 7-day long earth quakes ending with another
calendar earth axis wobble**

17 September 2015

Watch!

Three birthday parties:

Jesus Christ

The Saints

Israel Reborn

**The Feast of Tabernacles is linked to a Sun-Moon eclipse
synchronized crossing over the Milky Way coordinates**

15 Tishri 5776

Starting the Cosmos clock - Third Civilization on Earth

28 September 2015

(A Femtosecond time pulse perspective of the 7th Babushka egg Concept book)

ELOHIM JOD SUPREME COURT

A Micro-Babushka Egg Concept Booklet #10

The Science-Metaphysics Kaleidoscope Reveals Jonah's Big FISH

By Herbert R. Stollorz

ELOHIM
JOD SUPREME COURT
A Micro-Babushka Egg Concept Booklet #10

and

The Ultimate Renewable Energy Engine
Harnessing "Magnetic Gravity"

Babushka Egg Concept Book #9
Third Edition September 2012

By Herbert R. Stollorz

Use of Photos and Illustrations

All astronomical photos on the cover and in the text were sourced from nasa.gov and edited to adapt to publication design. Unless noted otherwise, all tables, illustrations and images were created by the author.

Use of Scripture

All Scripture quotations, unless otherwise noted, are from the HOLY BIBLE: NEW INTERNATIONAL VERSION®. NIV®. Copyright © 1973, 1978, 1984 by International Bible Society. Used by permission of Zondervan Publishing House. All rights reserved.

Scripture quotations marked NRSV are from the NEW REVISED STANDARD VERSION BIBLE. Copyright © 1989, Division of Christian Education of the National Council of the Churches of Christ in the United States of America. Used by permission. All rights reserved.

Faith in the Future Foundation
P O Box 6384
Minneapolis, MN 55406
www.apocalypse2008-2015.com

ISBN: 1439241171
EIN13: 978-1439241172

Printed on demand in the USA by
BOOKSURGE LLC, North Charleston, SC

ELOHIM
JOD SUPREME COURT

The US President Obama and Israeli Knesset
received in 2011 a courtesy Babushka Publication #9 about

FREE ELECTRICITY FOR THIS CENTURY

The Ultimate in Renewable Energy

Since there was NO RESPONSE, I ask:

WHY are TESLA's DISCOVERIES SUPPRESSED?

Why are they ignored by churches, universities & governments?

This 21st Century Generation must be the most depraved civilization!

WILL YOU to help the EARTH and MANKIND to survive and:

- Obsolete Criminal Nuclear Power Cartels.
- Obsolete dirty gross polluting Coal Power.
- Obsolete the monopolistic power grid system.
- Obsolete polluting gasoline - drive a car with green energy.
- Obsolete Corrupt Bureaucrats and Immoral Corporations.

?

Will You continue to live with:

- High cost of gasoline? Polluting the environment?

- Poisoned dirty AIR - Expensive automobiles?

- Water shortage - Food shortage

- Dirty coal - Diminishing oil resources?

A Precursor

Announcing the discovery of free ELECTRICITY, I expected live TV coverage or a visit from our government. Instead, total silence. WHY?

Across Western Civilization a deep-rooted worldwide apathy dominates as conditions on earth sinking faster than the Titanic. Many politicians have discovered what Hitler-Goebbels pioneered with devastating results throughout Europe: a new science, a powerful force useful in MIND-Control to manipulate people from kindergarten to retirement aided by an atheistic professor priesthood infesting every university. Being deceitful, both suppress TRUTH instead of being open-minded enough to welcome challenging dialog, thus they have become totally destructive and evil.

Because of MIND-CONTROL, the NEW WORLD ORDER has skidded down a slippery path. It changed the fundamental values that enabled mankind to survive for 6000 years and distorted the divine law defining Good and Evil. The modern world lives by Relativism with all of its confusion and conflict. Relativism is destroying eternal values in conflict with the Creator ELOHIM, who will soon interfere with mankind's mismanagement.

Watching worldwide NEWS, not a single government has a program to stop the escalating destruction of our environment. The public does not seem to care and is mostly comatose ignoring logical conclusions of what can be observed with common sense by a child. Mankind has become so ineffective, paralyzed in thought manipulated by the illogical atheistic evolution theory. Unscientific theories can only exist if fuelled by lucrative government grants. Teachers brainwash every child in irrational speculations, thus corrupting the human spirit into atheism denying the Creator and challenged his commandments clearly identified in his Bible. Degraded minds are hell bent to destroy HIS planet. But the Creator is still around; He has declared another asteroid judgment is waiting on the horizon to start mankind over again. The ELOHIM will end this last Civilization know historically as the most Evil generation. This is Jonah's Last WARNING: your LIFE may be over after 17 September 2015 - 2017 depending on God's Grace.

Mankind's dilemma comes from denying a creator and the creation story, replaced by evolution fairy tales. Science rejects the idea that LIFE must be protected both in the physical and the metaphysical realms. Modern society does not accept the fallen sinful state of human nature in need of a savior. Everyone born, no exemption, must answer the question, "Was Jesus human or divine?" The answer will determine your future and mine, guaranteed.

This last generation has implemented high technology to end most LIFE on earth. Instead of maintaining Life in procreation, modern GMO science is causing massive extinction, an irreversible Death affecting all LIFE. The genetic modification of original seeds and animal life is totally evil and screws up the unique intelligence information of every living thing that has survived the last 6000 years. That challenge of a

modern civilization will be answered by the ELOHIM with the prophesied APOCALYPSE, now dated 2008-2015 - God's WRATH.

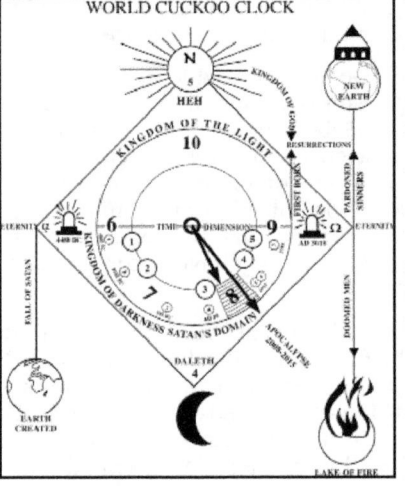

We are now in the middle of Apocalypse, which will end with an asteroid strike in 2015-2017 AD. The Bible prophesied that mankind would perish when it became totally evil by violating the ELOHIM contract.

Once more God's WRATH will be poured out on seven billion people. It repeats Noah's boat story from the ancient times reminiscent of a technically advanced Atlantis Civilization that totally vanished and was judged on 5 February 2287 BC by an asteroid strike resulting in a flood denied in our universities controlled by the evolution religion priesthood.

If you base your values on the brain-dead Atheistic-Evolution-Theory enforced in every school, you may have a problem reading my Babushka egg concept books. If you are offended, consider: an Infinite Energy big bang can never evolve from a NOTHING as energy must be controlled by intelligence, the precursor for anything in existence. Just track thermodynamic entropy laws in physics. That should convince you that a monkey could not possibly evolve and become president of the USA, because a cuckoo clock needs a designer. Denying logic is only useful for La-La-land and makes it difficult to understand my Jod Supreme Court story, so ask ELOHIM to turn the lights "ON".

God's Plan for Mankind represented in a Hebrew clock across eons of time 4488 BC - AD 3018

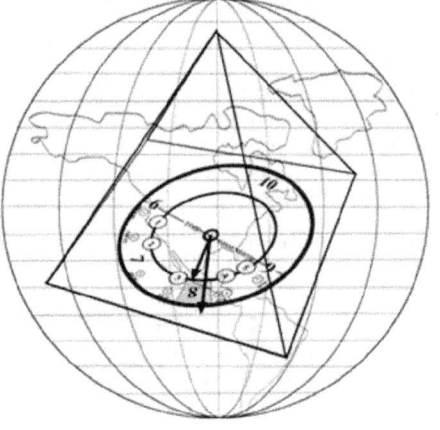

Legal Disclaimer

The social, cultural, theological and spiritual commentaries represent the best efforts of the authors to express their views and perspectives on the Bible in the context of historic and prophetic time, but the information presented in these books is not intended to constitute and should not be taken as any type of financial or legal advice.

First of all, we are only human and may be wrong when it comes to the specific facts and dates that someone might use in financial speculation, for example. Our methods of researching and understanding of the biblical texts are founded on some basic assumptions that are either only discernable spiritually at this time or may turn out to be incomplete.

Table of Contents

LAST WARNING - *Jod*

A historic ELOHIM Court Case from 21 December 2012

The ELOHIM appointed as prosecutor the inventor-scientist Herbert R. Stollorz.

The evidence is presented in ten Babushka Egg Concept Books for a global jury.

Ignoring Jonah's Warnings on the Internet caused the creator ELOHIM to make a final court case against the United Nations World Council to confront the entire 21st Century World Community of gross violation and desecration of the divine cosmic contract made with the representatives of Mankind in ancient times - Adam and Noah. He appointed a qualified technical prosecution to introduce the evidence.

His divine court case will end with serious consequences against this generation, here and now accused of High Treason in committing the greatest crimes in violation of God's Creation on a path ending with the extermination of all LIFE on Earth. A 6000-year old COVENANT has been desecrated.

The divine covenant spelled out the penalty of God's WRATH, and if repeated once more, the ELOHIM will implement the same terms as demonstrated in history by the destruction of a hi-tech, advanced society resembling the Atlantis Civilization. Billions of people were executed who violated that contract. They were judged totally evil and consequently disappeared as the result of an asteroid strike on 5 February 2287 BC. Only Noah's family of eight, being found righteous, survived.

Most people do not realize that much of science taught in universities is evil in God's sight. It no longer follows sound logic in physics, being muddied up by lies to pervert true science, and worse, to promote corruption by misusing modern computer technology in opposition to ELOHIM. It switched our collective mindset back to the Dark Ages and subjected every child worldwide to mind-control hostile to God. The ELOHIM of the Bible is thrown out as irrelevant. If you want to survive God's WRATH and live a little longer, choose wisely who you will serve:

ELOHIM or Satan

The message of Jod #10 is addressed to the 21st CENTURY GENERATION, which focuses on God's divine WRATH scheduled in the Apocalypse 2008-2015, ignored in the public town square in spite being confirmed by science.

For the skeptics, I will provide witnesses that God will once more employ the same methods previously used to execute his WRATH, this time by a 52 km asteroid publicized in various science magazines. It was projected to arrive 2015-2017-2020, depending on which 825-day orbit that gigantic rock will hit our earth, in a manner similar to what happened in Russia on February 15, 2013. The exact timing is wholly conditional on ELOHIM excising his unlimited grace according to his calendar as confirmed by the ancient Bible story of Jonah warning Nineveh.

Babushka egg books were meant for the next generation as a testimony that true science still existed at the beginning of the 21st Century.

The big JONAH FISH has vomited out a last warning for 2017!

Babushka egg concept books were designed to collect science and metaphysical data across the 21st Century to reflect on our collective culture and explain why God-the-ELOHIM will totally destroy MANKIND on earth again. They give the reason why this Second Civilization will perish like the ATLANTIS generation did on 5 February 2287 BC?

Our civilization can be compared to the Atlantis Civilization, which also used hi-tech science to transform God's creation, violating his contract by genetic perversion that modified the hereditary blueprint of all LIFE.

Notice: when we repeat the same evils from ancient times, we will get the same catastrophic result. Corrupted by wealth, world central bankers FRB cartel financed World Economic Forum (WEF), an exclusive international elite, now force every government and university to accept their atheistic, corrupt belief system or cut off their funding.

The FRB controlling the World Currency continues to print massive amounts of fiat money to keep a global economy afloat. Aided by modern computers, they pay off global institutions with generous bonuses. But it is stealing real wealth from the common people to force them to obey a one-world-centered business plan. That will make this global generation once more in conflict with ELOHIM.

It started in 1976 by appointing atheistic judges ordered to throw out the Bible as the foundation of the Supreme Court of America. They changed the US Constitution that has been the envy of the world. Modern Christian culture can now easily be manipulated like the Islamic spring nations. Both enforce an unscientific atheistic evil worldview creating chaos, which is the fuel for more death and destruction. The fundamental laws given by ELOHIM to MANKIND, which made civilization survive for 6000 years, have been totally altered or lost to human consciousness.

The ancient laws of Good and Evil were morphed into relativism based on an unscientific evolution theory now turned into a religion that caused the MIND to be controlled by what Hitler-Goebbels pioneered. The result is witnessed by our grossly polluted environment: fishless oceans, all vegetable and animal food are genetically screwed up, down to my grandkid's level of observation: he can no longer see life crawling under rocks as experienced by his grandfather.

Babushka concept eggs were designed to tell future generations why scientific knowledge was so screwed up and grossly misapplied to endanger all LIFE on earth, making it possible to become extinct in one generation.

Why were 50% of Sir Isaac Newton metaphysical writings suppressed? They link to Herbert R. Stollorz's discoveries that Newton's gravity is magnetic, changing science upside down into totally new theories presented in 9 Babushka egg books free on the Internet.

They reflect a 4300-year-old history and reveal the many counterfeit science theories developed by an atheistic mindset denying TRUTH. The essence of mankind's knowledge was crystallized into a complete concept from philosophy across the bandwidth of science linked to the metaphysics denied by this last culture due to belief in evolution fairy tales.

The sum total was projected in a 360⁰ vision exposing why it is so difficult to understand TRUTH. Being brainwashed by political MIND-CONTROL, worldwide every establishment enforces many lies and pays off people with fiat money, the root of all EVIL. Logically, it becomes obvious why our generation will collapse in 2015 or 2017!

The future generations surviving the global Apocalypse will greatly appreciate that someone collected and preserved mankind's knowledge from the eve of the APOCALYPSE.

Read about a new Donut Atom theory, free electricity - what is it and where it comes from. Discover a new perspective of the universe linked to space-magnetism and fuelled by the invisible infinite Alpha(+ONE) force.

$$\infty E = m(+\infty C/-\infty C)^2$$

How and where is life implanted on the atomic level?

Why only on earth?

And what is the purpose of MANKIND?

A kaleidoscope of unusual concepts across ancient history, prehistoric bronze-gold clocks, pyramids mathematics linked to a better Bible interpretation ending theological confusion. It will prove Jonah's Apocalypse Dating for the skeptic.

A Science - Metaphysic Kaleidoscope Reveals the Jonah's Big FISH

Give ear, O my people, to my teaching;
incline your ears to the words of my mouth.
I will open my mouth in a parable;
I will utter dark sayings from of old,
things that we have heard and known,
that our ancestors have told us.

We will not hide them from their children;

We will tell to the coming generation the glorious
deeds of the LORD, and his might, and the wonders
that he has done.

He established a decree in Jacob, and appointed a
law in Israel, which he commanded our ancestors to
teach to their children;

That the next generation might know them, the
children yet unborn, and rise up and tell them to
their children, so that they should set their hope in
God, and not forget the works of God, but keep his
commandments;

And that they should not be like their ancestors a
stubborn and rebellious generation a generation
whose heart was not steadfast, whose spirit was not
faithful to God. (Psalms 78:1-8, NRSV).

A footnote: As a scientist, I like to use the illustration of a kaleidoscope.
When it is turned, the student sees different color patterns, but on the
inside, nothing is changed, except the rearranged pieces of colored glass
to reflect a mirrored pattern. It highlights exposed different perspectives useful to teach principles of life. That could broaden our minds'
horizon needed for survival of what was prophesied only in the Bible.

When studying God's Word and searching for its logical conclusion, we
should include true science to better understand the APOCALYPSE. Being retired, I have had more time for research, which have become the
best years in my life, isolated in barn, surrounded by a vineyard fenced
in by redwoods. It took some time to gather all the information in these
nine Babushka Eggs. I hope you will benefit from my journey of linking
true science to forbidden metaphysics.

Herbert R. Stollorz,
Philosopher, Author, Vintner, Hi-tech Inventor
Founder and Director of Faith in the Future Foundation, an
educational charity to further biblical knowledge linked to true science

1. Introduction: Why God's Wrath?

When an effect is caused or linked to science, even Isaac Newton thought that it could fine hone the dating of the Apocalypse. The Great Pyramid in Giza was built by Enoch seven generations from Adam. (Pearl #224) With fantastic mathematics embedded in it, the Great Pyramid is really extraordinary but needs a little digging to explain a number of hidden suggestions. The pyramid's design aligns with a few discovered Bible projections. One is a future asteroid projected with two different impacts to change the earth's axis.

That could explain why some projected events on my Babushka egg web site were clouded a little. These issues can be rationalized later as explained in other Pearls still following my inclination digging deeper to add more Babushka eggs, if we assume two or three asteroids in matching Bible prophecy. Science projections of one or two orbits around the sun could change conclusions that will change the dating of the Apocalypse and its three WOES. (Revelation 9:12)

Dating the Apocalypse is a red-hot iron. Many got burnt trying, but most theologians do not even know why an apocalypse was prophesied to totally destroy our civilization. Being a Christian inventor-scientist has the advantage investigating what is invisible - especially when God appointed a warning for this civilization. That process taught me a little here, a little there, like Isaiah experienced. I learned how to uncover suppressed TRUTH linked to forbidden metaphysics.

> Whom will he teach knowledge, and to whom will he explain the message? For it is precept upon precept, precept upon precept, line upon line, line upon line, here a little, there a little.
>
> Therefore hear the word of the LORD, you scoffers who rule this people in Jerusalem. Because you have said,
>
> We have made a covenant with death, and with Sheol we have an agreement; when the overwhelming scourge passes through it will not come to us; for we have made lies our refuge, and in falsehood we have taken shelter; therefore thus says the Lord GOD...
> (Isaiah 28:9-29 NRSV)

It is not by accident that we are confronted by the Prophet Isaiah divinely inspired to direct a forecast toward the closing of our civilization in a projected end-time apocalypse scenario.

His WARNINGS have been preserved for 2600 years. They can only apply to our modern world and is addressed to the **Antichrist Global United Nations World Ruler**. It uses an ancient metaphor at the center of our worldwide civilization never experienced before on earth.

Projected...

> **25 July 2015**, "We have made a covenant with death signed in Hell an agreement...and have made lies our refuge and in falsehood we have taken shelter."

2. God's ELOHIM Covenant Violated

Isaiah describes the Last Evil Generation on Earth before the Kingdom of Light, which can now be dated by science fulfilling Bible prophecy to end in 2015. But God's Grace and Mercy cannot be measured with our mortal intellect; therefore, it could be open-ended, perhaps extended one more asteroid orbit to 2017?

The skeptic usually is not educated in metaphysics and needs more proof. If you have a logical MIND, you might discover confirmation. This Babushka story, Pearl #250, is written and available free on the Internet for easy access. Hopefully, it would widen horizons to 360⁰ in the midst of a total foundational collapse of the whole science establishment now exposed, irreversibly damaged by lies and false theories. Watch the many YouTube renegade true scientists who can no longer be silenced. But turning a metaphysic kaleidoscope will give us a better overview of what is happening in the disturbed science forum.

Reflecting on the smallest egg of (9) nine Babushka egg concept books linked to dating the Apocalypse. It can now be recalculated more accurately with the earth axis wobble at rest on 21 December 2012. That date was globally celebrated. Many visited pyramid tourist attractions in Mexico. They mistakenly thought that a 2012 date came from the Mayan culture based on falsified data.

For a much better real science story, check out how Julius Caesar changed our calendar in 54 BC by inserting 62 days, and Pope Gregory III added 14 days in the Middle Ages. But what surprised me greatly came from investigating the mathematics of Enoch's Great Pyramid in Gaza. I discovered that our earth axis 23½° when overlaid on the Aztec clock calendar date linked to 2012 could extend a knowledge horizon.

I discovered that once more the earth axis possibly could be disturbed again by an asteroid strike from calculating the messages from embedded design of the Enoch Pyramid who lived before 2288 BC. The Bible is not the only place dealing with prophecy. We are daily surrounded with miracles expecting the sun to rise exactly to the minute but do not know why. We take for granted many other incidents during the day.

What will shock many are the witnesses in science cross referenced to the Bible that point to a future, second earth axis adjustment towards a greater angle tilt than 23½° that will cause gigantic tectonic changes like in Noah's time. If it does not match science and prophecy cross-linked, forget about it. Jesus and Daniel forecast that another asteroid will plunge into the earth to repeat Noah's apocalypse, but this time a 52 km asteroid has been observed by telescopes to be coming our way. But luckily, that rock is moving in the direction parallel to earth's daily rotation.

That strike will not cause a flood again as promised by ELOHIM, but it will still disturb the earth axis wobble once more and rip apart tectonic plates dislocated in massive earthquakes lasting 10 days and over 10 on the Richter scale. Trembling mountains will break apart creating valleys and ocean size sinkholes.

America will never be the same. Half could be covered with water as the dome collapsed from a pumped out underground aquifer lake irrigating the Midwest for the last hundred years. Globally, every skyscraper will be gone guaranteed. The detailed effects have been reported in the Bible. Forget theologians' still speculating from the Middle Ages Bible interpretation as prophecy only makes sense if compared to true, accumulated science knowledge.

Theologians do not know that the second asteroid is reported in Daniel 2:34 and can be linked to various science magazines. The asteroid was planned before creation. It is connected to an ELOHIM covenant linked to the history and purpose of MANKIND only recorded in the Bible. Adam and Eve were placed on this planet and told to make an inventory of the exceeding LIFE that Adam categorized. Everything was given a Hebrew name coded on a much higher level. I hope we learned some of it before it is disappearing changed by evil GMO technology.

> So out of the ground the Lord God formed every animal of the field and every bird of the air, and brought them to the man to see what he would call them; and whatever the man called every living creature, that was its name.
>
> The man gave names to all cattle, and to the birds of the air, and to every animal of the field; but for the man there was not found a helper as his partner. (Genesis 2:15-20, NRSV)

The original contract with Adam (4068 BC) was ratified with Noah (2288 BC). He agreed to adhere to the agreement spelled out about managing God's Garden, the only place in the universe where LIFE existed. If mismanaged so that mankind's world system became totally corrupt and degenerate, then humans will experience extinction from poisoning the environment, which is evil in God's sight and fulfills the terms spelled out in the contract.

GOD's WRATH was experienced the first time around eliminated totally a technical highly advanced Atlantis civilization. Only Noah's family survived by building a boat. Mankind once more started anew with the Bronze Age. It is analogous to a second baby of three being born. The third baby after 2018 AD is totally different, with a given name - God's Kingdom on Earth.

Compare the histories of all three human civilizations. A baby must be educated to maintain LIFE. A new civilization must start over and be re-educated, beginning simply, like with a Bronze Age. But the earth has been seriously mismanaged twice, so ELOHIM-Jesus the Creator will soon return and personally supervise the last administration of earth instead of Satan. That will for the first time establish peace to last thousands of years into eternity.

In nature the principle of the placenta represents the rebellious, evil and bloody world system. It will be judged and totally perish in a Great APOCALYPSE according to the covenant. But some can escape if righteous, which is - the GOOD NEWS!

See, I am laying in Zion a foundation stone, a tested stone, a precious cornerstone, a sure foundation: "One who trusts will not panic." And I will make **justice** the line, and **righteousness** the plummet; hail will sweep away the refuge of lies, and waters will overwhelm the shelter.

Then your covenant with death will be annulled, and your agreement with Sheol will not stand; when the overwhelming scourge passes through you will be beaten down by it.

As often as it **passes** through, it will take you; for morning by morning it will pass through, by day and by night; and it will be sheer terror to understand the **message**. For the bed is too short to stretch oneself on it, and the covering too narrow to wrap oneself in it. [Cannot hide anymore, fearful what will be happening. Matthew 24:22]

For the LORD will rise up as on Mount Perazim, he will **rage** as in the valley of Gibeon to do his deed - strange is his deed! - and to work his work - alien is his work! Now therefore do not scoff, or your bonds will be made stronger;

for I have heard a decree of destruction from the Lord GOD of hosts upon the whole land.

Listen, and hear my voice; Pay attention, and hear my speech. (Isaiah 28:17-29)

The ELOHIM Prosecutor Qualifications

Representing the Jod Cosmos Supreme Court, a well qualified Prosecutor Herbert R. Stollorz was chosen to prove the evidence in Court that the Elohim Covenant made with Mankind was grossly violated therefore remind everybody that in any higher court of law, two or three witnesses are required to establish TRUTH.

He will present documents to establish that MANKIND, being blessed by God the Creator, was promised uninterrupted harvest seasons, the sun always shining, guaranteed. In return, God requires mankind to be a good steward to maintain LIFE on this planet, as agreed by the first representatives of humanity, ADAM and NOAH.

A special educated inventor-scientist was picked for the task to present a case in a full 360^0 science horizon documented in nine (9) Babushka egg concept books, for a Global Jury to examine, free on the Internet.

God appointed a German-born US citizen, non-denominational Christian, qualified by studying many years the Bible and the Torah utilizing a unique endowment of investigating nature learned in a lifetime. An inborn habits being curious and scientifically trained could be useful explaining the metaphysic higher-level laws.

Applying his tendency being inquisitive picked up as a kid while taking a walk with the family turn rocks over to see what was crawling under them. That started his science education, became an inventor

benefitting mankind by creating thousands of jobs. Because of his unusual true science and metaphysical perspective, he was appointed to present the ELOHIM's court case against modern civilization, that the Eternal Covenant of ELOHIM was grossly violated by global poisoning of the environment toward extinction. Much of LIFE is in the process of ending presented by a prosecutor and witnessed by his grandkids testifies to establish facts.

The Prosecution Indictment

The ELOHIM Supreme Court accused this vastly advanced 21st Century Generation violating His ancient Covenant now being charged with High Treason. Mankind once more has broken the eternal divine contract. Again civilization has been grossly mismanaged and unleashed destructive forces causing permanent extinction against the environment. A world wide destructive evil MINDSET driven by greed and crimes is on a path to total LIFE EXTINCTION designed to snuff out ELOHIM's creation.

The world's governments and educational institutions have ignored his divine authority, thrown out his Bible rules and changed the ancient laws of Good and Evil, being morphed into Relativism. They have once more endangered all of LIFE on this planet for the next generation, the only place where LIFE exists. They have violated the original Cosmic LAW Agreement, which required the DEATH PENALTY to any civilization. Since the Eternal Covenant twice ratified is again abused and exchanged for a Contract with SATAN, the consequences become more serious:

> ...we have made a covenant with death signed in Hell an agreement ... and have made lies our refuge and in falsehood we have taken shelter...

A modern scientist-inventor was appointed by the divine prosecution to present a case with three major witnesses among other qualified on YouTube to tell a **United Atheistic World Civilization** that time is up on God's Clock! His patience has run out.

The first witness presented is a newly enforced hellish genetic science development (GMO) and will prove without doubt that the Eternal Covenant was grossly violated and will forecast some dates when the terms of His settlement will be executed according to the tenets recorded in the Bible, the oldest book of mankind.

The evil agreement made with Hell proven by three science witnesses will be annulled and resolve for balance one notch higher the metaphysics perspective causing this generation to be executed with another asteroid strike on 17 September 2015. Still depending on God's Grace perhaps could give us extra time to get ready for the court's final decision feasible 2017 AD?

The Birthing of the 3rd Civilization!

The Apocalypse
A Cosmos-Epitaph Perspective (Daleth-Zayin)
2008-2015

The Quintessence Apocalypse
A Time Dimension Perspective (Heh-Chet)

4 January 2012 – 25 July 2015

God's Wrath
(Like a Clock-Escapement controlling the Cosmos)

**Starting 7th month of the 12th cycle
exploding Asteroid Impact**

10 September 2015

**Plus 7-day long earth quakes ending with another
calendar earth axis wobble**

17 September 2015

Watch!

Three birthday parties:

Jesus Christ
The Saints
Israel Reborn

**The Feast of Tabernacles is linked to a Sun-Moon eclipse
synchronized crossing over the Milky Way coordinates**

15 Tishri 5776

Starting the Cosmos clock - Third Civilization on Earth

28 September 2015

(A Femtosecond time pulse perspective of the 7th Babushka egg Concept book)

3. GMO Reflections:
The Divine Court calls its First Witness

To start with the proof of the ELOHIM global case, let's visit the Insectarium at the Montreal, Quebec museum to check the world largest insect-bug collection as a witness, exhibiting a massive assortment of original living species, which lived on this earth 50 years ago now disappeared after having lasted 4300 years since 2288 BC as dated from Julius Caesar's calendar change to 2012 AD. An evil generation of MANKIND, once more since 1976 caused much genetic extinction repeating a gross violation of Noah's contract.

Estimating the many missing bugs expanded to fishless rivers and dead oceans, wildlife disappearing, we may soon witness the possible extinction of the environment as we have known it, unless the Elohim can overwrite Mankind's evil. God can preserve any species. He only needs two of every original species, not modified, a male and female hiding somewhere globally to start over again no longer having a flood promised.

However, God has already preserved 500,000 unmodified food seeds, stored deep inside a mountain surrounded with ice. When the Apocalypse started in 2008, this inaccessible vault's steel doors were shut on the small island of Svalbard near Norway. It was constructed under UN supervision and meant to preserve non-modified seeds for the next generation. It became a concrete Noah's boat.

My Next Major Witness Called

...is known as the "Angel of Death". The powerful global Monsanto conglomerate cartel originally made obscene money as the inventor of Agent Orange. Forgotten history shows the effects of such horribly painful poison murdering millions of families with children in Vietnam, raining death from airplanes and applied to the many other wars, still suppressed by a corrupt US Government rewriting history.

Now the same cartel favored by US military establishment is making more profit manufacturing every year tons of lethal poison for military purpose among ROUNDUP just another name of the same deadly poison ten times worse than DDT[1], which caused more life to disappear, Silent Spring forgotten.

Have you ever wondered why so many pollinating honeybees disappeared? Brain-dead professors paid off by the cartel do not know the reason. Simple houseflies vanish together with wildlife.

I recently saw birds by the thousands falling from the sky on TV, along with literally thousands of beached fish, small and large whales, porpoises, and tuna mysteriously stranded. There were too many to count but still not enough death to wake up comatose people. The general public is unaware what is done worldwide to assault nature in a massive way.

---

[1] DDT (dichloro diphenyl trichloroethane) is an organochlorine insecticide, which is a white, crystalline solid, tasteless and almost odorless chemical compound...

Why do we see massive detergent foam so far away on the beaches in Australia killing everything in the ocean never investigated by corrupt governments protecting the oil cartel hiding the effect to cover-up oil leaks?

TV shows document how they avoid environmental laws and secretly spray detergent around leaking coastal oil wells. That becomes an invisible oil-gunk poisoning vast areas, even spreading submerged over the ocean-sea floor to exterminate all life. Experience a Vietnam movie again, and apply it across the oceans from the sky, this time massively killing marine life for more profit.

Check the price of fish this year. They have tripled, and I cannot find my favorite anchovies and herring. They have disappeared in my neighborhood. This is all ignored by corrupt bureaucrat scientists believing in the unscientific evolution religion as enforced in our universities. They do not understand nature. How can they advise senators to make good laws protecting the environment?

It gets worse. If you are scientifically inclined a little and do not have a closed MIND, follow reason and check the shelves of fruits and vegetables at the food market. Have you wondered why they now glue sticker label-numbers, which was meant for lawyers to collect gene patent fees or if someone gets sick to alert the vendor to remove that product from the shelves.

Remember the spinach fiasco in California some years ago? This worldwide GMO cover-up started with 28 people killed by salmonella. Since then, hundreds and thousands got deadly sick, but it was suppressed by the governments. The death path followed a full circle thousands died in India and hundreds of farmers committed suicide, ruined from poisoned harvest.

A global trail of death suppressed across the USA to Texas Peanut-butter- Tomatoes- Jalapeños- Corn- Tea billion dollar losses, all the way to San Francisco my neighborhood too numerous to count. Farmers deeply troubled and confused by Salmonella invested harvests do not get any help from a corrupt FDA that covers up facts to protect the GMO cartel, which caused the destruction of most original food genetics by allowing modifying genes to permanently alter the original embedded intelligence.

When original gene information is no longer available to procreate, it will be permanently shut down as every generation of seeds gets messed up due to damaged intelligence, receiving injured or missing genes until extinction occurs. It will come in a very short time according to entropy laws. Once you run out of the original to start the GMO process over for another cycle will make that species permanently extinct. Damaged GMO intelligence grossly violates the **Adam-Noah contract of ELOHIM.**

This raises the question, "Why won't the FDA enforce labeling laws to reveal the lie to make it obvious to the public what is inside the fruit, vegetables and Franken fish - especially meat, no longer genetic original pure?" Look at the beef displayed in the market now only shows red fat. It should have a white marbled color, does no longer taste beef which is

the indicator of foreign genes, no longer report Mad cow diseases secretly ground up in hamburger only time delayed, screwing up your brain, too.[2]

Perhaps they inserted crocodile or elephant genes to make a bigger bull? Jewish people desiring to eat only kosher are tricked into eating forbidden meat violating the Torah. Worse, GMO tasteless chicken produced a **million** eggs unavoidably contaminated with deadly salmonella and recalled in 2012. I wonder where they vanished. Perhaps given to charity for a tax deduction?

Investigating so many suppressed facts leads me to believe a government conspiracy exists exposed by much evidence. The many lies on TV designed to keep the public uninformed paying off the media to protect the invisible evil brotherhood Death-Angel corporate consortium known as the Monsanto cartel and their deadly sisters. Both are plotting for mutual benefit to force the consumer to buy exclusively only patent modified GMO food to make obscene profits to hide in offshore banks.

What ticks me off as a German is that I can no longer buy original potatoes to make my ethnic Klöße-dumplings. Even pancakes fall apart with the starch spliced out. Worse are potatoes showing evidence of death turning green in about one month?

Poor farmers are being screwed. They no longer can store potatoes without expensive bleaching gas and paying extra lawyer fees, and I am too irritated, being forced to inspect looking closer for green color. That tells me that the potatoes should no longer be eaten if you want to live a little longer fearing salmonella multiplying when the poison gas is fading.

What is not told to the public prevented by the criminal Food and Drug Administration (FDA) not compliant to the US content label laws being corrupted to the bone making many decision in favor of the cartel being paid off under the table.

Check YouTube. Somebody noticed a relationship of FDA executives being paid back for favors changing jobs later becoming a vice-president of the Monsanto cartel to collect huge bonuses perfectly legal millions deposited in untraceable off shore tax-free accounts.

But when we follow the deadly money trail could reveal the path to permanent extinction of all LIFE ultimately will end in the **Apocalypse** prophesied as a covenant for MANKIND from ELOHIM is still binding. The comatose public is kept totally ignorant by a controlled media that silenced many strange diseases. Medical science is bewildered by cancer going through the roof. This generation of children is much sicker, obese, diabetic and mentally impaired, and many are permanently on expensive drugs.

All can be linked to a new gene technology when the original gene immune system policeman was spliced out by a global cartel. We have forgotten that every food plant-animal genes were reproduced for thousands years by an embedded original blueprint saved in Noah's ark story told in the ancient Bible.

---

[2] Biomedicine-Mad Cow Disease might linger longer, N. S., Science News, July 15, 2006, www.sciencenews.org.

For those who cannot buy a forbidden Bible political prevented by global governments, I will print out a few verses so important not allowed in our schools either to find out why God in heaven = ELOHIM is horribly angry and infuriated at this generation of mankind.

Why are they screwing up his creation meant to last another 1000 years?

That will force God to return to earth and inspect the vast destruction of his despoiled garden and once more teach mankind what is **good and evil** and replace **relativism**, destroying so many cultures and all LIFE meant to last thousands of years.

> Now the earth was corrupt in God's sight, and the earth was filled with violence. The LORD saw that the wickedness of humankind was great in the earth, and that every inclination of the thoughts of their hearts was only evil continually.

> And the LORD was sorry that he had made humankind on the earth, and it grieved him to his heart. So the LORD said, "I will blot out from the earth the human beings I have created—people together with animals and creeping things and birds of the air, for I am sorry that I have made them." But Noah found favor in the sight of the LORD. (Genesis 6:5-11)

Why do we repeat what occurred in Noah's time? Let me explain the GMO process once more for my grandkid.

It started when criminal super wealthy global bankers financed worldwide universities with big fiat money bankrolling the process in hundred laboratories. Kept secret, air tight with negative pressure not allowing modified dangerous pollen to escape into the outside environment causing massive extinction. Those hybrid laboratories, some deep inside mountains, were closed to the public. Experimenting at high speed to splice out the reproduction imprinted DNA-RNA genes now altered and modified by E-coli bacteria.

That simple bacterium, cheaply reproduced, is outfitted with an embedded feature like a drill on its head, can penetrate and bore into thick hard gene-walls, the perfect toxic tool for GMO to wreck cell intelligence.

Pay attention! The deadly E-coli bacteria is able to picky-back any foreign genes of a specie aiding to enter a host gene but will be destroying the policeman protecting the species family. God the inventor has provided a guarding on the cell level from preventing unauthorized entering the inner cell-life-center and when spliced out replaced with cloning to prevent salmonella still will become extinct in the next generation guaranteed by the divine Laws of Fertility.

Applying deadly bacteria has more consequences. They not only destroy the protecting immune system of the interior cell intelligence but also open wide the door for salmonella, unhindered, to enter the center on the cell level. It now permanently alters the original blueprint of the cell, which makes it easy for salmonella to multiply and kill its host, perhaps you. The Bible prophesied diseases recorded the eyes will rot in their sockets and their tongues in their mouth sounds like flesh eating bacteria military science created. (Zechariah 14:12)

Do not wonder anymore much is now cloned, but it will cause more honeybees to disappear poisoned the nectar on the GMO cell level guaranteed to cause food shortages that will elevate food prices astronomically. A hundred dollar bill does not buy much in groceries now - a bad bargain for consumers.

When corrupt governments enforce genetic modified foods generated by evil criminal cartels, it raises the cost by paying off expensive lawyers who control every food store will punish anyone with big lawsuits making obscene money. If a farmer grows clean, non-modified, real organic food, avoiding residues of antibiotic and pesticides, he will now be persecuted with frivolous lawsuits anywhere in the world. He will be financially ruined by paid off judges favoring the international Monsanto cartel, as demonstrated in Canada by the Perry Schmeiser case, which violated his Canadian constitutional rights and the many neighboring farmers, now bankrupt in the same boat.

Organic dairies are being sued by the Monsanto cartel manufacturing dangerous r**BST** to produce more milk. Inoculated cows create a lot of puss in milk now also overloaded with antibiotics to counter the infections. This causes horrible stress on the animals, but it is approved by the corrupt FDA favoring the cartel. Worse, the FDA recently permitted the nuclear radiating of food. Perhaps too many salmonella outbreaks were killing people. If you want to live a little longer, read that again, or tell the government to stop that evil. It is their job to protect the public, not make it sick. Check the evidence collected in the 4th Babushka egg concept book - **Genetic Modification Exposed!** free on the Internet.

Turn once more the kaleidoscope. It gets worse. You will gain another insight watching recent Nobel Prize speeches revealing who is a braindead PhD professor. When you can no longer buy true organic food, you should tell what it is forbidden by the FDA. Like in India where hundreds of farmers committed suicide because only 40% of the rice harvested in 2010 could be sold, so they could not pay back the lender. Even worse, they lost thousands of sheep and water buffalo that died after grazing on fields planted with Monsanto GMO rice.

That is no fun to watch, especially after being honored with a Nobel Prize in Sweden, the Nobel laureate professor derided America for being the only nation having some renegade scientists not believing in evolution. He did not realize he was a brain-dead victim of the Hitler - Goebbels pioneered MIND-CONTROL culture. He has spent his life in a closed university cage that provided a high salary and fat pension to keep him compliant. My dog shows more intelligence, but he is not dressed up in a suit and tie to look for Judah Iscariot money in Sweden after he hanged himself.

What is amazing to me that the Agent Orange cartels and tobacco industry are able to distribute their poison globally to every store from remote tiny, inaccessible village drugstores to skyscrapers? If the poison is globally distributed in every store, under the sink and in most households' garages, heaven forbid gets into the environment over a wide area will eventually kill millions of people by starving them to death as GMO is no longer reproducing on top void of LIFE nutrition.

It is time for us to be educated. Hopefully you are motivated, or see a war-movie from Vietnam once more to see how to kill on a big scale and should learn from history. However, to become truly life essential educated should not forget ancient GMO technology. Let's follow a historic path to add something to our general education, like my grandkid turning rocks over in hope of finding something alive.

Pay attention. **Your grandchildren will die - all of them.** Being the next to the last generation as nature now is nearly totally screwed up, every original food gene will soon disappear according to entropy laws.

Get the proof. Your nursery flower plants die soon after planting being cloned. They can no longer reproduce the now genetic mortified, damaged seed stamped with a patent number on the outside package or skin changed to make obscene money even sanctioned by our corrupted FDA government. Being a favored class of privilege should for God's sake protect the consumer paying their fat salaries and big pension. Notice the many angry scientists on YouTube who were misled by universities. This will cause a total collapse of what has been taught for 150 years as the many lies are now exposed.

Lastly, Shoreline Aquatic Park in Mountain View was set aside to protect wildlife on the inside edges of San Francisco Bay. It is an important stopover for migrating birds every year. Recently, I sat in an outdoor restaurant and noticed not a single sparrow or ground squirrel around begging for french-fries. Even the hardy pigeon has disappeared.

I discovered a box labeled POISON next to the children's playground, even translated in Spanish. That explains the mystery why most wildlife has disappeared. Ducks and so many other birds no longer come there. A lonely Canadian goose was surrounded with piled garbage looking at my face, and looks do tell. "Why have you turned off the water leaving only burned grass, no more food for all my friends starving to death?"

A big expensive sign announced, "No Fishing" at a dead lake now poisoned with ROUNDUP, tons bought by corrupt, evil bureaucrats getting 6 digit salaries and fat pensions. What a joke! They never do their job to visit or check the park. They believe that kindergarten kids can read the POISON label boxes surrounded with invisible ROUNDUP soaked into the dirt all around the whole lake killing everything.

If we just fired one lazy upstairs bureaucrat lacking solid education, we could save some life and would have enough money left over to improve the whole Aquatic Park. Applied to minimum wage unemployed Mexican boys lingering around, they could collect the garbage and feed the stressed wildlife and turn the water "on". It would greatly benefit the next generation of kids seeing life again, provided those curious kids do not eat the poison bonbons if their parents are not watching.

Stupid, highly paid bureaucrats are protected by unethical bosses. This will lead to an Apocalypse affecting a huge population starving to DEATH worldwide. We are next to disappear as demonstrated in the Park. Do not blame God of coming disasters but rather be informed of Jonah's WARNING revealing forbidden prophecy on my web site. Turn the kaleidoscope again and watch God's WRATH the next asteroid al-

ready on its way similar to Noah (2288 BC) coming by, or compare it to the asteroid seen in Russia as shown on TV 15 February 2013.

Hundreds of people were injured and many buildings collapsed from the asteroid's shock wave. Billions of people were warned and will die being found in violation of his covenant. God's Kingdom on Earth will start over again another civilization with free energy no longer polluting and poison globally the environment prophesied to last thousand years and will not let puny mortals trained by Satan destroying all living things on earth once more. Why not be safe asked the ELOHIM how do I survive? Read Pearl #126 on the reason for the population spike and the Bible verses on page 7.

4. Free Electricity - Second Witness

Turn the Kaleidoscope once more while tracing forbidden GMO needs ENERGY to see the parallel-described Babushka egg concepts revealing deadly secrets hidden from the public. I wonder why President Obama never acknowledged my courtesy letter to inform the US government that electricity can be extracted free from Gravity-Water and Air.

Instead, pay attention to make my point. Recently CNN TV showed President Obama visiting a South Korea University in Soul (3/15/2012) where he reaffirmed nuclear power once more. The students listening had scared looks on their faces upon hearing that another lethal nuclear power station was to be started in America without the American people agreeing. Guaranteed it will end in massive numbers of future deaths not mentioned. It is obvious the US government protects the oil cartel and is not listening to the United Nation Energy Department warnings.

Germany, usually a leader in technology, is the first country to realize nuclear power's extreme and irreversible danger to society; therefore, it announced that it would shut down every nuclear-power-station under their control. In 2012 they discovered underground drinking water exposed to radiation seepage near to a nuclear plant.

It got worse when they realized that not a single container was designed to safely store nuclear rods. The storage tanks were not designed to use the original water pool 50 years. It now has cracks in the cement liner revealing rusting iron that had deteriorated much faster with hot radiation. The corrupt American government flouts dangerous facts and conceals the NEWS to keep the public deadly ignorant. Their nuclear garbage is sent to Oak Ridge, TN in America to be burned with the radiation ashes returned. It makes lucrative profit for a private corporation that is shared with the governor and their friends under the table.

They staged a short NEWS clip only seen in Germany. Two old women were in a coffee shop. One said, "We need the jobs." The other replied, "It is cheap energy." In the next scene by a big river, a man saying, "We can no longer eat the fish." The dirty, radiant hot air is not mentioned. The citizens never gave permission to the state government, which made a lucrative deal without them. A quote from a local, "A public who chooses to be uninformed deserves to eat radioactive eggs for breakfast. Where is the money? Cheap energy?"

Tell that to the Japanese people not yet educated. Most still trust their government's lies despite dirty hands under the table grabbing illegal tax money. They seem to need another accident to convince them of permanent, unavoidable Death. The sun not having set, they will not escape the next time around.

China suffers from horribly thick, polluted AIR, and I cannot understand why so many governments still suppress clean free energy for mankind invented 100 years ago by **Nicola Tesla**. Why is free electricity still silenced, although it is demonstrated on YouTube by many young renegade witnesses, people excited to show motor-generators that can run perpetually. They have found something denied in universities around the world avoid teaching it, not understanding what electricity really is.

No wonder global governments are stupid, too! In my neighborhood, why would an immoral US Government Energy Agency suppress the evidence and dictate to the Patent Department preventing free energy. I wonder why bureaucrats and the US Congress protect the dirty OIL cartel and the nuclear coalition club? Why do they allow environmental poisoning worldwide by deadly nuclear radiation killing all LIFE for 100,000 years? Highly paid scientists cannot stop deadly radiation from leaking out willfully lying, degenerate to being criminals guaranteed will fry in Hell for not shutting down nuclear power?

In spite of the evidence on YouTube, energy policies still have not changed even today. They squash unlimited available energy, free **ELECTRICITY** for eighty years. How evil can you get?

Free electric energy has been confirmed by many science witnesses in YouTube videos, but discussion of their discoveries is prevented in the global universities. Sadly, these degenerate lowlifes have been paid off with lucrative Judah Iscariot grants from corrupted governments. Perhaps Tesla was too advanced for his time and needed a new Babushka Donut Atom theory to explain it, but that has been suppressed, too.

Tell me for heavens sake, why is the boundless electricity invented by Nicola Tesla and documented in his 700 patents still forbidden? Why is that technology censored and kept buried in the US Patent Department? Why would a US State Department confiscate so many energy patent applications perceived as military sensitive? Some inventors were also mysteriously murdered to discredit the evidence.

Why does the public allow a criminal energy cartel to silence and deny that ELECTRICITY could be totally free? Even worse they deny that energy can be extracted from embedded magnetic Gravity, super-magnets or air-atoms. They point to upside down physics following an infinite energy path. Just check my 16 new UREE generator inventions: they work, if you are educated enough to test them.

The Chinese government is not controlled by America, yet I really do not understand why they go on suffering from thick polluted sulfuric air when clean electricity is freely available? That proves that a coal-cartel is more powerful than the Chinese government.

I could go on. Why not, for an example, become a billionaire and apply what was testified in the ELOHIM trial case as more witnesses lined up

to establish TRUTH can no longer deny the concept of The Ultimate Renewable Energy Engine now postulated on the free internet town-square.

The new light bulb sold in America is just a coiled up light pipe demonstrated in 1920 by Nicola Tesla wireless connected if grounded. Why is it suppressed? Our computers, telephone, Internet already is wireless. Why not go further and obsolete ugly wires on the pole and drive cars without gasoline and get clean air in the bargain? Universities should not deny Babushka books that trace where the energy comes from.

Universities have the money and faculties to prove or disprove Internet facts. The public should wake up now because they can watch many YouTube movies revealing the censored reality. Perhaps they are afraid of exposing evolution lies, or is corrupting money still the driver?

TRUTH cannot be forever silenced. Perhaps students watching YouTube will no longer being swindled out of an education exchanged for $200,000 in tuition and expenses. Ripped off students are being forced to go deep in debt, a poor investment wasted as many jobs disappeared by following a brain-dead doctrine based on evolution fairy tales. The cost of university education has increased 10 fold and surged 1120% since 1978 compared to the price for food 144%.

Do you wonder where our hi-tech society is heading by denying free energy, not bothered that millions were killed in the last century from useless wars over oil. It seems every other year the same story is repeated around the world. Suppressing free energy will make it obvious who compromised with evil, linked to the Military Complex Cartel, forever building new equipment and smarter bombs to recycle DEATH.

American corrupt politics appoint Pro-Fracking, Pro-Big Oil Scientist to Head the U.S. Department of Energy (DOE)?

Word on the street is that President Obama will nominate Dr. Ernest Moniz, Big Oil and fracking cheerleader, to head up the U.S. Department of Energy (DOE). If you think blasting toxic chemicals into the same ground that gives us the food we eat and the water we drink, is dangerous, if you think allowing fracking to destroy our farmland, contaminate our groundwater and endanger our health sounds like a bad idea, you're part of a growing movement that is determined to ban fracking across the U.S. and move the country toward a sustainable future of clean energy and organic food and farming.

Mr. Moniz is poised to leave his perch as director of MIT's Energy Institute, which boasts such Big Oil financial backers as BP, Chevron and Saudi Aramco. Think former Monsanto lawyer, Michael Taylor, appointed deputy commissioner of the FDA. And Tom Vilsack, another former Monsanto shill, named head of the USDA. (Like fox in the henhouse) Our voices may fall on deaf ears at the federal level....we owe it to our farms and our future to educate, speak out, mobilize - and oppose this appointment. (Organic Bytes a publication Organic Consumers Association 2/28/13)

One more example, many new technology applications are seen weekly in **Popular Science** magazine. In March 2013 page 8-9, the author speculated to keep the consumer ignorant, along the vein of the world's largest windmill turbine manufacturer in Germany. Recently, Siemens built 300 dinosaur windmills for the UK costing the usual billions for taxpayers. Each of its three rotor blades measures 246 feet - nearly the wingspan of an airbus A380. It will produce only six mega watts of electricity. But the well-known company favored by the energy cartel suppressing free energy will not look at modern electric generators producing free electricity. They build outdated designs invented 50 years ago all for higher profit.

Why are YouTube experts demonstrating free electricity oppressed? Or prove that just one of my 16 UREE generators do not work! They could give 1000 times more electric power a lot cheaper, guaranteed if you understand what magnetism is.

Discovering that clean electricity is available totally free was first so preposterous to me, too, which urged me to investigate where it starts. I looked at the universe with thousands of galaxies shining brightly with energy and uncovered physics proving that Gravity is magnetic. This source for electricity is summed up in Babushka egg concept book #9, The **Ultimate Renewable Energy Engine**, meant for a future generation.

It explains how infinite energy is embedded in atoms and easily converted to 16 UREE electric generator inventions. One is enough to fly a Concord-jet with totally free energy. That needed a different Donut Atom theory postulated as the old one from smashing atoms for 40 years is really pretty stupid and needed to be replaced with a much more logical theory linked to Babushka egg concepts books, without an unscientific evolution religion doctrine to make more sense.

In the 2012 elections for president in the USA, politicians raised two billion dollars to be reelected, but no money was appropriated to the millions starving around this globe. Many drought stricken areas do not have water anymore and could be helped by water wells, cost as little as $4,800 to drill a hole. [Contact Life Outreach International based in Texas.] Converted, it means 400,000 water wells could give clean fresh water to 40 million people in 1000 villages.

On TV we see kids walk five miles to a dirty slimy mosquito-infested waterhole surrounded by a collapsing local economy. We should tell our US President to make the next Presidential Decree and become famous, forcing the media to give free voting coverage and apply the collected money to stop global hunger. The public airways belong to the people. It is pathetic to watch little children walk miles for water, hungry and barefoot, to help the family survive. The overabundance in one country should alleviate global hunger. The UN official statistics shows that our food outlets throw away 50% of the food wasted as GMO is spoiling sooner, not aware that God in heaven sees it.

From my perspective it does not matter anymore. Mankind has become totally evil. Our culture based on relativism is evil to the core, which got God the ELOHIM's attention to send a Jonah to the global town square

to announce time is up on God's clock. The judgment is now dated, and it gives us very little time to repent or think about Jonah's WARNINGS.

Why is true science forbidden by world governments although demonstrated by so many witnesses on YouTube? This free Internet witness has hundreds of witnesses demonstrating free clean energy and exposing many lies. Surely our 21st Generation will end in chaos. We should not blame God.

Personally, be informed. Whatever we do is recorded in books opened at the White Throne on the last day. We will appear naked with a record of our deeds on earth opened to the divine judge to be evaluated.

> Then I saw a great white throne and the one who sat on it; the earth and the heaven fled from his presence, and no place was found for them. And I saw the dead, great and small, standing before the throne, and books were opened.
>
> Also another book was opened, the book of life. And the dead were judged according to their works, as recorded in the books.
>
> And the sea gave up the dead that were in it, Death and Hades gave up the dead that were in them, and all were judged according to what they had done. Then Death and Hades were thrown into the lake of fire. This is the second death, the lake of fire; and anyone whose name was not found written in the book of life was thrown into the lake of fire. (Revelation 20:11-15 NRSV)

Summed up:

The ELOHIM has presented two major witnesses from recent science, plus ONE, your and my grandkid. Proven even on a kid level, investigating nature can no longer find bugs crawling under every rock in your neighborhood indicating death toward extinction.

Your food market points to massive GMO destruction by charging increasing, astronomical higher prices every year, which is making billion dollar bonuses for the Death-angel cartel. Lastly, the energy supplier is suppressing free ENERGY proven by hundreds of witnesses on YouTube, still raising the energy cost.

Check the latest gasoline price hike. Look at how the environment is polluted globally with dirty coal poisoning the air and worse deadly nuclear radiation leaking over vast areas, no longer managing the cracked cement-lined pool not designed for 50 years to be a safe nuclear storage container to protect all LIFE.

This evil generation is grossly violating the ELOHIM's contract. On top of being totally evil, its leaders forbid free ELECTRICITY for mankind, still suppressed. This deadly cancer has gone too far to be healed.

Global universities should have perused true physics but now will witness an irreversible revolution storm turning upside down a 100-year old atheistic counterfeit science forum on a free YouTube. It will be

reminiscent of the Islamic Spring and spread uncontrollably to expose the evolution religion's hatred bent to destroy Christianity.

Against the many criminal lies rooted in an unscientific evolution religion taught globally in the academic world, it will gain speed similar to a gigantic snow avalanche to bury obsoleted, irrational theories in an icy balled up mass getting bigger.

Even scheming to close the Internet can no longer suppress TRUTH! It will grow stronger ending with divinely directed consequences, unstoppable and will surely cause a response from the Creator ELOHIM with an Apocalypse PENALTY for breaking his covenant to allow LIFE to continue for the next generation.

Watch, announced now greatly angered, furious, extremely incensed because mankind has messed up his garden genetically and grossly polluted the earth causing extinction. Condemned are those who deny free energy for mankind, ripped off by corrupt governments protecting wicked cartels.

The evil tenants will be thrown out to burn in Hell forever. They never paid their rent to acknowledge and honor the divine benefactor in heaven who gave us LIFE, nor were they gracious to his children living in utter poverty. But one more witness for the skeptic, one notch higher linked to the metaphysic ancient history dated to seal the ELOHIM prosecution case against a corrupt MANKIND destroying all LIFE on earth.

5. Enoch Pyramid - Third Witness

When a civilization once more screwed up GMO and denied free energy, it will also suppress a forgotten ancient story. Our godless global atheistic system has rejected ancient wisdom and thrown out most of ELOHIM's Bible wisdom instruction censored in every court and schoolroom. They replaced the ancient laws that made society survive for 6000 years by knowing the difference between what is good and evil.

It got replaced with relativism, guaranteed to make ELOHIM angrier the second time around. Perhaps, if you live to 2015-2017 now dated with science, you might witness from the sky God's WRATH.

Another asteroid will repeat God's judgment to end our civilization. It will do the same to the modern world as when the ancient but hi-tech Atlantis civilization disappeared on 5 February 2287 BC, totally without a trace, destroyed by a bigger asteroid. The YouTube TV clip of the Russian asteroid splitting up in February 2013 gets you the idea postulated.

When EVIL is morphed into relativism, now taught to every child in kindergarten, and they no longer know the difference between Good and Evil and falls under the atheistic mind-control to accept the extinction of all LIFE on this earth. We will see a repeat of what happened in 2288 BC. Satan is well organized to the smallest detail repeating what worked so well before.

Predicting an Apocalypse using a kaleidoscope makes different pattern visible will widen a knowledge horizon. Most people will not recognize that our modern science we are so proud of pre-existed 4300 years ago.

Mankind's prehistoric knowledge was far more advanced during the ancient Atlantis Civilization, which disappeared submerged in the Bay of Mexico caused by a huge asteroid fire from the sky like Sodom and Gomorrah, not 60 million fairy tale years ago as speculated by an atheistic evolution religion.

To get a new perspective of true science, you must look at all the evidence and not just pick what is politically correct at the moment. A powerful super wealthy priesthood has pushed their unscientific atheistic evolution theories for hundreds years, controlling most universities. If you are truly educated in science, you will readily recognize revealed modern mathematics in some special studies of the great pyramid in Gaza-Egypt. We could not do without physics, but more is needed from the metaphysics embedded in the Bible the oldest book of mankind.

Enoch, the seventh generation from Adam, was the ancient architect-builder of the Great Pyramid in Gaza. It was constructed before Noah's time but also gave warnings to that generation of God's WRATH. They had become totally evil in GMO, too.

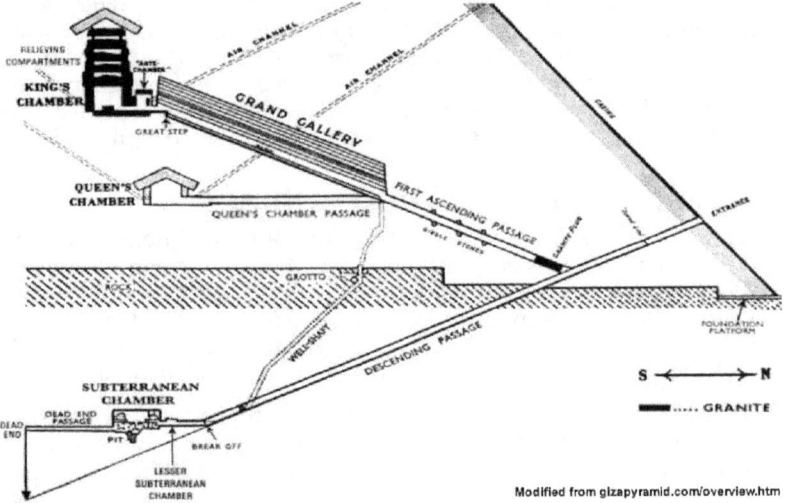

Modified from gizapyramid.com/overview.htm

Enoch was an exceptional math genius who embedded many cosmic secrets in his building design, which were only recently re-discovered. Buried in the makeup of his structure Enoch hid extraordinary mathematics and advanced physics rarely taught in universities any more. Modern education has forgotten the forbidden metaphysics now linked to my third Babushka egg concept book. That helped me to decipher the many ancient bronze gold clocks exhibited in museums still being suppressed.

The most important concept revealed by the study of pyramids around the world was the built-in physics mathematics that described the moving earth-tilt-angle. The only exception is the Methuselah - Enoch Gaza Pyramid built before 2288 BC. It does not match the earth axis wobble position of later designed pyramids.

Perhaps it predicts a future different degree of earth axis tilt forecasting the axis wobble projection of the upcoming asteroid impact. Most pyramids have math embedded that provides useful information in measuring historic time cycles on earth along a wobble declining calendar like the Mexico City outdoor museum has five (5) pyramids on top of each other measuring an axis wobble change.

The range is dated right after Noah in 2288 BC with an earth axis wobble declining very fast to Moses' time at 1400 BC, which made it obvious. The Aztec priests were confused in comparing their calendar from before Enoch's time. They must have had a reason to build 5 pyramids on top of each other as the Moon Pyramid on the inside has (7) pyramids. Why? Before 2288 BC one solstice sun orbit had 52-year short cycles, the sun going up on the west-east reversing east-west seven times in one cycle.

But the nearby Teotihuacán Sun and Moon pyramid were built later on a lake. I noticed the apex is now less steep conforming to an earth backwards wobble measured against zodiac star position therefore designed the next pyramid with a more flattened apex indicating a double earth axis wobble angle at that time. They later made it stepped to save some money in adjusting to a small shift of the wobble angle.

The adjustment was made four times on the bottom footprint instead of building another one on top. That is seen on other pyramids measuring an axis wobble over just 100 years but a little steeper. Others, built 1500 years later, ended with 2x23½ degrees. Somewhere in this process, one priest got a better idea: pyramids are expensive, so he made a calendar using balls indicating wobble cycles declining.

In essence he invented an abacus. Some ancient ball calculators have bigger balls made from stones like a bowling ball exhibited in the Mexico City's outdoor museum, piled up on a wall everybody could see. They were from far away but found nearby. I assembled the same calculator principle as an instrument maker in 1960 as a mechanical Bendix military ball-resolver for rockets. It had three turning output axes measuring automatically sine-cosine math equations needed for space travel. It could have been adapted perfectly to an Aztec calendar wobble system.

Check my third Babushka clock egg book to learn more about ancient history ending with prehistoric clocks never postulated before why is it still suppressed by the establishment. Study the Antikythera clock with 32 bronze gears over 2000 years old. It was fished out from the ocean 100 years ago and exhibited in Athens Greek museum collecting dust. It now is deciphered but still suppressed by the brain-dead scientists who cannot figure it out. It is not possible for them because they are being fooled by an evolution religion, so they remain perplexed seeing so many other stone-gold clocks in other museums, still a mystery to them.

Tracing the Mexican Aztec civilization history from Noah's children, I started from Babylon via China, always moving toward east. Their path is littered with pyramids getting steeper. Many still are buried under jungle growth, but investigating pyramids globally surprised me. I even learned some exist in China along the same route, but only seen

from satellites. Compare that on your paper-globe horizontal traveling across China floating down the main rivers to Shanghai being expert boat builders.

However, coming back to our Endtime scenario and examining the most ancient book, which is very unusual for us. The Bible can be proven as divinely inspired because it has the history of mankind written in advance, which can be traced backwards. Now pin down ancient Ezekiel's prophecy in the Bible. He revealed that the last pyramid build after 2018 AD is a Jewish Temple, now only a semi-cubic shaped structure. The last pyramid built on earth is rising from the Temple center surrounded like a city square Hilton donut hotel (Pearl #224 and Pearl #174).

Investigating the future earth axis tilt angle / wobble, I noticed that Ezekiel's last pyramid apex can be calculated from Scripture. Compare it with Enoch's first pyramid apex, and the differences reveal a future calendar. That explains the mystery of the only apex exception now logically defined and becomes a third witness projecting an upcoming Bible Second Asteroid impact angle. This prophecy is made using Enoch's mathematics as defined with forbidden metaphysics.

Summed up, averaging various pyramid axis wobble positions provides useful information to calculate a future earth-axis wobble as prophesied if compared to Enoch-Ezekiel pyramid to determine the future earth axis wobble. A bigger tilt axis is important for climate change because the bible promises a little more tropical climate for the next 1000 years to have super thriving food-supply feeding 70 billion people.

Look at a map of the earth expanding across unpolluted areas like northern Siberia. Trace it horizontally to China across the Canada's North Territory above 45^0 to 60^0 latitude opening new wide virgin agricultural zones with thousands of lakes but much warmer to grow food. Once more the ocean will be filled with fish, the Antarctic colder again and running rivers in the African desert opening the underground aquifers.

The Bible foretells that the hills in ISRAEL will be flowing with milk and honey once more. Plenty of water will be moved by tectonic dislocation opening Jerusalem to a wide valley to become a global world city, the capital of all nations. (Zechariah 14:1-11) Horrendous earthquakes will fulfill prophecy expanding the royal city the foot stool of Jesus Christ King of kings polarized around a global worship center with a raised temple area to a Sacred High Mountain set aside as an altar for Elohim's footstool and set with a pyramid temple on top.

Underneath, the foundation becomes the source of several springs with healing power growing into a new huge river filling the salt sea now sweet water brimming with fish. Just read the many Bible predictions written history in advance make now much more sense overlaid with true science linked to studying the cause of the second asteroid.

Notice in every country. Just observe the many embedded, tilted up layers of broken mountain ranges and think a little in logic to trace the last effects of a historic asteroid impact 2287 BC only taught in Christian schools.

It is Coming, One Way Another!

10 September 2015

Since a second asteroid was prophesied in Daniel 2:34, I am not surprised to find it identified in a science magazine's 2003-2006 front page picture of a rock 52 km in size in orbit around the sun. It was advertised as projected to hit our earth in 2020, now forgotten.

But a Russian scientist in the meantime re-calculated its orbit around the sun postulating that if one side is heated more than the other, it will change its projected path. If we allow metaphysical pyramid calculations, then we can conclude that it will terminate our civilization. The date originally calculated of 2015 is perhaps delayed to 2017, like one more orbit. Still, it all matches a divinely inspired Jewish New Year date system and all the other Hebrew feast-fast holidays embedded in a celestial schedule.

Enoch's mathematics connect with science magazine projections, save for God's Grace. Remember Nineveh experienced Grace. The asteroid schedule could be modified from a projected 2015 end scenario to 2017. This is similar to the owner of the Titanic, who had rights over the pilot to run his ship according to schedule. But if the owner warns the pilot of an oncoming impact danger, and if the pilot is ignoring advanced warning, not put on alert, who will not prepare the crew and passengers?

For sure history will be repeated like Noah's relatives never made it to a closing door to be saved in the only boat available; therefore, they perished. Do not put your faith in stupid unscientific atheistic evolution religion. It is not proven with real science, and is in fact already has sunk like a Titanic with too many holes if you have the intelligence and watch YouTube exposing the many lies taught in universities. But if you are a professional, you cannot deny what was seen by honest scientists looking through a telescope. The atheistic science priesthood soft-peddle do not worry the asteroid no longer is visible, disappeared back into the cosmos. Was it to appease a scared public no longer hearing anything about it, business as usual, like denying free energy?

But not being ignorant of true physics while applying mathematics learned from Enoch's Pyramid still around could project a gravity calculation tracing a rock to follow its velocity projection still destined to hit the earth in 2020, which made big headlines in the first place to get our attention! But further investigating with true science reveals a sooner hit. Pay attention if you want to survive, or would you prefer to follow Noah's relatives to a nameless death?

Checking the Bible record, three civilizations were originally planned and destroyed by asteroids, which can be dated with an ancient Hebrew calendar schedule. But the timing could be modified by God's GRACE overriding a schedule. Being an off-beat inventor-scientist applying modern science and measuring the apex of pyramids designed with an impact axis wobble apex, perhaps I could re-calculate along a

corrected Julius Caesar - Pope Gregory calendar axis wobble ending in 21 December 2012.

An asteroid is God's method of divine Judgment. It is even recognized by a same-sex couples that know about a prehistoric Sodom and Gomorrah example conveniently suppressed linked to when a society degenerated and became grossly evil, will follow entropy laws. They will become more violent in causing the environment to disintegrate and put all future LIFE in danger of extinction. Demonstrating the same lifestyle in our time linked to corrupt governments that allow many evil scientists and corporations to pollute and screw up genetically our food supply heading toward extinction, gets the same divine reception once more - guaranteed.

The Atlantis civilization ended in God's Judgment with the first asteroid plunging on 5 February 2287 BC to create Mexico Bay with continents drifting apart raising mountains 20,000 ft. high like in Chili and North America destroying an ancient civilization, their ruins now under water after polar ice melted.

Ignorant scientists brainwashed in an unscientific evolution religion postulate that this documented asteroid event happened 60 million fairy tale years ago. But they have no witnesses to prove it. They should read the oldest book of mankind, Job, instead. It described some survived aquatic dinosaurs or Jonah's big fish 4000 years ago recorded in the Bible. A 200-ton whale is twice as big as the largest dinosaurs. Their study should include mystery bronze-gold clocks exhibited globally in many museums, now encoded to understand what the nature of time is.

Still following the pyramid trail linked to forbidden GMO, take just another turn of the kaleidoscope to give us notice why God once more planned to destroy our civilization, as in 2288 BC so in AD 2018. Next time around, the Apocalypse will be similar to the destruction of the Atlantis civilization, which became totally corrupt in Noah's life, full of absolute evil including devastating GMO linked to LIFE extinction.

For the skeptic, to prove that GMO existed in ancient times requires an extra witness (+ONE) for my grandkid would finalize the case of ELOHIM. Why do we find so many paintings in museums of animal-human hybrids? Look in your garden-nursery for a GMO mermaid fish-human figure, or visit the Great Pyramid in front a 150 ft. stone cat monument with a human head buried anciently in silted sand covered by a wobble axis flood.

The math of Enoch's Pyramid provides extra bonuses to reveal more. Does that make sense? Understanding that GMO verified even by the ancient civilization once more is connected to an asteroid apocalypse God's WRATH? Atheistic scientists will deny anything reported in the Bible but logical tell me how big where the waves in Noah's flood?

I invite you to look up to the Gaza-Pyramid on top, notice a washed out break. Now measures from the polished cover plates visible for thousands of years to the bottom Sphinx footing. It is very obvious. Everything is rough below where water pressure waves could have flushed off the outer massive cover veneer plate and buried a stone monument sphinx 150 ft. deep.

"All Gizah Pyramids" from en.wikipedia.org

Why do we see a high-technology man-animal gene modification applied during the Atlantis civilization demonstrated in a proud monument, which had advanced science, too? It could not have been done without electricity and computers as verified by many objects exhibited globally in museums and seen on many TV programs made by an atheistic **National Geographic Magazine**. If you want to know more, watch the many YouTube discoveries. It will surprise you: not all are fake deceptions sent from the atheist priesthood, pushing their unscientific evolution lies.

Enoch's pyramid embedded mathematics was linked to ancient clocks in museums now deciphered is just another witness. Clocks are designed to measure time. Ancient ones help to better analyzing future prophecy DATES. No wonder I was previously a little off when compared to a calendar earth axis "wobble".

Now the earth axis is no longer moving in math exponential difficult to understand its logic if you are common educated. But since 2012, the axis stopped. It will get better data aligned. No one should quickly reject Bible projections to find the reason but should not forget investigate physics linked to the metaphysics applying logic if your Mind is still opened to it.

6. Apocalypse Dating Adjustments from a Metaphysical Perspective

Back to fine honing the dating of the Apocalypse, I turned my kaleidoscope again wondering what would be the effect on a global perspective if God's GRACE should it be extended further from the divine schedule?

If you ask me, I would not give Grace. Lucky for you, I am not divine. But look with open eyes at the evil GMO in a race to modify permanently all seeds and domestic animals for food to extinction. Or look at politics driving mankind by high technology to the very edge of a nuclear catastrophe.

Ignorant theologians still wait for the apocalypse, insisting that the Daniel's seven-year prophecy is still future. Check out the statistics in science. Study the many signs of the Apocalypse Jesus mentioned in Matthew 24-25, now mostly fulfilled; however, where is the spark to blow up the powder-keg in ISRAEL, having passed the fearful 21 December 2012 date?

Keep watching NYC. The real 9/11 perpetrators want to rule globally. Like an omen, something will soon vaporize the big Islamic Satan to gain more power. I hope YouTube would be around to tell the real truth to the American people linked (Rev. 18:6) for future generations destroyed from within by absolute evil.

It moved me to double-check the prophetic date schedule. Perhaps I missed something to find only God's Grace could change a timetable, but I was still challenged to analyze some events not heard in church preaching a different scenario on Christian TV.

Following an old habit of mine of turning a stone to see what is under it, I discovered a gem linked to our changed historic Gregorian calendar. Many theologians wrongly quote a verse, and consequently they don't know that our calendar needed to be corrected by (2) two years. Why is it odd dating Jesus birthday 2 BC?

Dating the First Jewish Temple destruction reference consequently was not 586 BC but 588 BC, according to ancient bronze-gold clocks exhibited globally in many museums. Using that Bible-science information could delay God's WRATH prediction perhaps adding 2 years, too? I also applied it to Nineveh's generation that experienced God's Grace and lived a little longer. You are the judge of whether it makes sense.

I discovered too late, resembling what the prophet Isaiah experienced, that revealed truth comes in small portion here a little and there a little, and never all at once. It is too big for our mortal minds. (Isaiah 28:9-10) The concealed Hebrew math formula 7:5 ratio embedded in HANS can be applied once more to expand God's Plan for Mankind in a little investigating about God's extension of divine Grace.

A few years ago discovered a Hebrew code HANS first explained in the Second Babushka books, **Mystery of Tammuz 17** (page 319 - Chapter

7) describing Daniel's obscured Prophecy Rosetta Stone [7:5 Ratio]. In a previous dating opinion, everybody expected that something unusual worldwide would happen on 21 December 2012. I thought it was time for Satan to be cast out to earth. (Rev. 12:9-12) It was a natural reflex but logical because the war in heaven is fought without a time dimension and difficult to date if you are mortal. Since Satan did not show up according to my scheduled opinion, Antichrist is still invisible and roaming the earth like before told in Job.

> The LORD said to Satan, "Where have you come from?" Satan answered the LORD..."From going to and fro on the earth, and from walking up and down on it." (Job 1:7-8 NRSV).

Consequently, I concluded that Satan is not yet mortal. I learned we can only approximate some dating proceedings by comparing it to our earthly mortal clock events because we are living in a time dimension. The first indication starting the middle of the Apocalypse, which is the Chet or 8th Hebrew cycle, meaning corralled in / a new beginning and linked to a Chinese New Year of the Snake.

The Vatican announced German born Pope is resigning on February 28, 2013. It shocked everybody, but many see it as prophesied by Saint Malachy (1094-1148 AD) that a future #16 Benedict Pope will be the last before the world will end.

Now the corral Hebrew HANS doors are opened for a serpent to appear on the world stage symbolized in Satan. I am persuaded that the Antichrist will appear in 2013 in the same year. Benedict #16 was the last Christian Pope leaving a sinking boat as the next Pope is the true Benedict Shepherd coming back from heaven linked to the Hebrew number #17 but prophesied in the Bible Jesus Christ binding Satan in the abyss ending Satan's school of teaching us evil.

However, one more last counterfeit atheistic pope must appear being linked to Satan like so many in history. He will be adored by the secular world that he will promise to unite globally. All religions will come together in a global World for Peace Party Movement as the answer to worldwide unrest. He will preach that all religions must come together joining Islam and Christianity (Chrislam) to solve the ISRAEL conflict and to obsolete the many wars on earth. Watch out!

Read the Bible proclaiming this biggest deception. History will be repeated like Goebbels, a prophet politician during World War II who stirred up evil by announcing his National Socialist Party would last 1000 years. I remember it as a kid living in Germany.

Likewise the greatest deception Jesus said has arrived will be the most terrible time on earth when Satan really angry now cast out from heaven and knows little time is left to rule physical globally on earth. Now Hell will be visible transferred to our mortal domain. Watch massive demon-possessions affecting the class of privilege who benefitted from a free sponsored UN Visa Card prepaid with fiat money as the rest of a world population is starving run out of energy embroiled in wars as usually.

Only the elite will have food and credit to buy gasoline. The Bible warns us never have the Visa-Beast-number embedded in your right hand or forehead for easy identification to get rationed food and keep your job belonging to the class of privilege. This is mentioned in Revelation 14:9. The salvation destines one for eternal LIFE, but becomes impossible once sold to a Satan enforced global Peace Party system of privilege. Even lukewarm Christians when starving will be misled and sifted like wheat from tares, as God's harvest has started.

Once more to make the middle of the Apocalypse a reality and to expand our vision, we need to apply the obscure HANS 7:5 ratio as a help in dating the final events of Satan appearing first in an assassinated body of a world leader. When an angel is cast out from heaven to earth it can only become visible if permitted to steal a dead body.

Satan will be resurrected in an assassinated world leader body to enter the Jewish Tabernacle Temple as a final act of rebellion against ELO-HIM, proclaiming he is the divine SON, being resurrected as proof. Therefore, desecrate the ancient covenant holy symbols given to ISRA-EL. The center-hub for the Endtime prophecy surfaced when the UN allowed a replica of a temporal tent tabernacle being built on the Jewish Temple Mount next to the Islamic Mosque.

Articulating a biblical appointment, we can attempt to mathematically date that major confrontation and conflict that will defile and violate the ancient Holy Place. I believe it is still possible to look ahead. But be careful. Our limited MIND perception can fool us, and we need to investigate more Bible history written in advance to align it with simple logic. You be the judge.

However, I am strongly convinced that the most major event linked to Satan is anchored firmly grounded in mathematics as too many witnesses merge like many spokes supporting a wheel all focus to a target center of 25 July 2015.

It is even dated from other systems matching HANS and measured from a corrected fixed Julius Caesar calendar with a 2012 perspective cross linked with Enoch Pyramid projections, the oldest information. (Check Pearl #224.) But the possibility of applying God's Grace could shift the main center event by one more asteroid orbit, which is the final Armageddon global war concluding the Second Civilization on a Hebrew holiday calendar schedule 17 September 2017.

Analyzing the APOCALYPSE dating system as adjusted from metaphysical discoveries, I find that many previous dates are still right on, but perhaps need the final Armageddon closing death event expanded by two years to be correlated with a newly discovered science perspective. Let's investigate once more the timing of the coming 52 km asteroid destined for GOD'S final WRATH.

Remember, I read in various science magazines published in 2003 and 2006 that announced on the front cover the possible strike of an incoming 52 km asteroid rock. It was on its way, orbiting around the sun in 825 days, as discovered and traced a little by the Hubble Telescope now forgotten. (Daniel 2:34)

An exponential curve is difficult to visualize looking against the sun's profile. It is much too bright to see a small object, and we can only see it contrasted in the night-sky silhouette on the side. If astronomers have forgotten what was published years ago in science magazines, they could miss that rock, as it is only visible a few hours when passing between sun and earth. It could go either way and even hit the sun.

I believe that solar flares continually observed are really asteroids hitting the sun from the asteroid belt source as the sun is a much bigger magnetic attraction like the size of a hot air balloon compared to our cherry-size earth a tiny magnet. Only a horizontal rock projection is dangerous to earth with the likelihood to happen nearly zero having nine planet-shields in between and the moon protecting a special designed earth home for humans, unless caused by a Creator who invented the solar system.

Investigating science closer could open the horizon redefining false teaching in global universities. Since Galileo looked through a primitive telescope to prove his sun religion, the atheistic science establishment teaches contrary to the Bible viewpoint that the sun is fixed and all the other planets are moving around which is not correct. If the sun is fixed, we would not experience gravity on our body. Not understanding that made other science theories wrong, too. Please check:

http://www.youtube.com/watch?v=o6jBK1ZV-qs

A Russian scientist made a movie on YouTube showing a totally new concept, which was proved years ago in my theory and explained last year in my 9th UREE Babushka egg book. Namely, that Galileo was wrong to assume the sun is fixed.

Now based on a thousand photographs from space, which leads to an illusion all is stationary. I wrote **Pearl #225 - Is Gravity Magnetic?**, which explains a chain carousel. When seated, we feel gravity and compare it to our universe turning with the lights on in galaxies.

That same allusion we get seeing the asteroid belt and think that all those rocks and fragments are just motionless. But our satellites, obeying physics, must move; otherwise we could not have a photograph from space. Therefore, it is not possible on one side to trail an incoming asteroid in danger to collide with earth, but in looking at the asteroid belt I conclude that a billion possibilities exist.

On the other hand, if a moving sun is the leader in the pack with chasing planets, it will behave like a huge vacuum cleaner sucking in magnetically-gravity wise any oncoming asteroids on the way, which makes the earth the safest place in the universe being behind. That is consistent when you plan for LIFE to have a safe place in a hostile universe. Look through the telescope and explain why a shrinking time dimension is so dangerous in physics. Measure the energy exchange of an atom bomb will prove a theory?

When we use the 4th creation day perspective, the earth trailing behind the sun makes good sense, like driving behind very close to a truck shielding us from sand storm debris damaging the windshield. I postulated the daily explosions flare-outs visibly observed on the sun are

impacting asteroids. Why not? Watch a rock hitting stationary pond. It creates a water spike, but a fluid surface sun moving would produce an arc magnetically returned as seen photographed.

By the way, that proves why my 16 UREE electric generators work. They could give us free ELECTRICITY from magnetic gravity expelled from the sun. A one-sided science-view based on Galileo's sun religion observing the planetary system should be re-examined after 400 years and compared to our modern technology to get another look. An unscientific 150-year-old evolution theory should be discarded when it is in conflict with physics. Never mind the metaphysics, only understood when your MIND is not brainwashed by the establishment when it comes to scientific evidence established by experiment and careful observation.

7. The Final Big FISH Dating Proof

Similar to what happened in Russia on February 15, 2013, an asteroid passed by, which will be repeated again being the greatest spectacle possibility of applying God's Grace dated with science. On **17 September 2015** AD a 52 km asteroid rock will pass by the earth and be seen globally by everybody's eyes big time on TV meant as a last warning of God's WRATH, like the biggest JONAH FISH to appear.

Previously, I thought that Jonah's big FISH was **"free energy"** for the next civilization. But logic changed my mind from what was around for hundreds of years as demonstrated by many rebel scientists rarely would be extraordinary a big fish. But a huge asteroid in the sky seen by everybody passing the earth is truly a bigger FISH terrible scary unable to prevent it.

Let's look again if we can project an asteroid from the metaphysic perspective. If one orbit is extended from the 2015 schedule by God's Grace could mean a delay of the impact by 2.2 years (825:365 days = 2.2) still within the timetable of 2018 AD. But forecasting dating cycles must be linked to the Hebrew calendar gears being offset by 8 months to the Gregorian calendar. That would date the ending of the apocalypse (**2015+2 = 2017 AD**) on two entropy levels. But on the other hand, when a rock orbit is moving exponentially, it will get consequently a shorter time, too, making up the difference. But the metaphysical projection dovetailed more accurate than the previously dated **17 September 2015.**

A little exponentially shorter time like 8 months must match the previous date perfectly and start the countdown of Daniel's prophecy on **5 February 2012**, now expanded using the 7:5 HANS overlay by five more years to 2017. The **September 17** date is still linked to Jesus' birthday on 18 September and 10 days later on 28 September ending with Hebrew Holidays of Yom Kippur and Atonement followed by the Feast of Tabernacles and centered on the most important date of inaugurating the newly appointed Jewish King as linked to an old Hebrew moon calendar New-Year festival.

All must match. In the case of clock gears, as little as half a tooth will stop any time projection. Check the Hebrew calendar system of the 19-year cycle. The cycles must mesh like gears inserting 7 times an extra

Addar month to synchronize a Julius Caesar **21 December 2012** solstice calendar correction. Amazingly, moon cycles seem only three Jewish calendar days off looking at **2017**. That is still astounding. I calculated the accumulating wobble fraction in math across 2000 years in the UREE Energy Babushka Egg book #9 if you are familiar with true science linked to prophecy. It must be perfectly aligned by seven Jewish feast and five fast days matching a divine HANS 7:5 ratio logic proving these dating opinions.

First Witness: the Metaphysics Bible

One more proof: start counting the 7-year Apocalypse from **21 December 2012**, which is also the middle of twelve cycles, like 7:5 or the seventh (7) Hebrew cycle and add Daniel's prophetic 1260 days to get to **17 September 2017**. However, we must convert Daniel's wobble axis calendar of shorter backwards days.

His ratio is (7:5=1.4). If we multiply it by 1260, we get a future adjusted time cycle that is wobble factor corrected to our calendar to be **4.8297** years (1260 x 1.4/364.24 = 4.8297) added (2012.9698828 + 4.829) = 2017.799582 Jerusalem time, with a leftover fraction to arrive at **18 September, Jesus Birthday** coming back to start his Kingdom on Earth – a perfect match again.

Theologians are not familiar with a calendar change around Julius Caesar's time in 54 BC. It means that any date from Moses, or 1400 BC (last wobble), must be multiplied by 1.4 to correct to the Gregorian calendar.

Second Witness from Science

Much information is available on the Internet that make universities obsolete. You can get a better education without paying a penny.

Check the address of the IDA asteroid mystery 52 km in size, which I pick as a candidate that must match the mathematics of physics with the metaphysics on a higher level now added the Enoch Pyramid information as a third witness combined with logic. The Hebrew Hand-rule 4+1=5 overlaid with the 7:5 Daniel ratio (#4) of ancient clocks measuring time (#5) should prove it to a brain-dead skeptic believing in an atheistic evolution religion. The 52 km IDA asteroid's orbit was given 4.84, and if we calculate with Babushka egg method, we get the same date: a coincidence once more doubled. 21 December 2012 = 2012.972620742 + 4.84 = **2017.81262** fraction = SEPTEMBER **22.8**

Again 4 days off, as the scientists do not know our calendar needs to be corrected again since the Middle Ages by 4 days to match the Jewish moon calendar. (Check the 9th Babushka concept book.)

Seeing a monster asteroid fizzing by in **2015**, atheistic scientists will postulate that a rock passing between earth and sun would more likely fall into the sun in order to reduce popular anxiety. However one more orbit is scientifically possible being invisible behind the sun, and it could also exponentially turning still to incline and plunge into the earth in **2017** instead as prophesied in Daniel 2:34, Revelation 16:**17**, Isaiah 24 and other Bible places. You decide.

Remember the number **17** overlay from the metaphysic perspective. A strong case could be made for 2017 in HANS. It means (Beth) 2 = house on a higher level (reaching hand =20) now linked to the **Hebrew 17 = Victory** Jesus appearing in the sky starting God's Kingdom on earth eliminating EVIL as expressed in 10+7=17 (Jod-Zayin), as **Zayin** = sword linked to **Jod** on a higher level ending the Armageddon war. To learn more, read Revelation 11:15, 17:1-18.

But skeptics need more witnesses. The first became the ancient Aztec stone clock exhibited in the Mexico City outdoor museum having 52 x 7 years oriented toward solstice positions, but measured in short cycles reversed from our fixed axis projection.

It was demonstrated when hit by the first asteroid from Noah's time in 2288 BC. That date was **17** September recalculated on Enoch's clock mathematics? Why does it overlay with a recent date **17 September** repeating? A triple witness number **17** was postulated 6 years ago published, but check this pearl has the number 17 in Bible verses buried, too. What a coincidence?

Believing in an unscientific evolution religion was designed to keep your MIND brain-dead loosing the ability to think logically and should never be considered next to a Bible-science book. It only creates confusion. Our American education has sunk to #36 on the world scale. It used to be on top in 1945. Primitive bush people have jungle religious opinions too, but are not exclusively enforced on a Western Civilization.

Why impose a "evolution religion" violating science in every school system? It is criminal to force my grandkid to believe in religious evolution fairy tales not even logical and push an agenda purposely designed to keep the masses compliant and ignorant to accept corruption so prevalent in global governments.

The footprint of the sun-pyramid in Teotihuacán near Mexico City is identical to an inch with that of the Pyramid in Gaza - why?

But the apex measurements represent a different earth axis wobble being built later in time. Why is Enoch Pyramid math design showing the next earth tilt axis similar to a future Ezekiel Temple pyramid and even embedded in ancient stone-bronze clocks? A different earth axis tilt would prove the prophecy projection of a major climate change to bring back what was before Enoch's times, a land dripping milk and honey. High up in Israel's sacred mountain, the footstool of the Elohim will have plenty of water running to fill the Dead Sea basin once more, alive with massive fish again like before.

Every child and PhD regardless of education will be confronted with reality when they see a 52 km asteroid flashing by fire in the sky. My first thought was this asteroid is much too big could destroy the whole earth but coming by twice being first close touching the air atmosphere breaking up in a fire rain even the Aztec mentioned it check it out in the 3rd Babushka book. Traveling one more orbit became a little smaller still is a big hammer. That matches Revelation 8:10 dated 2015, the third trumpet of falling debris poisoned the water like Wormwood many died from drinking it, followed by the fourth trumpet so much dust in the sky to dim the sun and moon one third now called out three (3) more

WOES which is one more orbit coming back ending God's WRATH matching Bible prophecy! (Revelation 9)

Now everybody will understand the nature of God's anger to take it seriously but still will be debated by foolish brain-dead PhD and theologians speculating with unproven fabrication watching a disturbed TV. Or become educated in (9) nine Babushka egg concept book, and if not accepted: ...will perish like Noah's relative guaranteed. Watch

17 September 2015
21 September 2017

This is what the Lord Almighty says:

> "Look! Disaster is spreading from nation to nation; a mighty storm is rising from the ends of the earth." At that time those slain by the Lord will be everywhere — from one end of the earth to the other. They will not be mourned or gathered up or buried, but will be like refuse lying on the ground.

> Weep and wail, you shepherds; roll in the dust, you leaders of the flock. For your time to be slaughtered has come; you will fall and be shattered like fine pottery. The peaceful meadows will be laid waste because of the fierce anger of the Lord. (Jeremiah 25:32-37)

8. Why did I Survive?

Nine (9) free Babushka egg concept books present much information never preached in any church or postulated in universities. God took me aside and said write a Jonah WARNING over my inner protest.

We are all born on this planet with a MIND, seems naturally to think of an infinite God, being conscious defined with five senses functioning inside the box controlled by an intelligent SPIRIT. But outside the box looking up at night see so many lights wonder how is it formed burning with energy, the circle of life and death the finality of it. We all wonder is that all?

I started with that question as a kid in East Germany occupied by the Russian Army. We were mostly starving, living on the wrong side in Germany. Always being so hungry, the oldest ran away from home and tried to cross the military East-West border to the food on the other side. Sneaking up to the border, ignorant of its danger I sprinted at the proper time like a rabbit making a dash to the other side.

But I encountered an unscheduled party of two soldiers patrolling and inspecting the military trenches, a highly decorated officer accompanied by his bodyguard. The bodyguard looked at me, aged 17 I remember, but now ordered to shoot as no one was allowed to cross the boundary supervised by military force.

But God said, "NOT this ONE." His machine gun got stuck, useless, to give me life a little longer. Now 81 years later, I write about God's infinite GRACE once more, never was forsaken and never been hungry again, should read more (Grace bathed in mortality - Pearl #227).

In closing this important Pearl, I want to leave you with a thought like a little Kabala in a nutshell as an Apocalypse only makes sense outside the box. We were created by God who has many names. A few were defined in Babushka egg concept books trying to define a higher intelligence. Anyway we could accept that the infinite God is the energy source, and we are the receptors designed with five senses inside the box.

God designed a box with purpose like an inventor. He miniaturized and replicated himself to exist forever, "Let us make mankind in our image." Perhaps fun to watch having a free will not being a robot but meant for important application in God's administration. Read the Bible to get a different view.

My electricity energy and new Donut Atom egg book defines a little what energy is seeing galaxies brightly shining in the night sky. Try to think of them on the metaphysic imagination as created shiny angels from another dimension. Each is different but seeing it from a creation perspective, a universe so many stars why do they exist and shining like angels? The other spectrum we cannot see in the sky many invisible planets on a lower domain, cold, and no longer light energy emanating from inside, empty decayed the final death stage of God's purpose. That is our domain, being mortal also part of a universe but no longer shining.

The mortal box was designed with five receptors our five senses to convert energy to fuel our MIND therefore being aware of our existence. Originally, Adam was created similar to an angel-body brightly shining like a radiant 1000-watt light bulb cannot see the inside invisible like our sun. Mankind became a new invention not yet final, God said there are two trees to give our parents a choice what you want to be with the light on or off? But do not eat the forbidden fruit, if you do, the lights in your body will be off.

Now Adam's children live in darkness made the wrong choice and ever since our body is decaying like a cold planet no longer shining following physics ending in death according to entropy laws, but still have the box receptors functioning with five senses receiving invisible energy we call life to live a little longer.

The design of the box has hidden possibilities to move some receptors a little to reflect energy back where it came from like a mirror now becoming a reflector as a contributor of energy sent via our environment we call Christian compassion, reflecting the ∞ energy light.

Some of our five physical senses now reflecting ∞ energy could morph outwardly into the metaphysic sixth sense that works in addition as demonstrated in charity activities. Now we – mankind, children of Adam, can mysteriously function again according to the original design, transforming the built-in receptor of energy only turned a little like a filter becoming part of a reflector-conductor of a higher level energy source to pass on God's Grace of Blessings.

This is how humans become suitable to reflect light, shining energy back to a lower environment poverty, despair, hunger and help the sick. We can shine brightly in a dark hateful evil world, looking like angels of light becoming visible again. The modified box is now once more united and linked to the ELOHIM Creator closing an infinite energy

loop to be connected to the source to match the original purpose that started with Adam.

By recycling his infinite energy once more we become visible even shining on a dark earth matching the original design purpose. It is like an inventor building an electric generator running from an outside source will give us light passing energy from our five receptors for the benefit of our surrounding environment. Or when the lights are turned "on" will make the theater stage curtain rise to make everybody happy who came to learn more or be entertained.

Please take the opportunity to turn the lights "on" like Adam to experience God's GRACE being designed with receptors but could make a difference turning and become reflectors passing on God's Grace. Checking the Gospel reflectors, you could discover that Jesus will make the difference in your life, too. Mortals were created to be redeemed to live forever but first need to be inoculated with a good dose of evil learned in Satan's school, thus building up an immune system against evil never to get corrupted again - even if surrounded with hellish fire like a sun shining again.

Being infinitely linked to the outside energy source invited to move my reflectors to redirect the energy helped a little by Jesus Christ - the WORD who loved us much, did not mind to come to earth and became flesh like mortals to let me find my hidden light switch now turned "on" reflecting his love. Now we have the chance to become a one notch higher - SAINTS, shining in heaven like the angels and invited to the entourage of the King of Kings, Lord of Lords being privileged by intimate fellowship with Jesus Christ. The nature of light, like God's Word, will illuminate anything outside the box.

Seeing more, like Adam originally and the angels, they communicated back and forth with reflectors to be united with the energy source to live forever in total safety never to live in darkness ever again totally immune from evil, being inoculated while mortal.

Lucky for me, rising from the cocoon by God's Grace, I will become a beautiful butterfly in a world now open to expand my MIND. It will still have the receptors brought along to the next dimension to investigate my new environment that is promised to be without pain, tears and death no more.

I could not keep quiet about God's GRACE, though even being unpopular, so it became my Grandmother Babushka egg story.

The Evidence Trail Continued -
Nine Babushka Egg Concept Books
(Printable free on the Internet)

1. The 1st Babushka egg concept book tells the story of God's Plan for Mankind, which has never been preached in church. If you want to know what happens after death and what our destiny is, which is not answered by the many confused theologians, here is a simple book with profound consequences. My promise is that you will never be the same. **Apocalypse Prophesied 2008-2015, From Eden to New Jerusalem.** It is translated into German, Spanish and Arabic.

2. The 2nd Babushka egg concept book tells the story about what will happen during the Apocalypse. All events are dated. **Mystery of Tammuz 17, Ancient Hebrew Keys to Dating the Apocalypse,** also translated into German, Spanish and Arabic.

3. The 3rd Babushka egg concept book deciphers a number of ancient mystery clocks, which are exhibited in museums globally. I am sure it will become included in the canon because it connects three civilizations with different earth axis wobble calendars. All it needed was science and mathematics, linked to ancient prophecy and history. **Asteroid Answers to Ancient Calendar Mysteries, Aztec-Mayan Calendar, Antikythera & Other Ancient Clocks Confirm 21 December 2012 in Prophecy.**

4. The 4th Babushka egg concept book is about dangerous **Genetic Modification Exposed!** Atheistic, criminal scientists do it again. Just like the Atlantis civilization, they incur God's wrath, as they genetically destroy God's creation. Check out the consequence in physics linked to metaphysics and why we have an apocalypse in our time.

5. The 5th Babushka egg concept book corrects prevailing false opinions: **Reflections on Global Warming** answers the question, Why. It destroys myths postulated in many universities, which are only interested in grants for higher salaries and do not pay attention to causes, since they reject Bible information linked to the moon's gravity.

6. The 6th Babushka egg concept book demonstrates a new representation in physics: **A Donut Atom Nuclear Story, How the Universe was Created.** A totally different atom theory; linked to galaxies and never before postulated. Gain an understanding of how nature works and how it is connected to the metaphysical. For the first time, it will explain how life is embedded inside the atom. It will also explain what gravity, magnetism and electricity are and explain why the Platinum Standard Kilogram PIS and its 6 sister copies have changed after 135 years; and how it is linked to the ancient mystery stone-bronze-gold clocks?

7. The 7th Babushka egg concept book expands a new atom theory, now proven with more scientific facts linked to Genesis 1:1: **What is the Time Dimension? An Invisible Galactic-Atomic Force Connecting Gravity with Magnetism Linked to the Strong - Weak Force.** It is still a mystery, as the missing links could define a New Model of Structural Physics. It became a special synopsis that explains what a Time Dimension is, which is associated to our mortality.

8. The 8th Babushka book, **New Research Pearls,** is a collection of unusual questions on the Internet. Requested by readers and answered from both physical and metaphysical perspectives. In the HANS code, the Hebrew letter Chet = #8 and corresponds to meaning, "corralled in – a new beginning".

9. The 9th Babushka egg concept book describes a new invention The Ultimate Renewable Energy Engine (UREE) that creates a free electricity profit. If it works, it will make nuclear, coal and oil energy obsolete, whereby we could be driving an electric automobile perpetually free. The energy is extracted from Gravity and will prove that Elohim controls the infinite energy cosmos. The Hebrew letter Teth = 9 "New Life" – **nishmat ha-avir** – (air-breath) will start a new civilization with cheap ∞ fuel.